The Order of Nature in Aristotle's Physics

In this book Helen S. Lang enters into the point of view of the ancient world to explain how they saw the world and to show what arguments were used by Aristotle to support this view. Lang demonstrates a new method for reading the texts of Aristotle by revealing a continuous line of argument running from the *Physics* to *De Caelo*. The author analyzes a group of arguments that are almost always treated in isolation from one another and reveals their elegance and coherence. She concludes by asking why these arguments remain interesting even though we now believe they are absolutely wrong and have been replaced by better ones.

The author establishes that we must rethink our approach to Aristotle's physical science and Aristotelian texts. In so doing, her book will provide debate and stimulate new thinking among philosophers, classicists, and historians of science.

Helen S. Lang is Professor of Philosophy at Trinity College, Connecticut.

The Order of Nature in Aristotle's Physics

Place and the Elements

HELEN S. LANG

CAMBRIDGE
UNIVERSITY PRESS

CAMBRIDGE UNIVERSITY PRESS
Cambridge, New York, Melbourne, Madrid, Cape Town,
Singapore, São Paulo, Delhi, Mexico City

Cambridge University Press
The Edinburgh Building, Cambridge CB2 8RU, UK

Published in the United States of America by Cambridge University Press, New York

www.cambridge.org
Information on this title: www.cambridge.org/9780521624534

First published 1998

A catalogue record for this publication is available from the British Library

Library of Congress Cataloguing in Publication Data
Lang, Helen S., 1947–
The order of nature in Aristotle's physics : place and the
elements / Helen S. Lang.
p.   cm.
Includes bibliographical references.
ISBN 0-521-62453-3 (hb)
1. Aristotle.   Physics.   2. Science, Ancient.   3. Physics–Early
works to 1800.   4. Philosophy of nature – Early works to 1800.
I.   Title.
Q151.L36   1998
509'.38 – dc21   97-51317
CIP

ISBN 978-0-521-62453-4 Hardback
ISBN 978-0-521-04229-1 Paperback

To
Berel, Ariella, and Jessica
places in the heart

Contents

Acknowledgments

This work has taken more than five years to complete. In that time, I have acquired many debts of various kinds. Indeed, without considerable help it would not have been possible for me to finish this book at all. It is therefore with pleasure that I acknowledge these debts, take this occasion to thank those who have so generously helped me, and recognize that any mistakes must be my responsibility.

Grants from the National Endowment for the Humanities and Trinity College made this work possible. In addition, I spent a wonderful year as a Visiting Scholar at the University of Pennsylvania and would like especially to thank Gary Hatfield for making all things possible in the Philosophy Department and library.

My colleagues at Trinity College, Howard DeLong, W. Miller Brown, and Drew Hyland, generously answered my questions, discussed systematic issues raised by this work, and offered criticisms of my views. I owe a special debt to my colleague in Classics, A. D. Macro, who was always available, interested, and informative about problems of language. The library staff at Trinity is one of the college's great resources, and their help was invaluable for this project; I want especially to thank Patricia Bunker, who never failed to find anything no matter how confused my starting point.

Paul J. W. Miller was my first teacher of Aristotle and remains my best Aristotelian friend and reader. He read a draft of the manuscript and offered many insightful comments. The final form of this work owes much to his criticisms. An early draft was also read by Joe Gould, Philosophy Department, SUNY Albany, who was an astute and most helpful critic. David Konstan, Classics Department, Brown University, has spoken with me over the years with energy and enthusiasm and I owe much to both the spirit and content of his reflections. Robert W. Sharples, University College, London, provided a number of telling criticisms of my account of place, as well as sharing his own work on the subject.

Some of the material in this book has been published, albeit in a quite different form, as "Aristotle's *Physics* IV, 8: A Vexed Argument in the

History of Ideas," in *The Journal of the History of Ideas* 56, 1995, pp. 353–376 and "Why the Elements Imitate the Heavens: *Metaphysics* IX, 8, 1050b28–34," in *Ancient Philosophy* 14, 1994, pp. 335–354. Anonymous readers of both these articles offered important criticisms. And no author has ever been luckier in anonymous readers than I have been in the three readers for Cambridge University Press. Their criticisms were detailed, thoughtful, and intellectually generous in the highest degree.

Aristotle argues that in the process of development what is best comes last. Last thanks go to my husband, Berel, who has been reader, critic, moral support, and more. His enthusiasm for this project gave me the courage to go on in its darker moments. Indeed, he has suggested with some energy that I dedicate this work to Aristotle because without him it would not have been possible.

Abbreviations

Citations to several texts or translations have been abbreviated as follows:

Ackrill
Aristotle's Categories and De Interpretatione. Trans. with notes by J. L. Ackrill. Oxford: Clarendon Press, 1963.

Apostle
Aristotle's Physics. Trans. with commentaries and glossary by H. G. Apostle. Bloomington: Indiana University Press, 1969.

Barnes
The Complete Works of Aristotle: The Revised Oxford Translation. Ed. J. Barnes. 2 vols. Princeton: Princeton University Press, Bollingen Series LXXI-2, 1984.

Carteron
Aristote: Physique. Text établi et traduit par Henri Carteron. Vol. I. 2d ed. Paris: Budé, 1956.

Charlton
Aristotle's Physics: Books I, II. Trans. with introduction and notes by William Charlton. Oxford: Clarendon Press, 1970.

Couloubaritsis
Aristote: Sur la nature (Physique II). Introduction, traduction, et commentaire par L. Couloubaritsis. Paris: Vrin, 1991.

Didot
Aristotelis Opera Omnia. Ed. A. F. Didot. Vol. 2, in Greek and Latin. Paris: 1874.

Guthrie
On the Heavens. Greek-English. Trans. with introduction by W. K. C. Guthrie. The Loeb Classical Library. Cambridge: Harvard University Press, 1939.

Hardie and Gaye   *Physica*. Trans. R. P. Hardie and R. K. Gaye.
                         Oxford: Clarendon Press, 1930.

Hussey   *Aristotle's Physics: Books III and IV*. Trans. with
                         notes by Edward Hussey. Oxford: Clarendon
                         Press, 1983.

Moraux   *Aristote: Du Ciel*. Texte établi et traduit par
                         Paul Moraux. Paris: Budé, 1965.

Nussbaum   *Aristotle's De motu Animalium: Text with
                         Translation, Commentary, and Interpretive
                         Essays*. Ed., trans., and with commentary by
                         M. Nussbaum. Princeton: Princeton
                         University Press, 1978.

Ross   *Aristotle's Physics*. Revised text, with
                         introduction and commentary by W. D.
                         Ross. Oxford: Clarendon Press, 1936.

Stocks   *De Caelo*. Trans. J. L. Stocks. Oxford:
                         Clarendon Press, 1922.

Wicksteed and Cornford   *The Physics*. Trans. Philip H. Wicksteed and
                         Frances M. Cornford. Vol. I. The Loeb
                         Classical Library. Cambridge: Harvard
                         University Press, 1934.

Part I
Place

1

Aristotle's Physics and the Problem of Nature

"Nature is everywhere a cause of order."[1] This claim, together with the evidence that Aristotle marshals to support it, forms a consistent theme throughout his entire corpus. There has been and continues to be considerable disagreement about Aristotle's various arguments – their topics, what they say, if they are valid. But there can be neither doubt nor quarrel that within his philosophy as a whole this claim is central: nature is everywhere a cause of order.

As Mansion argues, nature gives the world not only order, but intelligibility.[2] I shall conclude that Aristotle's claim that nature is everywhere a cause of order constitutes a first principle that informs his physics as a science, and consequently, the particular problems and solutions within physics. Beyond his physics, it also appears at work within his metaphysics. Indeed, it constitutes one of his most important philosophic commitments.

As a first principle, this claim is never proven and so is not derived by or within physics – or any other science. Rather, as I shall argue, the topics, proofs, and arguments of Aristotle's physics and metaphysics presuppose and work in terms of the claim that nature is everywhere a cause of order. Hence, we see this claim at work because we see the physics that assumes it as a first principle. Because first principles cannot be proven directly (if they could, they would not be first),[3] Aristotle's topics and arguments provide the only, albeit *ex post facto*, evidence for his claim that nature is everywhere a cause of order. And these topics and arguments constitute the object of this study.

---

1 Aristotle, *Physics* VIII, 1, 252a12, 17; on this and other closely related claims, e.g., nature does nothing in vain, cf. *Physics* VIII, 6, 259a11; *De Caelo* I, 4, 271a33; II, 11, 291b14; III, 2, 301a6; *De An.* III, 9, 432b22; *Parts An.* II, 13, 658a9; III, 1, 661b24; *Gen. An.* I, 1, 715b14–16; II, 5, 741b5; 6, 744a36; III, 10, 760a31. All translations are my own, unless otherwise noted.
2 Mansion, *Introduction à la physique aristotelicíenne*, 92.
3 Aristotle, *Metaphysics* IV, 4, 1006a5–11; *Physics* I, 2, 185a1–3, 12–17.

In short, I shall not take up Aristotle's "philosophy of science," e.g., his concerns with the methodology and/or the status of science, including physics, but shall consider his actual practice of physics and the results of that practice insofar as they relate to his metaphysics and conception of the world at large.[4] Garber describes his book *Descartes' Metaphysical Physics* as "a kind of handbook of Cartesian physics, a general introduction to the mechanical philosophy as Descartes or a sympathetic but not uncritical contemporary of his might have understood it."[5] So I propose a "sympathetic but not uncritical" analysis of arguments at the heart of Aristotle's teleological physics.

Speaking of Kant, Friedman observes that "there has been a marked tendency to downplay and even to dismiss the philosophical relevance of Kant's engagement with contemporary science, particularly among twentieth-century English-language commentators."[6] And the reason for this tendency is clear: the mathematics and science of Kant's time, essentially those of Euclid and Newton, have long since been replaced; hence if Kant's achievement is to be thought significant it must be understood insofar as it transcends "the details of his scientific context."[7] If that point holds for Kant, how much more it applies to the pre-Euclidean, pre-Copernican mathematics and physics of Aristotle! Indeed, it has become common practice to read Aristotle in terms of Newtonian physics and to find his physics engaging just insofar as it appears to anticipate features of Newton's physics.[8] If the goal of analyzing Kant's philosophy has been "to transcend" his science, that of analyzing Aristotle's has been "to anticipate" a science entirely different from his own.

This study constitutes a *de facto* rejection of this view. Against it, I argue that to analyze Aristotle's physics in terms formulated in later scientific contexts and language is not only to diminish the scope and importance of his work but also, and more importantly, to skew the meaning of concepts central to his physics – in short, to misrepresent it. Thus I shall present a detailed analysis of an important set of arguments insofar as they reveal Aristotle's physics and his practice of physics as a science. The fact that in terms of post-Copernican science Aristotle's physics turns out to be wrong diminishes it neither as an extraordinary accomplishment of its time nor as a subtle and sophisticated set of arguments. Aristotle's physics

4 For analogous projects concerning Descartes and Kant, cf. Garber, *Descartes' Metaphysical Physics,* and Friedman, *Kant and the Exact Sciences;* for an argument concerning the importance of such a project, cf. Hatfield, "Review Essay," 1–26.
5 Garber, *Descartes' Metaphysical Physics,* 3.
6 Friedman, *Kant and the Exact Sciences,* xi.
7 Friedman, *Kant and the Exact Sciences,* xii.
8 Like many of his contemporaries, Newton himself thought of his ideas as known in antiquity; such claims added authority to science, and in this sense, current work in this vein continues a venerable tradition. For one example, cf. Hine, "Inertia and Scientific Law in Sixteenth-Century Commentaries on Lucretius," 728–741, which includes a useful bibliography on this topic. Newton forms a theme throughout Hussey; for one example, cf. 130.

constitutes a full engagement with the problems of physics as they were defined by both his predecessors and his contemporaries. As such, he provides invaluable evidence concerning the practice of physics and the arguments constituting it as a science – a physics remarkable first in its own achievements and then as central to the history of philosophy and science in the Hellenistic world, the Byzantine world, the Islamic world, and in Europe from the thirteenth to the seventeenth centuries.

To compare Aristotle's physics with that of Newton (or others) is neither impossible nor unimportant. But such comparison cannot be conducted on a "point-by-point" basis when the language, concepts, and context of physics are so radically different. In my conclusion, I sketch a program for comparison. In short, I suggest that each view must be understood fully in its own terms before it can be meaningfully compared with another. Comparisons and evaluations must rest on the presuppositions underlying the physics, the problems that the physics is designed to explain, and the internal coherence of the arguments that purport to solve these problems. The continuing interest of Aristotle's physics derives, at least in part, from this last point: his arguments are remarkably coherent. But his presupposition – that nature is everywhere a cause of order – turns out to be wrong in important ways, and the problems that he sets out to solve are not the most important or productive for physics. And as a consequence, his enormously powerful arguments serve a project that is wrong about everything.

There is no way of knowing this outcome, however, until the starting point has been spelled out and its implications made clear. And there is no way of evaluating either Aristotle's physics or the history of science until we understand how well Aristotle does the job. This is the task of this study: to understand the conception of nature as everywhere a cause of order and the definition of problems whose solutions exhibit this presupposition. And this task is accomplished by an analysis of the arguments – their structure, language, and logic – that solve the problems defined by Aristotle as central to physics. Finally, this task is both an end in itself, i.e., understanding a complex set of arguments, and a first step in a larger project, evaluating the history of physics.

As I argue, Aristotle's physics engages then-current problems and is conducted by someone in full control of those contemporary issues. His arguments present a clearly defined structure, which in its turn reveals the practice and outcome of physics as he understands it. Furthermore, analysis of these arguments establishes not only his solutions to various problems of physics but also his conception of the problems themselves, both his own and those of opponents whose views he rejects.[9]

The text of Aristotle's *Physics* served as the starting point for the physics (and astronomy) of the so-called Aristotelian system which "continued to

---

9 Although I shall later criticize details of their account, for an example of Aristotle's physics treated in this way, cf. Matthen and Hankinson, "Aristotle's Universe," 417–435.

hold the allegiance of the overwhelming majority of the educated classes
in the seventeenth century, its final century as a credible system."[10] As I
show in examples throughout this study, translating Aristotle's physics
into modern terms, e.g., those of Newtonian physics, at once falsifies his
position as well as the history of philosophy and physics more generally.
It falsifies his position by translating it into intellectual concepts and commitments entirely foreign to his own, and it falsifies the history of ideas
by treating this history as teleological insofar as Aristotle is made into an
anticipation of later ideas. Such falsification suppresses the sense in which
choices have been made not with the benefit of hindsight but because of
immediate pressures of local intellectual contexts, commitments, and
problems. Sometimes such choices, like those Aristotle made, have in fact
turned out to be flatly wrong. But translating Aristotle into an anticipation of Newton neither "saves" this physics nor adds any weight to Newton's. Rather, it falsifies the histories of science and philosophy, impedes
our present understanding of them, and, finally, prevents us from grasping an important possibility: the conjunction of false starting points with
powerful systematic arguments. This conjunction should give us more
pause for thought than is often the case – or than *can* be the case if Aristotle is read simply as an anticipation of post-Copernican ideas.

The history of Aristotle's physics bears witness not to the achievement
of Newton's physics, but to the power of Aristotle's ideas and arguments.
Indeed, as the literature on them testifies, they fascinate their readers
still. Herein lies the mystery for contemporary readers. From our post-
Copernican perspective, Aristotle's views are often wrong both in their
larger claims (e.g., nature is not necessarily a principle of order) and in
their details (e.g., the earth is not in the center of the cosmos with the
stars going around it). And when Aristotle is "right," as when he argues
that the earth is a sphere, often he is "right" for the wrong reasons; for
example, he claims that all individual pieces of earth are oriented toward
the center of the cosmos and so cohere together there, making the earth
a sphere (*De Caelo* II, 14, 297a8–297b20). Perhaps worst of all, Aristotle
offers explanations that seem positively feeble for problems central to
post-Copernican physics; e.g., he apparently says that the motion of a projectile, such as a stone, after it has left the hand of the thrower, is caused
by the air that rushes in behind it.[11] All in all, it is a mystery not only why
anyone would be interested in these arguments now, but why anyone
would ever have found them persuasive or engaging.[12]

---

10 Grant, *In Defence of the Earth's Centrality and Immobility*, 3.
11 Aristotle, *Physics* VIII, 10, 266b27–267a10, and *De Caelo* III, 3, 301b17–30, are often read
     in this way. I shall suggest a different reading later.
12 O'Brien, "Aristotle's Theory of Movement," argues that Aristotle's physics is now of only
     antiquarian interest: "Does Aristotle's theory of movement, does his theory of weight,
     have any more than an antiquarian interest, today? Quite frankly, no," 78.

But the arguments constituting Aristotle's physics were and are gripping: "The general excellence of the 'secondary literature' on Aristotle, extending from the Greek commentators to the present, is unrivaled (as far as I know) by what has been written on any other philosopher."[13] Why? Unlike later physics (and much philosophy), Aristotle's arguments respond immediately and directly to experience. We experience objects as heavy and as always going down when they are left "to do their own thing"; the stars appear to move while we feel ourselves to be stationary in the center. Indeed, the range of Aristotle's examples, consistently appealing to everyday experience and common sense, may be unequaled in the history of science or philosophy (except perhaps by Plato). By answering to experience, these arguments speak and have always spoken to their readers. And so they speak also to us – modern readers who inhabit a profoundly different world from that of Aristotle.

The appeal to experience raises an important issue. Aristotle's claim that "nature is everywhere a cause of order" is a starting point that cannot be proven directly; but he marshals impressive *ex post facto* evidence and arguments for it. Consequently, it presents the problem of any genuine starting point: it was not obviously false when Aristotle assumed it, nor is it obviously false for anyone now setting out to examine the presentation of nature in everyday experience. And Aristotle's analysis, using this starting point, often seems to penetrate to the very heart of natural phenomena as it appears to any observer. In this sense, he offers a starting point that is both powerful and promising.

That Aristotle's position on central issues is now universally recognized as wrong is itself the outcome of a long history involving complex sets of choices. Sometimes these choices arise out of criticisms of Aristotle's physics (and various interpretations of it) and would have been impossible without the full articulation of his "false" view, resting as it does on a "false" starting point. And herein lies a second important issue.

Aristotle's starting point, that nature is everywhere a cause of order, lies at the heart of his practice of physics, which defines the problems to be solved by the physicist and divides phenomena in such a way that those problems can be solved successfully. I shall argue that much of the cogency of his arguments lies in his success in conducting physics as a project given his first principle and the problems defined by it. That is, he defines the problems that his starting point is best able to solve; he divides phenomena in the most useful way, given the problems he wishes to solve; and he develops powerful arguments to solve these problems. So, for example, I shall argue that his definition of "nature" is intimately linked to his definition of "motion" and that these two together define the terms required by his account of nature; "place" is such a term, and his analysis

---

13 Irwin, *Aristotle's First Principles*, vii.

of it is designed to support his definitions of motion and, ultimately, nature. In this way, he constructs a coherent science of physics that he uses not only to account for a wide range of phenomena but also to present strong claims for a universal principle embracing a remarkable set of details.

Thus according to Aristotle, several facts bear witness to the claim that "nature is everywhere a cause of order": (1) The same cause, assuming it is in the same condition and acting in the same way, always produces the same effect; (2) nature is everywhere opposed to chance – chance presents rare events whereas nature presents those that occur always or for the most part.[14] Furthermore, (3) as a cause of order, nature everywhere "flees" the infinite;[15] hence (4) nature always desires what is best.[16] For these reasons, nature cannot but remind us of god and what is divine. Like god, nature does nothing in vain or uselessly.[17] And the absence of what is vain or useless in nature clearly returns us to the view that nature desires what is best and is thus everywhere a cause of order.

The history of choices and criticism that has proven Aristotle's account wrong – wrong from its universal principles to its treatment of details – has neither merely criticized details nor grandly rejected universal principles. That history is itself very complex, and this complexity arises not only from a progressively more sophisticated analysis of Aristotle's arguments but also from other sources. Physicists have taken up problems, such as a creating God as the first cause of the universe, originating outside Aristotle's physics as a science; they have redefined the problems within Aristotle's physics (e.g., the problem of projectile motion), and they have devised new answers to those problems (e.g., impetus theory). And we are bound to lose sight of this important and complex history if we judge the details of Aristotle's physics only, or even mainly, as anticipating or failing to anticipate later ideas.

Furthermore, we also lose a full sense of *modern* physics as it represents our own intellectual commitments, if we fail to see that the problems and starting points of physics are not "given" but are themselves constructed within a historical process called "science."[18] So, as I argue later, the problem of how to account for the motion of a body, such as a stone after it

---

14 Aristotle, *De Gen. et Corr.* II, 6, 333b5; 9, 336a27; *Physics* II, 8, 198b35; *De Caelo* III, 2, 301a7.
15 Aristotle, *Gen. An.* I, 1, 715b14; *Physics* VIII, 6, 259a11.
16 Aristotle, *Physics* VIII, 6, 259a11; 7, 260b23; *Gen. An.* I, 1, 715b14; *De Gen. et Corr.* II, 10, 336b28.
17 Aristotle, *De Caelo* I, 4, 271a33; II, 11, 291b13; *De Anima* III, 12, 434a31; *Parts of Animals* II, 13, 658a8; III, 1, 661b24; IV, 6, 683a24; 11, 691b4; 12, 694a15; 13, 695b19; *Progression of Animals* 2, 704b15; 8, 708a9; 12, 711a18; *Gen. An.* II, 4, 739b19; 5, 741b4; 6, 744a37; V, 8, 788b20.
18 Cf. Hatfield, "Review Essay," who criticizes Garber (*Descartes' Metaphysical Physics*) because he "does not inquire into the origin of the concept of physics assumed in the book itself," and such inquiry "might have brought the difference between Cartesian and Newtonian physics into stronger relief," 9.

leaves the thrower's hand, appears to us as central to physics because, standing on this side of its history, we live with a physics that defines it as such. But there is nothing *necessary* about this supposed centrality. Aristotle's physics defines *its* problems in a way that renders this problem marginal, and *consequently,* he never treats it directly but always "in parentheses." Thus – and here is the crucial issue for the history of physics and philosophy – his "parenthetical" treatment occurs *not* because he cannot solve the problem and so chooses to sidestep it, but because as a problem it is of little interest within his physics. The failure to understand this background in Aristotle's work is simultaneously the failure to understand the project *we* call physics and how it defines its characteristic problems, significant phenomena, and solutions.

Here is the methodological crux of my thesis. We do not understand Aristotle's position until we understand its starting points, the definition of problems that it presents, and the solutions to those problems as they are presented through his actual practice of physics. Furthermore, if we fail to understand Aristotle's position, we also fail to understand the history of choices that have led to our own physics, and so we fail to understand our own scientific surroundings as shaped by the history of choices that ultimately rejects Aristotle's physics.

The world that Aristotle constructs is, as it must be, a world very different from the world that appears within the constructs of contemporary physics. Aristotle presents a cosmos with the earth stationary at its center, the stars going around eternally, and each element – earth, air, fire, and water – is moved always toward its natural place. Because it is like form, place renders this cosmos determinate in respect of direction, i.e., up, down, left, right, front, and back; and all things within the world are determined in respect of place. In this world, nature operates always and everywhere as a cause of order.

That this world is so strange and so different from our own – and yet so coherent and responsive in its own right – tells us something important about the nature of conceptual starting points, the definition of problems, and the power of arguments. Among other things, it tells us that a position constructed from a starting point with extraordinary appeal and developed through coherent, even elegant, arguments may be entirely wrong. It also tells us something about the nature of criticism and history – that history has a way of making our own world and its familiar problems look real and absolute ("given as a fact") rather than the outcome of a set of historical choices, which, as historical, may be compared and evaluated.

These points taken generally seem neither surprising nor profound. Yet they have not in their particulars been appreciated in the history of physics, perhaps because as a science physics claims to yield not a moral or a legal world, but the "real" world. A recent essay, which falls within a long-standing tradition of analysis of Aristotle's physics, makes the latter claim explicit: "Aristotle's ideas so far appear to us as hardly more than cardboard cut-outs, because, so far, I have said nothing about the origins of Aristotle's theory. I began instead (as most histories of science do begin) with the facts."[19] It is against this background that I present a quite different account of Aristotle's analysis of nature as always and everywhere a source of order: a conception of physics as the construction of a world complete with its starting point, the definition of its problems, and the development of solutions to those problems – and all the choices entailed thereby.

My analysis focuses on the two fundamental features of Aristotle's account of nature: "place," which I examine in Part I, and the elements, along with their respective "inclinations," which occupy Part II. At the conclusion of the examination of the elements, I shall also examine the account of potency and actuality in *Metaphysics* IX insofar as it bears upon the elements. Taken together, I shall conclude, place and the elements constitute nature, and so an examination of them exhibits nature as everywhere a cause of order.

Place, according to Aristotle, resembles form and is the first limit of the containing body (*Physics* IV, 4, 211b10–13; 212a20). Hence, the cosmos is intrinsically directional: "up," "down," "left," "right," "front," and "back" are not just relative to us but are given in the cosmos itself. I shall argue that place is the formal constitutive principle that renders the cosmos directional in this sense and so constitutes all place within the cosmos as "up," "down," and so forth.

Within the cosmos all things, whether made or natural, animate or inanimate, are made up of the elements – and there is nothing material outside the cosmos. Each element possesses a specific nature, its "inclination": the active orientation of each element toward its proper place, e.g., up or down. Consequently, place and inclination work together: place constitutes the formal limit and directionality of the world – the "where things are" – while each element is limited and possesses inclination toward its proper place. As a result of this partnership, natural motion in things (in the absence of hindrance) is completely regular: it exhibits the order of nature and the account of nature as itself orderly.

The account of place found in *Physics* IV and the account of the elements and their relation to place found in the *De Caelo* yield not only regular natural motion, but the cosmos itself. Because (as Aristotle argues) there can be only one cosmos within which the elements comprise both natural things and artifacts, an account of place and the elements yields a universal account of nature and of all things that either are by nature or, like artifacts, are composed of natural things. In *Metaphysics* IX, Aris-

---

19 O'Brien, "Aristotle's Theory of Movement," 73.

totle examines the notions of potency and actuality (the terms that define motion), claiming that these terms are important not only for becoming but also for being; he concludes that the elements imitate the heaven because they too are ever active (*Metaphysics* IX, 8, 1050b28–30). In short, by examining the accounts of place and the elements as they appear in these texts, we examine the constitution of the natural world itself insofar as it is limited, determinate, and orderly in respect to both its becoming and its being.

The sense in which the cosmos and all things within it are constituted by place and the elements comprises the heart of this study. I here anticipate my analysis more fully, turning first to the method of analysis utilized within this study and then to a brief outline of its substantive parts.

Methodological Assumptions/Methodological Restrictions

Any study of Aristotle must address the method of analysis to be used. Several are available. The "genetic" or developmental method informed much of the scholarship on Aristotle's work earlier in this century and has recently been revived. This method argues in effect that one cannot analyze any given argument in Aristotle's corpus without first establishing a chronology for the development of his thought across the entire corpus.

In contrast to the genetic method is the "acontextual" method: any given argument in Aristotle may be analyzed without referring to the corpus as a whole or even to the local context of the argument. On this method, all arguments are, as arguments, subject to the same analysis and thus comparable to one another. Taking a problem such as motion in a void, one can ask how – or if – Aristotle solves it in a given argument and how his argument compares to any other, e.g., that of Philoponus or of Newton.

I shall propose a third method, a "method of subordination." It is generally agreed that Aristotle wrote *pragmateiai*, or *logoi* – what are now referred to as "books" (e.g., *Physics* I, *Physics* II, *Metaphysics* IX), and that these were later bound into treatises, for example, the *Physics* and the *Metaphysics,* which now comprise the corpus traditionally termed "Aristotelian."[20] I shall argue that at or near the beginning of each *logos* Aristotle announces a problem or topic and then develops arguments that define and/or solve that topic or problem. Replies to objections, criticisms of alternate views, and so on follow. Thus arguments are subordinated to the particular topic or problem that they address. With this method, no decision need be made about the corpus as a whole to understand an

---

20 For a brief but helpful account of the issues surrounding the Aristotelian corpus, cf. Barnes, "Life and Work," 6–15.

individual treatise (unlike the genetic method); but because arguments are subordinated to a specific problem or topic, they cannot be read independently of their relation to this topic or problem (unlike the acontextual method). Before I give a fuller account of this method, it will be useful to consider the genetic and acontextual methods and the methodological problems posed by them.

The genetic method originated in the discovery of the chronology of Plato's dialogues, which revolutionized Platonic studies early in this century. It seemed natural to hope for similar success from a chronology of Aristotle's *logoi*, and the task for any analysis of Aristotle's texts, or arguments, came to be defined first as establishing a series of chronologically defined periods in which the development of his thought occurred.[21] The most influential study of this sort was and remains Jaeger's *Aristotle: Fundamentals of the History of His Development*.[22] This study and its method, i.e., construing the substance of Aristotle's philosophy via an account of his development, dominated studies of Aristotle's corpus for upwards of three decades.[23]

However, virtually every page of Jaeger's study has been criticized; in response to his thesis, diverse, even contradictory, accounts of "Aristotle's development" appeared. Solmsen, for example, argues that the *Timaeus* represents a new turn for Plato's academy and is a starting point of sorts for Aristotle's physics – while de Vogel argues, in the same volume, that Aristotle did not start out as a Platonist, but moved toward a Platonism.[24] In a more radical critique of Jaeger, Frank rejects the entire thesis of an early Platonizing Aristotle, arguing that several of Aristotle's central concepts could not possibly have originated in Plato's views.[25] (Such claims raise two further problems, first what constitutes Platonic or anti-Platonic positions and then whether Plato's thought evolved.[26]) Recently, the entire situation has been reviewed by Rist, while Irwin concludes: "I see no

---

21 The suggestion has recently been made that this project was first defined by Thomas Case in "Aristotle," 501–522, and "The Development of Aristotle," 80–86. On Case, cf. Wians, "Introduction," ix, and Chroust, "The First Thirty Years of Modern Aristotelian Scholarship (1912–1942)," 41. This point is also made by Ross, "The Development of Aristotle's Thought," who connects the discovery of a chronology for Aristotle to the success experienced in Platonic studies, 1–17; Irwin, *Aristotle's First Principles*, 11.

22 The importance of Jaeger's work is acknowledged in virtually every essay in Wians, ed., *Aristotle's Philosophical Development*.

23 For one example (of many), Guthrie criticizes the particulars of Jaeger's view, but wholly accepts the project as Jaeger defines it in "The Development of Aristotle's Theology," *passim*. A recent study refers to Jaeger as presenting the "dominant paradigm" for theories of the development of Aristotle; see Bos, *Cosmic and Meta-Cosmic Theology in Aristotle's Lost Dialogues*, 97–112.

24 Solmsen, "Platonic Influences in the Formation of Aristotle's Physical System," 213, and de Vogel, "The Legend of the Platonizing Aristotle," 248–256.

25 Frank, "The Fundamental Opposition of Plato and Aristotle," 34–53; 166–185.

26 Owen, "The Platonism of Aristotle," points out that "The catchword 'Platonism' . . . is too often taken on trust and too riddled with ambiguity to be trusted," 200. The evolu-

good reason to believe that he [Aristotle] spent most of his time deciding whether to agree or disagree with Plato, and hence I doubt if attention to debates with Plato or Platonism is likely to explain his philosophical development."[27]

In the face of these problems, recent practitioners of the genetic
method sometimes abandon the Platonic/anti-Platonic dichotomy and
define "Aristotle's development" by identifying issues on which Aristotle "changes his mind." Irwin, for example, suggests that Aristotle changed
his mind about the possibility of a universal science, and Graham finds
a profound change signaled by the presence or absence of the word
"matter."[28] Indeed, in his introduction to a recent collection of essays
on the problem of "Aristotle's development," Wians suggests that the
work of Graham and Rist "may signal a renewed interest in develop-
mentalism."[29]

It seems to me, however, that the contradictory substantive results produced by the various applications of the genetic method are symptomatic
of serious problems with the method itself. Expressed generally, a genetic
method requires a sufficiently unified chronological origin for each book
within Aristotle's corpus because such an origin is presupposed by the notion of a chronological order among all the books. This method was ef-
fective for Plato's dialogues precisely because of their remarkable character as individual compositions.[30] Thus if a chronology of individual
books could be established with certainty, then it would reveal first the sequence of Aristotle's ideas and then the substantive implications of that
sequence.

The assumption of a specific chronological origin for each book cannot be sustained, however, because these books are not compositions in
the sense of being written at a certain fixed date. Indeed, as Moraux emphasizes, historically they were never thought of as literary units.[31] Rather,
they are almost universally thought to be teaching texts that were revised
and/or extended frequently – perhaps throughout Aristotle's teaching
life. Consequently, they are not fixed "datable" literary compositions, but

tion of Plato's ideas and its effect on the analysis of Aristotle's development is raised by
Pellegrin, "The Platonic Parts of Aristotle's *Politics*," 348.

27 Rist, *The Mind of Aristotle;* Irwin, *Aristotle's First Principles,* 12.

28 Irwin, *Aristotle's First Principles,* 13; Graham, *Aristotle's Two Systems,* vii, 15, *passim.* For a
"topology" of views of Aristotle's development, cf. Witt, "The Evolution of Developmental Interpretations of Aristotle," 67–82.

29 Wians, Introduction, ix; the claim is also found in Graham, "The Development of Aristotle's Concept of Actuality," 551–564; cf. Menn, "The Origin of Aristotle's Concept of
Ἐνέργεια," 73–114. Both Menn and Graham are criticized by Blair, "Unfortunately It Is
a Bit More Complex," 565–579.

30 On the character of Plato's writing, cf. B. Lang, "Presentation and Representation in
Plato's Dialogues," 224–240; Hyland, "Why Plato Wrote Dialogues," 38–50.

31 Moraux, *D'Aristote à Bessarion,* 67–94. This point is granted by Graham, *Aristotle's Two Systems,* although he claims to be able to overcome it, 4–14.

ongoing discussions that incorporate shifts in interests, newly devised (as well as older) objections, new questions, and so forth.[32]

Because it seeks to establish a chronology and each *logos* contains parts (re)written at different "periods," the genetic method divides each *logos* into parts with different historical origins.[33] Far from establishing the historical sequence of Aristotle's writings and so revealing substantive relations among his ideas, this method leaves the reader with a pastiche of writings that derive from different "historical periods" and are unintelligible as arguments. Grene sums up the problem: "The whole procedure finally issues in a sort of Heraclitean flux: from one page to the next one is never reading the same Aristotle, and finally there is no Aristotle left to be read at all."[34]

To address the problem of reconstructing the arguments from this pastiche, chronological assumptions require further systematic or substantive assumptions.[35] In their turn, these substantive assumptions often lead to a reevaluation of the dates assigned to different arguments within each *logos*.[36] But at this point, the situation exactly reverses the position from which proponents of the genetic method set out nearly a century ago: rather than a firm chronological foundation establishing the sequence of Aristotle's ideas, substantive decisions determine the chronology of texts.

The problem endemic to all treatments of Aristotle by the genetic method lies in the fact that its proponents face two related but different questions: (1) the chronological question, i.e., (assuming that he did write them) when did Aristotle write the *logoi* comprising his corpus? and (2) the substantive question, i.e., considered in terms of their content, which of Aristotle's ideas are early and which late? If there were some in-

---

32 On Aristotle's school as the origin of these teaching texts, cf. Lynch, *Aristotle's School.*

33 For evidence on how Aristotle's writings were viewed in ancient times, cf. Lynch, *Aristotle's School,* 89, and Düring, "Notes on the History of the Transmission of Aristotle's Writing," 37–70.

34 Grene, *Portrait of Aristotle,* 27–28. Graham, *Aristotle's Two Systems,* makes the same point: "Furthermore, there seems to be in principle no terminus to the successive subdivision of Aristotle," 6.

35 For two examples of the problem here, cf. Guthrie's introduction to *On the Heaven.* In an earlier article, "The Development of Aristotle's Theology," Guthrie proposes a view of "pure materialism" (169) but later revises his view, saying that the term "materialism" is "too positive in expression" (*On the Heavens,* xxxi n. a; xxxiv-xxxv). In both articles, he claims that Aristotle has difficulties reconciling various aspects of his views ("The Development of Aristotle's Theology," 98; introduction to *On the Heavens,* xxxiv). Elders, *Aristotle's Cosmology,* refers throughout to serious problems created by bungling editors.

36 This widespread practice is impossible to detail in a note. It permeates the commentary on the *De Caelo* by Elders. In Jaeger, *Aristotle,* for one example among a myriad, cf. 219–221 where Jaeger argues that Aristotle held a Platonic view of metaphysics in *De Philosophia* and in *Metaphysics* XII; also 140, 151 ff. Jaeger has been very ably criticized by a number of scholars. For one example, cf. Mansion, "La genèse de l'oeuvre d'Aristote d'apres les travaux récents," 307–341; 423–466. For a critique of Jaeger on philological grounds, cf. Von Staden, "Jaeger's 'Skandalon der historischen Vernunft,'" 227–265.

dependent evidence, such as internal historical references, on which to base a chronology of Aristotle's writings, then we could deduce which positions are early and which late; conversely, if there were some independent evidence, such as testimonia by his students, as to which positions he held early on and later rejected, then we could deduce which writings are early and which late. But there is no known evidence for either question. Hence proponents of the genetic method *per force* work with two unknowns. Jaeger, for example, argues that *Metaphysics* XII, 8, must be early because of its apparent polytheism; therefore, he removes it from *Metaphysics* XII so that *Metaphysics* XII, 9, may connect directly to *Metaphysics* XII, 7.[37] Wolfson argues that it may be Jaeger, not Aristotle, for whom polytheism is "early" – and that there is no substantive ground, other than Jaeger's own prejudice, for dating this chapter as early.[38] In effect, utilizing one unknown to establish another unknown cannot produce a known result – only an arbitrary one.

The conjunction of claims that mix chronology and substantive assumptions when neither can be established independently of the other, produces a second, equally serious, problem for the genetic method: acknowledgment, assessment, or defense of substantive assumptions is suppressed. The genetic method takes chronology as its primary and explicit criterion for analysis of Aristotle's arguments but then of necessity resorts to substantive assumptions. These assumptions, however, remain largely unjustified, in some measure *because* of the initial emphasis on chronology. Consequently, when the relation between chronological and substantive elements is reversed, unjustified substantive assumptions determine the chronological "period" of Aristotle's writing and hence the sequence of ideas that presumably represents his development. And this situation is intolerable: unjustified substantive assumptions in effect produce the moments that constitute "Aristotle's positions."

The views of Graham are interesting because he explicitly claims to use both chronological and substantive assumptions. Rejecting the Platonic/anti-Platonic distinction, he says he will argue for two distinct systems (based on the presence or absence of a notion of matter) and that these systems stand in a genetic relation.[39] At the end of his first "methodological"

---

37 Jaeger, *Aristotle*, 345 ff. Graham criticizes practitioners of the genetic method (Jaeger in particular) because "they almost invariably omit to give any *philosophical* reasons for the change" (*Aristotle's Two Systems*, 6, italics in original). Such a claim is remarkable; cf. Barnes's comments on Jaeger's philosophical analysis, "Life and Work," 16–17.

38 Wolfson, "The Plurality of Immovable Movers in Aristotle, Averroes, and St. Thomas," 1–21. I have argued that the logic of *Metaphysics* XII requires that XII, 8, be left where it is, H. Lang, "The Structure and Subject of *Metaphysics* Λ," 264–267. For another view, including Jaeger's claim that it is early, cf. Modrak, "Aristotle's Epistemology," 167–168.

39 Graham, *Aristotle's Two Systems*, 15, 19. Graham is criticized by Wildberg, "Two Systems in Aristotle?" and Goldin, "Problems with Graham's Two-Systems Hypothesis"; Graham replies to these criticisms in "Two Systems in Aristotle."

chapter (pp. 18–19 of more than 300), he asserts: "The study of Aristotle's two systems raises fundamental problems concerning his philosophy: its unity, its development, its overall coherence. There are problems that confront us on page one of the *Categories* and which should vex us to page 1462 of the *Poetics*." Graham begins his next page, and chapter, with a distinction between the two systems. In short, "two systems" are not the conclusion of the study, but its starting point: a starting point, I would suggest, that produces the "fundamental problems" that "confront us" on virtually every page of Aristotle's corpus.

Here is the heart of the matter: analysis and evaluation of arguments require that the method of the analysis and the criteria for the evaluation of the arguments be prior to any substantive decisions. And insofar as the genetic method fails to justify and/or to evaluate substantive assumptions, it fails as a method for establishing and analyzing Aristotle's arguments.

Setting aside the genetic method, we may consider the problem itself of analyzing Aristotle's arguments. The analysis of arguments and the prior task of providing criteria for such analysis do not require – are quite independent of – determinate chronological origins. In short, Aristotle's texts and their arguments must be considered as such, i.e., as arguments. The problem, of course, is how to understand these texts as arguments without producing in another guise the same arbitrariness that the genetic method does.

If ever a method could claim to consider Aristotle's arguments as such, the "acontextual method" would seem entitled to. On this method, which now largely dominates Anglo-American Aristotelian studies, any given argument in Aristotle can be independently analyzed, evaluated, and compared to other arguments. After so analyzing Aristotle's arguments, Waterlow, for example, concludes that there is a major inconsistency between *Physics* II, 1, and *Physics* VIII, 4;[40] Hussey concludes that Aristotle's first objection to the persistence of motion in a void received devastating criticism from Galileo whereas Aristotle's second objection contains an observation that accords with Newton's First Law.[41]

Like all methods, the acontextual method begins with an assumption in order to analyze the arguments; namely, that the context in which an argument is formulated is irrelevant to it as an argument and so may be disregarded. Consequently, practitioners of this method first analyze "Aristotle's arguments" in abstraction from the initial problem posed by him and then evaluate the "success" or "failure" of these arguments in terms of various interests. Such analysis and evaluation lie at the heart of this method.

40 Waterlow, *Nature, Change, and Agency in Aristotle's Physics*, 193, 240, and chap. 5, 204–257; for a similar argument, cf. Charlton, 92.
41 Hussey, 130. Another comparison of Aristotle and Galileo appears in O'Brien, "Aristotle's Theory of Movement," 47–86. Ostensibly arguing about Aristotle's development, Cleary raises the problem of "the mechanical motion in the universe" in "Mathematics and Cosmology in Aristotle's Development," 227.

Yet this assumption presents its own problems and in some ways produces results not unlike those of the genetic method. Treating any particular argument independently of the context in which it occurs first (1) views as given, or obvious, the formal structure that constitutes an argument and then (2) abstracts that argument from any immediate relation to other arguments. But there are corresponding objections to each of these. (1) Aristotle's arguments are often exceedingly complex in their formal structure; for example, he may pause to provide a proof of a premise before proceeding with the larger argument in which that premise operates. An undefended assumption of what constitutes an argument in effect construes this structure without justification. And such a construal determines substantive features of the argument. Consequently, substantive features of an argument result from undefended assumptions.

(2) The use of abstraction, i.e., treating an argument as independent of its context to facilitate formal analysis, has its origins in logic. Logic abstracts arguments from specific conceptual content of the argument as a condition for determining validity, which is entirely formal, i.e., content free. As a tool for formal analysis, abstraction is enormously powerful. But it achieves this power by eliminating all reference to meaning, because the meaning of arguments rests on the domain of the premises as well as on the concepts at work within them; and these are irrelevant, even obstructive, to logic, which produces an analysis of the formal relations among the terms of an argument to determine validity.

As applied to the arguments of Aristotle, however, the acontextual method pursues abstraction not to determine validity alone but to analyze the content of particular arguments. In so doing, this method would extend the power achieved by logical analysis *via* abstraction from the purely formal realm to that of meaning. But such an extension fails to recognize both the potential and the limits of abstraction: its strength derives from excluding content that is irrelevant to its task; consequently, abstraction is limited to the important but not exhaustive issue of analyzing arguments in regard to their validity.

Here we reach the heart of the matter: if an argument is to have meaning, it must address a topic or solve a problem. Because the acontextual method abstracts individual arguments from their larger context and the problems that can be identified there, the arguments stand in need of a topic or problem. This need is met *per force* by the commentator. For Hussey, the problems of physics derive from Newton, for others from Philoponus or someone else.[42] Once an argument is treated independently of

---

42 For an example of reading Aristotle in terms of problems defined by Philoponus, cf. De Groot, *Aristotle and Philoponus on Light*. I criticize her treatment of Aristotle in my review of this book, 190–192. For an example of reading Aristotle in terms of problems defined by Galileo, cf. O'Brien, "Aristotle's Theory of Movement," 47–86, and Lewis, "Commentary on O'Brien," 87–100.

its context in Aristotle's *logos,* a wide range of problems may provide the terms of its analysis.

The acontextual method claims to treat Aristotle's arguments as such, i.e., as independent arguments. But if "independent" means independent of a problem or topic, then the claim is false because all arguments produce a conclusion that (except for the entirely formal question of its validity) must bear upon a topic or problem. And if "independent" means independent only of Aristotle's writing but dependent on any question that can be posed, then, as with Jaeger and the genetic method, however astute the questions posed may be, they bear no necessary relation to Aristotle's arguments.

Here we find the arbitrariness of the acontextual method. Problems such as motion in a void or projectile motion are not formal, but substantive. When posed for Aristotle's arguments, the assumption is that the arguments must in some fashion address these problems. But such problems are assumed prior to the analysis of the arguments that presumably address them and, as prior to analysis, they are not themselves justified. Because all problems are equally possible once an argument is analyzed apart from its larger context, there can in principle be no justification for posing one problem rather than another. As substantive assumptions, the problems posed by this method are no less arbitrary than are the substantive assumptions of the genetic method.

The difficulties with the acontextual method do not end here. Abstraction of individual arguments produces a pastiche of arguments not unlike that produced by the genetic method. When Hussey finds in *Physics* IV, 8, the object of Galileo's criticism followed by an anticipation of Newton, any sense of a coherent position has disappeared. And the same may be said for, among others, Waterlow: on her reading, it is difficult to find a coherent position in Aristotle's arguments. Presupposing a problem in order to evaluate an argument conjoins that argument to the concerns presented by that problem, e.g., Newtonian or modern concerns. Thus, the argument is not treated "acontextually"; there *is* a context – but it turns out to be that of the reader, not of Aristotle's *logos.* In short, the acontextual method (1) presupposes an abstraction that is incompatible with analysis of specific meaning, (2) poses problems that present unjustified substantive assumptions, and (3) recontextualizes these arguments without justification.

My own "method of subordination," I would claim, avoids the major problems of both the genetic and the acontextual methods. It too begins with an assumption, which, although expressed *vis-à-vis* the Aristotelian corpus, I would propose for *any* analysis of texts (and arguments) in the history of philosophy: the interpreter must assume at the outset that the apparent coherence of Aristotle's arguments (and writing) is in fact their primary characteristic; the conclusion that arguments (or texts) fail to be coherent can only be a last resort and must rest on analysis of the arguments themselves as presented within the texts.[43] Indeed, I shall conclude on the basis of the analysis that I shall elaborate that Aristotle's arguments about nature, motion, place, void, and inclination are remarkably consistent and present a fully developed account of nature.

Furthermore, I shall argue that the methodological assumptions that I utilize throughout this study – and to which I now turn – are both minimal and warranted by the texts themselves. They solve a number of difficulties traditionally associated with these arguments that on other accounts appear hopeless. And readers can evaluate my account directly on the basis of a principle of coherence – the coherence of my own assumptions and the coherence of the account of Aristotle's position at which I arrive by means of those assumptions.

Coherence is a general philosophic criterion to which virtually all systematic philosophers lay claim. Those who reject it, like Nietzsche, do so precisely because it is so intimately identified with philosophy as systematic and because they wish to challenge that conception of philosophy. And the main outlines of coherence as a general philosophic criterion are not difficult. Within arguments, conclusions must follow from their premises. Across arguments, concepts at work and conclusions based on them must be compatible; they must solve the problem at hand and, ideally, will do so completely. Taken at its broadest moment, a coherent position presents the world itself (whether the starry world without or the moral world within) as a whole within which each part is accounted for with a minimal number of compatible starting points. Aristotle, I shall conclude, when he takes nature as a cause of order, produces a remarkably coherent set of arguments – and a remarkably coherent world.

Even assuming a general notion of coherence, we must recognize that coherence and a notion of being systematic are spelled out differently by different philosophers. The task of establishing the exact meaning of "coherence" for any moment in the history of philosophy must be prior to any analysis of that moment. Although I focus on Aristotle, I would suggest that many of the points in the theory of interpretation I apply here would apply equally to Plato and reveal the conception of philosophy unique to Plato and Aristotle (and perhaps the Presocratics).

For Aristotle (and Plato) the task of philosophy is to address particular topics and the problems that they represent. In Plato's dialogues individuals pose problems to one another; in Aristotle's books the opening

---

43 Grene, *Portrait of Aristotle*, 33, expresses the point well: "The Aristotelian corpus must have, in its main outline, something like the order it says it has. Secondly, even if scholars can ferret out earlier lines of organization, the corpus as we have it does, as I have just said, represent, in most subjects at least, the lecture course at the Lyceum as Aristotle conceived it after the definitive period of biological research. And so we are entitled to study it as such, whatever its genesis."

line announces the problem at hand and thereby defines the conversation or lecture for which the arguments that follow are developed. And the coherence of Aristotle's arguments and his position as constituted by them lies here: topics or problems are announced, and then arguments are developed solely to resolve them.

The way in which Aristotle is all business in his opening lines bears witness to the force of the lines in defining the issue at stake in the subsequent arguments. Perhaps for this reason many of these opening lines are memorable. "All men by nature desire to know . . . " (*Metaphysics* I, 1); "there is a science of being *qua* being . . . " (*Metaphysics* IV, 1); "substance is the subject of our inquiry" (*Metaphysics* XII, 1). The *Physics* presents the same economy: "Of things that are, some are by nature and some by other causes" (*Physics* II, 1); "Nature is a principle of motion and change and it is the subject of our inquiry" (*Physics* III, 1); "The physicist must have a knowledge of place, too, as well as of the infinite . . ." (*Physics* IV, 1); "Everything moved must be moved by something" (*Physics* VII, 1); or, the rhetorically extended:

Did motion come to be at some time before which it was not, and is it perishing again so that nothing is moved? Or is it neither becoming nor perishing, but always was and always will be? Is it in fact an immortal never-ceasing property of things that are, a sort of life as it were to all naturally constituted things? (*Physics* VIII, 1)

This rhetorical practice – announcing the topic immediately – appears regularly in Aristotle's *logoi*. (Plato's dialogues too present a topic announced early on, e.g., what is knowledge, what is justice, pleasure, courage, piety.)

After announcing his topic, Aristotle either pauses to indicate its importance or proceeds with his argument immediately. His pauses are very brief – in *Metaphysics* XII, 1, ten lines emphasize the importance of the investigation and in *Physics* III, 1, twelve lines.[44] In *Physics* III, 1, these lines do considerable work, indicating the importance of the topic at hand and listing the specific topics that occupy *Physics* III-VI. When Aristotle does not pause, the directness with which the argument begins can be startling. So *Physics* II, 1, opens:

Of things that are, some are by nature, some from other causes; for by nature are both animals and their parts and plants and the simple bodies, such as earth and fire and air and water, for we say these things and other such are by nature; for all of these seem to differ from those not constituted by nature.

Physics VII, 1, is even more striking: "Everything moved must be moved by something. For if it does not have the source of motion in itself, then it is evident that it is moved by something other than itself for there will be some other mover."

---

44 I have considered the rhetorical structure of *Metaphysics* XII in H. Lang, "The Structure and Subject of *Metaphysics* Λ," 257–280.

Often, although not always, the end of the initial argument follows in short order. For example, in *Physics* II, 1, having announced that some things are by nature whereas some are from other causes, Aristotle defines nature as a principle or cause of being moved and being at rest and takes up a possible counterexample to his definition, a doctor who heals himself. All of this occupies twenty-five lines, and he concludes "Nature then is what has been stated; things have a nature which have a principle of this kind." The remainder of the *logos* takes up further problems concerned with nature – e.g., is nature more properly identified with matter or with form? what is the science of things that are by nature? what are the causes of such things? are chance and spontaneity causes of these things?

As I shall later consider these arguments, another example may be useful. *Physics* III, 1, opens by announcing its topic: "Since nature is a source of motion and change and our investigation concerns nature, what motion is must not remain hidden" (200b12–14). Motion in its turn requires a number of universal and common terms, including "the continuous," "the infinite," "place," "void," and "time" (200b14–24). "Proper terms," Aristotle asserts, will be examined later (200b24–25). Fourteen lines after announcing the topic, he begins his analysis of motion, asserting its primary characteristics, defining it and criticizing alternative views. An examination of the infinite completes *Physics* III. The remaining common and universal terms form the explicit topics of *Physics* IV through VI. In each case, the same pattern emerges: an initial announcement of the problem to be examined followed immediately (or almost immediately) by the analysis itself.

Looked at methodologically, this point could hardly be more important. Aristotle raises the topic or problem that he will address at the outset of his arguments, and his arguments are then directed toward solving it. In some cases, e.g., *Physics* II, the topic announced at the outset (i.e., things that are by nature) dominates the entire book. In other cases, e.g., *Physics* III or *Physics* IV, one topic may succeed another, although in these cases, each topic is sharply delineated. For example, in *Physics* IV, after discussing place (*Physics* IV, 1–5), Aristotle begins his account of void (*Physics* IV, 6), and after rejecting the void, he turns to the problem of time (*Physics* IV, 10). Because each argument is defined by the problem at hand, Aristotle's arguments are topical in structure and substance. That is, they address the problem announced at the outset.

The topical character of Aristotle's arguments – and of his philosophy conceived more broadly – is central to the construal of his arguments. His announced topics define the logical units of his arguments because the announced topic (or problem) to be addressed (or solved) determines what is at stake in the argument or what the argument is designed to prove. In short, as systematic, Aristotle's arguments operate within the topics that open the discussion.

Furthermore, these topics or problems determine the concepts at work in the arguments subordinated to them. For example, in *Physics* II Aristotle defines nature as a source or cause of being moved and being at rest. He then turns to the question of whether nature is more properly identified with form or with matter. Nature is not undefined in this inquiry, but presupposes the definition just established. The argument makes a further determination of what nature is. After identifying nature primarily with form, although it must also include a reference to matter, Aristotle raises further problems that presuppose and follow the definition of nature and its specification as form plus a reference to matter. In this sense, the subsequent arguments may be called "subordinated" to the definition and specification of nature reached in *Physics* II, 1.

One may justly ask what evidence supports this claim of "subordination." The strongest evidence is presented, I believe, in the chapters that follow: it makes sense of Aristotle's arguments, and it solves a number of problems insoluble with other forms of analysis. But there is some *prima facie* evidence for this claim, although it is admittedly speculative.

In an oral tradition, the beginning provides the mnemonic key on which the rest hangs.[45] And as each point is proven, it provides another mnemonic marker for the listener. Rather than requiring that the listener remember a series of points while waiting for the conclusion that provides the logical relation among them, the determining point appears first; consequently, the listener immediately possesses the unifying principle for the argument that follows. Arguments do not build up to a main point, they follow from a main point. In this sense, rhetoric reflects the logical structure of argumentation.[46]

The difference between this and later procedures is telling. Hellenistic philosophy defines the task of philosophy as combining the multiple (often apparently conflicting) truths established by Plato and Aristotle. Thus philosophy no longer addresses topics or solves specific problems; now it forms coherent systems across problems and texts.[47] Byzantine commentators, such as Alexander of Aphrodisias, Philoponus, and Simplicius, further develop this conception of philosophy as systematic. In their commentaries, a single line from Aristotle can evoke a commentary of several pages with quotations from other arguments of Aristotle and/ or other authors. The conclusion, and the gratification that it brings, can

---

45 For the evidence concerning Aristotle's school and the practices there, cf. Lynch, *Aristotle's School*, 68 ff.
46 Cf. H. Lang, "Topics and Investigations," 416–435.
47 Pfeiffer, *History of Classical Scholarship*, 152, argues that the Hellenistic philosophers invented "scholarship"; Chroust argues that after Aristotle's death "the commentary in a way took the place of Aristotle the teacher," in "The Miraculous Disappearance and Recovery of the Corpus Aristotelicum," 55; I am suggesting a further point, namely, that the conception of philosophy itself was redefined.

be considerably, but not indefinitely, delayed because the reader has, quite literally, a text to hold on to.

When the rhetorical presentation of the argument changes, the logical role of elements within the argument also changes. That is, empirical observation, thought-experiments, and so on, as well as the relation of these elements to the premises of the argument, are transformed, because writing broadens both the rhetorical and the logical framework of thinking. Within a tradition of written philosophy, these changes allow the postponement of the main conclusion in a way that is impossible within the framework (both rhetorical and logical) of an oral tradition. Although the issue lies beyond the bounds of this study, it could be argued that the broader logical framework available in a written tradition is a necessary condition for the Hellenistic conception of philosophy as systematic across topics, texts, and authors.

The logical structure of Aristotle's arguments – an announced topic supported immediately by a primary definition and/or argument – tells us what constitutes coherence for his arguments. First and foremost, Aristotle's philosophy is constituted by an examination of topics – what is nature; what is place, void, time; what is substance; what is happiness, pleasure, friendship; and so forth. In short, the subordination of arguments to topics implies that propositions or arguments cannot be emancipated from or compared to one another without reference to the topic that defines them. As I shall argue throughout, Aristotle's arguments, defined by specific topics, are remarkably coherent. And I suggest in conclusion that different positions in the history of philosophy or science can be compared by establishing criteria for evaluating the importance of different problems, topics, and starting points.

Furthermore, because Aristotle addresses important objections, views to the contrary, counterexamples, and so on, his arguments are not only orderly, but as topical investigations they are often complete. This is not to claim that they answer every possible question that can be raised – such would be an impossible task (and a ridiculous claim) – but that they are complete as direct proofs of a primary thesis supported indirectly by resolutions of important objections. Indeed, I would suggest that much of our fascination with Aristotle's arguments lies not in their particulars – the view that the earth is stationary at the center of the cosmos or that god moves as an object of desire – but in our sense of their rigorous economy, order, and completeness.

I have, in effect, accused both the genetic and acontextual methods of importing substantive assumptions into Aristotle's texts without justification. Because these assumptions determine "Aristotle's position" in significant ways, the analysis conducted by practitioners of these methods is arbitrary. The method of subordination, I shall now argue, cannot be accused of the same difficulty.

The first starting point of any argument is always identifiable within the text itself. It is neither imported from the outside nor requires the construction of a new whole. In its role as a first starting point, the topic cannot be defended on prior philosophical grounds – such is the meaning of a starting point. But I have provided here indirect, i.e., rhetorical, evidence, and the harder evidence lies ahead: I shall argue that Aristotle's arguments are, on this construal, remarkably coherent, consistent, and economical.

Because the announced topic defines the domain of the arguments that follow, the answer to the question of whether a given argument proves its point lies in the analysis of that argument. The criteria posed for such analysis must rest on the announced topic and on the question of whether or not the argument addresses this topic and operates consistently utilizing concepts appropriate to its defined domain. Analysis of arguments on this method fully reflects the assumption just identified, namely, the coherence of Aristotle's arguments. And it does not reflect more than that.

This method, however, raises two serious questions that must be addressed. (1) If a given argument produces more than one conclusion, which conclusion(s) represents the position of the author? Does a philosophic position necessarily include every implication that follows from its concepts and arguments? If so, then Aristotle is responsible for every variety of Aristotelianism – even though the varieties often contradict each other. But if the answer is no, then we require a dividing line between what constitutes Aristotle's position and what does not.

The answer, on my view, is that the structure of Aristotle's arguments provides the dividing line. Because his subsequent arguments and theses are subordinate to his primary thesis, that primary thesis defines the domain for conclusions that can be drawn from subsequent arguments. Hence the first task for any analysis of Aristotle's arguments – my first task here – is to determine the primary thesis of Aristotle's arguments. This step is prior to analysis itself because this thesis (along the concepts used to establish it) defines what can be derived from the arguments that complete the *logos,* insofar as they represent the implications of the position being developed there. In effect, the thesis defines the domain of subsequent arguments, and one can never derive more from an argument than the work it does in the service of a thesis. Thus in reading Aristotle, this thesis must be identified and its identification defended; the domain defined by it cannot then be violated.

This method may seem excessively restrictive. For example, Aristotle's definition of place and his arguments against the void were used in a variety of productive ways for centuries and are still being evaluated in light of Galileo's and Newton's analysis of motion. Does not the fact that Aristotle reaches a definition of place or that he claims [valid] conclusions about the void entitle one to a usage beyond the limits set by the initial topic of the investigation? The answer to this question requires a distinction. Claims about Aristotle's definitions and conclusions are substantive and so open to substantive debate. However, a method of subordination rests on procedural, not substantive, grounds. If one develops a position that one attributes directly to the texts of Aristotle, one cannot emancipate a substantive definition or even conclusions (such as those concerning place or void) from the thesis that it serves and still call it "Aristotle's argument."

Here we reach the second, even more pressing, question raised by the claim that Aristotle's arguments are subordinated to topics and are coherent, indeed systematic, *as* subordinated. (2) How can one establish any sense of a larger position across topics, or across *logoi*? Actually, there are two questions here: how can one understand Aristotle as possessing a larger position across topics or *logoi*, and how can one understand our own impulse to systematize across topics and/or texts? I shall address these issues in the conclusion, i.e., after the analysis of topics. But some sense of my response can be anticipated here. First and foremost, systematic relations among topics never themselves form a topic in Aristotle's writings.[48] So, for example, he says that the examination of "proper terms" comes after the common and universal;[49] but although he provides examinations of attributes, he never compares these different examinations or his arguments about the different kinds of attributes.

Yet there are numerous cross-references throughout the corpus. What are these references if not evidence of a system? They are evidence that Aristotle is systematic, not in the sense of making a system, but in the sense that he always assumes that "nature is everywhere a cause of order." Because this principle underlies different topics and investigations, their results are at least compatible and often seem to work together in a stronger sense. They do so, however, not because a "system" is being built, but because individual topics are investigated under the auspices of the same starting point. Speaking of Aristotle's corpus, Barnes comments: "Our corpus is not a strongly systematic body of work. Nonetheless Aristotle had systematic thoughts about the nature of the enterprise to which he was contributing. And here and there in his work the system peeps through."[50] That is, the systematic character of Aristotle's philosophy "peeps through" as the *outcome* of the subordination of arguments to topics and investigations based on the same first principle.

However, the absence of a system outside of topics renders systematic interests neither illegitimate nor irrelevant to Aristotle's arguments. It

48 Barnes effectively criticizes "the traditional supposition that there was an Aristotelian system" in "Life and Work," 22–26.
49 Aristotle, *Physics* III, 1, 200b23–24; cf. also I, 7, 189b30–32.
50 Barnes, "Life and Work," 26.

indicates that these interests originate independently of Aristotle's arguments and so must be defined in themselves and as relevant to Aristotle.
As I have suggested, such interests originate in a later conception of philosophy and not Aristotle's topics. This later conception then informs our
impulse to systematize across topics and across texts. Hence, we must acknowledge and take responsibility for the construction of systematic relations across topics and texts of Aristotle. Only if we do so can our interests and the analysis based on them acquire full and legitimate
philosophic status.

I shall suggest in a moment a sense in which Aristotle's writings may
be understood as forming a system across topics, and I shall argue for this
view throughout my analysis of *Physics* IV, the *De Caelo*, and *Metaphysics* IX.
But such a project must acknowledge that only particular topics, e.g., the
definition of place, are directly analyzed in Aristotle's arguments. Hence,
this notion of Aristotle's position as forming a system is not found in a
particular statement or argument and so it is not, strictly speaking, Aristotle's. It results from the method of analysis that I am proposing, the
method of subordination, and is in this sense – for better and for worse –
my responsibility.

Identifying my responsibility brings me to a more general project: the
history of philosophy. Using the structure of philosophic writing allows
one to establish the questions appropriate to Aristotle not so as to deny
the validity of further questions, but to identify *their* origin, philosophi-
cally as well as historically. Thus, I shall argue, Aristotle has not "failed"
to answer the important questions raised by Galileo or Newton; he has
succeeded in defining, arguing, and defending his thesis. At the same
time, we see that Galileo and Newton succeed in raising different questions and posing them in a form of philosophic discourse native to a different generation. If we declare Aristotle a failure because he has not answered Newton's questions, we lose *both* the force of Aristotle's arguments
and the originality of later thinkers. As I am proposing that we read it, the
history of philosophy can be understood both as discrete moments and
as continuous: new questions posed within the context of long-standing
traditions. And this project in its turn gives us philosophy itself: an understanding of arguments (and the world presented by them), their coherence (its coherence), logic (intelligibility), and ultimately a basis for
evaluating them (and the world itself) – an issue I shall take up in the
conclusion, i.e., after examining Aristotle's account of nature as a cause
of order.

In this study, I consider several topics raised by Aristotle: nature, motion, place, and the elements. Combining these topics within a single
book is, of course, itself an act of construal across topics and one, obviously, of my own doing. I shall argue that my analysis of Aristotle's arguments on these topics is itself defined by, indeed subordinated to, the

topic at hand in each instance. I shall also make the case for the order and propriety of these particular topics and texts. But I should like here to anticipate in a brief form the topical framework of the analysis that follows.

Topics and Analysis

"Things that are by nature" forms the primary topic of physics. Aristotle defines nature as containing within itself "a source or cause of being moved and being at rest in that to which it belongs primarily in virtue of itself and not accidentally."[51] But what exactly is included within – and excluded from – nature as the primary topic of physics? Furthermore, how does nature relate to the other topics raised within the science of physics?

Outside of nature, according to Aristotle, are all unmoved things, e.g., god and mathematicals. Nature includes plants, animals, their parts, and the four elements. Artifacts too, such as a bed or a cloak, which are not themselves by nature, are composed of the elements, which are by nature. Hence, whatever is true of natural things is also true of artifacts, insofar as they are composed of the elements.[52] In short, there are no moved things that are not in some sense "by nature," and no natural thing can be unmoved.

Physics III, 1, opens with an explicit connection between nature and motion: "Since nature is a source of motion and change and our investigation concerns nature, what motion is must not remain hidden" (200b12–14). And the account of motion begins with an account of things in motion because there is no motion apart from things (200b25–33). In effect, all moved things are either natural or composed of elements that are natural, and motion cannot be independent of things. Consequently, the domain of nature and the domain of things in motion are one and the same.

Herein lies the broadest definition of the science of physics: to understand nature as a cause of order is to understand moved things and to understand them as moved. And because "things that are by nature" and "motion in things" form the primary topics of Aristotle's physics, he defines them first. But the problem of motion requires several terms, such as "place" and "void," "without which motion [in things] seems to be impossible" (200b20–21). These terms are common and universal, and analysis of them must come before consideration of proper terms. As I shall argue, because nature and motion constitute the domain of physics and are defined first, analysis of common and universal terms, which are

---

51 Aristotle, *Physics* II, 1, 192b13, 21; III, 1, 200b12; VIII, 3, 253b5; 4, 254b17; *De Caelo* I, 2, 268b16; *Metaphysics* V, 4, 1014b19; 1015a14; VI, 1, 1025b20; *N.E.* VI, 4, 1140a15.
52 Aristotle, *Physics* II, 1, 192b-23, the opening lines of this discussion.

somehow required by motion, presupposes their definitions. Analysis of proper terms, Aristotle adds, will come later – but neither here nor later does he specify what these terms are (200b23–24).

The analysis of place and void, like that of all the common and universal terms, presupposes the accounts of motion and nature. And it presupposes these accounts precisely because they are proposed as terms without which motion seems to be impossible – and motion must be understood in order to understand nature, the proper subject of physics. As we shall see, Aristotle affirms that "the where" of things is place, and he defines place as a limit resembling form and in an important sense causing motion; he rejects the void as an incoherent concept that in every way fails to serve as a cause of motion. As I shall argue, the full force of Aristotle's account of "the where," including both his definition of place and his rejection of the void, require and are best understood in terms of the assumption of his definitions of motion and nature.

Aristotle opens *Physics* IV by asserting that all things that are and are moved must be "somewhere" [πού] (208a27–30). The problem is thus posed: what is "the where" of things that are and are moved? Two answers are possible: place and void. Place, I shall conclude, acts as a cause of motion, although it is not one of Aristotle's four causes. Place is the first limit of the surrounding body and thereby causes motion by rendering the cosmos determinate in respect to "where," e.g., up or down, while (as we shall see later) each element is ordered toward its proper place. Thus, place is a term without which motion is impossible and causes motion because it acts as a principle of determination.

Because place is a limit, many commentators identify it as the outermost surface of the cosmos. But Aristotle himself never calls it a surface. I shall argue that whereas all surfaces are limits, not all limits are surfaces: place is a limit that is not a surface. As a limit, place is a constitutive principle that is in every way unlike surface: it is not itself in place, it does not divide the contained from what is outside the contained, and as a topic, it belongs to the science of physics rather than to the science of mathematics. Indeed, place is a unique constitutive principle.

However, place is not the *exclusive* answer to the question "what is 'the where'?" until its most important competitor, void [literally, "empty"], is rejected. Thus void must be examined and the same questions asked of it as are asked of place. Indeed, people who propose "void" as "the where" do so because they think of it as a kind of place – place with nothing in it.

Consequently, the void has an odd logical status. It is both a competitor of place, posed as an answer to the question "where are things," and proposed as a kind of place itself. Its status as a proposed universal and common term reflects the former view, whereas Aristotle's analysis of it reflects the latter. And as a kind of place, void presupposes the necessary features of place that have just been established. In this sense, analysis of the void is subordinated immediately to that of place and ultimately to that of motion and nature.

First, Aristotle argues, the void is an incoherent concept that on every account produces a contradiction. Void is itself indeterminate, but is proposed as place, which is a principle of determination. As I shall argue, this rejection of the void as an incoherent term constitutes Aristotle's primary argument. However, after concluding unambiguously that the void is conceptually incoherent, Aristotle presents several difficult and (in)famous refutations of the claim that the void is a term without which motion seems to be impossible.

These arguments, I propose, deliberately presuppose a false premise (that the void causes motion) to show that in every case this premise produces a contradiction. Aristotle ultimately concludes that the assumption of a void not only fails to show that a void causes motion, but would render motion impossible. And the void fails, finally, because it cannot render the cosmos determinate: it cannot serve as a principle of order and so cannot be a principle of nature.

I shall argue for two conclusions concerning the logical status of these arguments. (1) This construal of the arguments solves the most important difficulties traditionally associated with them. (2) Because they presuppose a false premise, these arguments reveal nothing about Aristotle's own views. Even premises other than the void may be assumed solely for the sake of these arguments and so cannot be taken as constructive moments within Aristotle's physics without independent confirmation – and such confirmation must itself be derived from a constructive argument. In conclusion, I suggest a sense in which these arguments make an indirect contribution to Aristotle's position concerning the construction of the cosmos. With the rejection of the void, place is left in sole possession of the category "where," and place, by rendering the cosmos determinate, explains why it is inherently directional.

Although Aristotle says that an examination of "proper terms" will come later, he never identifies them as a topic of examination or even specifies what they are. In Part II, I argue that the De Caelo constitutes this investigation. The evidence for this claim lies in *its* opening lines, which identify the investigation as a continuation of physics that takes up the parts of the cosmos. "Proper terms" include (but need not be limited to) the inclination unique to each of the elements, e.g., up for fire and down for earth. Inclination constitutes the nature of each element and is in some sense "a source of motion."

An examination of the arguments of the *De Caelo* and *Metaphysics* IX involving the motion of the elements and inclination occupies Part II of this study. There I shall argue that just as the accounts of place and void presuppose the definitions of motion and nature, so the account of inclination presupposes the definition of place that has preceded it in the investigations comprising physics. (The void disappears because it has been rejected.) Because inclination presupposes Aristotle's definition of place, it also presupposes the definitions of motion and nature. Thus the examination of common and universal terms comes before that of proper terms because the latter require the former. In this relation, I find the basis for speculation concerning how Aristotle's larger position can be constructed: nature as a cause of order emerges not only in individual topical arguments but in this relation among them.

In *De Caelo* I, the relation between the elements, their respective natural motions, and place is used to prove the necessity of aether, the fifth element and the first body. And an account of the first body follows immediately. In *Physics* IV, the account of place as a formal limit refers to what place is a limit of, i.e., the first containing body, but does not provide an account of this body. *Physics* IV, as I argue, cannot take up the first body because it concerns a common and universal term, place, whereas *De Caelo* I investigates a proper term, the first body. Its account of aether completes the account of place in *Physics* IV by examining the first body that is bounded by place and referred to in its definition. I conclude that taken together, place and aether constitute the cosmos as determinate and exhibit nature as a cause of order.

In *De Caelo* II, each element is by definition identified with its own specific inclination because inclination constitutes its very nature. And whereas nature is defined as a source of being moved and being at rest in *Physics* II, so in *De Caelo* II, inclination is nothing other than an intrinsic orientation of each element toward its proper place. Consequently, in the absence of hindrance, fire is always moved upward by nature, earth downward, and the middle elements, water and air, to their respective places in the middle. In short, proper place and inclination are causally related: place is a principle of determination for the entire cosmos and as such constitutes all place within the cosmos as "up," "down," etc.; each of the four elements possesses an active inclination toward and whenever possible must be moved toward its respective proper place within the cosmos. This inclination is further identified with being heavy, i.e., earth, or light, i.e., fire. And, for example, in *De Caelo* II, 4, the relation between natural place, being moved, and inclination downward founds an argument, which I shall examine in detail, that the earth must be a sphere.

As the intrinsic ability to be moved, inclination contrasts with force, which is an extrinsic source of motion, e.g., when a stone is thrown upward. Aristotle provides an account of motion as produced by extrinsic force that is often read in conjunction with the argument concerning extrinsic movers in *Physics* VIII, 10, because both are taken to account for "projectile motion." Hence they represent Aristotle's "mechanics of motion." I analyze these arguments insofar as they reflect Aristotle's account of nature, moved things, and the task of physics as a science. I argue that each of these arguments is defined by its own specific topic and that in neither case is this topic "projectile motion." In fact, projectile motion as a phenomenon *never* forms a topic in Aristotle's physics. I suggest a pathology of this "absence" and consider its implications for understanding Aristotle's "mechanics of motion."

Even on Aristotle's own terms, however, problems remain here. Neither the actualization nor the activities of the elements are clearly explained. Furthermore, both the actuality of each element and what moves it remain unclear. And at this point in Aristotle's account of the elements and their inclination, these problems are pressing.

They are taken up explicitly in *De Caelo* IV in the context of an important question: in what way are the elements generated? But this question reverts to a prior question: what differentiates each element, thereby rendering it unique? This question is prior because an account of the generation of the elements will be an account of the generation of these differences. Hence an account of what differentiates each element anticipates the account of their generation – they are generated from one another – by accounting for the nature of the elements.

After criticizing his predecessors, Aristotle gives a constructive account of the elements, and this account defines both elemental motion and the mutual relation between the elements and the cosmos. Aristotle first asks "on account of what" each element is moved naturally to its proper natural place. First, the mover is what generates an element. The generator moves the element by granting it potency, thus making it able to be moved to its proper place. Second, each element, when moved naturally to its proper place, cannot be moved by any chance thing but must be moved by actuality proper to the potency of the element. Each element is moved, as like to like, to its proper place. Here the relation between proper place and its respective element is defined: proper place appears as the form of the element. And in this relation both the order and the teleology of nature emerge clearly.

Each element is simple and possesses only one proper, i.e., natural, place, which serves as its actuality and form. When an element arrives at its proper place, it no longer becomes, but is: the activity of each element is to rest in its proper place. Having been moved naturally to this place, it can be dislodged from its place only by means of force applied from the outside. Indeed, in the case of the elements, their simplicity and possession of only one place that serves as actuality seem to imply that the light and the heavy more than other things possess an intrinsic source of motion "because their matter is closest to substance" (*De Caelo* III, 3, 310b31–311a12).

This remarkable claim, along with the account of elemental motion and place as a potency/act relation, raises in its strongest form the question of what relation obtains between place and the elements. Here I turn to two related accounts of elemental motion, that of *Physics* VIII, 4, where

Aristotle argues that, like all things, the elements must be moved by something, and that of *Metaphysics* IX, Aristotle's most extended account of potency/act relations, which explicitly refers to the elements. After examining these arguments as defined by *their* topics, I conclude that these accounts are fully consistent and give a remarkably coherent view of the motion of the elements as orderly and hence as natural. The cosmos itself emerges in *Metaphysics* IX when Aristotle concludes that "the elements imitate the heavens."

I shall propose that the larger structure and coherence of Aristotle's physics as a project emerge here. The topics comprising Aristotle's physics are neither strictly serial in the sense of one after another without prior commitment, nor are they progressive in the sense of leading up to some end. Rather, the domain of physics is defined first by nature and then by motion. The other topics of physics appear solely within this domain and are defined by it. Terms such as "place," "void," and "inclination" are examined for the sake of understanding motion, which itself must be understood if nature is not to remain hidden.

A conclusion, Part III, completes my analysis. I argue throughout that Aristotle's arguments, when read as topical, are remarkably coherent and produce a full account of nature as a cause of order. As I have already suggested, a pattern among the topics that constitute his investigations can be identified. The domains of nature and motion are identical, and the definitions of these terms are presupposed by the analysis of common and universal terms, including place and void. Analysis of proper terms presupposes the definitions first of common and universal terms and ultimately motion and nature as well.

The separation of Aristotle's analysis of common and universal terms from that of proper terms raises the problem of what these categories mean as well as the problem of the relation between them. I shall argue that "place" and "void" are common and universal terms not only in a logical sense but in an ontological sense: place is a single unique principle of determination for the cosmos as a whole. (The void is proposed as such a term but is rejected.) The inclinations of the four elements are proper terms because each specifies its particular element, and the elements may be considered as a group composed of these kinds. Consequently, I reject the frequently held view that in the *De Caelo* we have an account that applies the "common and universal" terms or conclusions of the *Physics*. The difference between the accounts does not lie in one being universal and the other applied. Rather, they are alike as arguments in the sense that each possesses its specific topic and is conducted as an investigation of that topic. And in this sense, both sets of arguments fall within the science of physics, i.e., the study of nature. They differ only insofar as their objects differ: place is a unique cause for all things in the cosmos because it renders the cosmos determinate in respect to "where," whereas inclination is an effect. These accounts work together because both presuppose the same first principle: nature is everywhere a cause of order.

However, Aristotle's account of place and the elements, as well as his rejection of the void, coherent though they may be, have not generally been thought to be true or even persuasive. And the question is why – or why not. I argue that coherence is an "internal" criterion of evaluation of arguments, i.e., it takes them on the terms defined as their domain; hence, it must be established first. Questions of extrinsic standards of evaluation can (and should) be established only after the specific content of the arguments themselves has been established. And they can be used to produce an evaluation of the importance of the problems that arguments are designed to solve as well as the character of the first principle at work in the development of a given position (in this case Aristotle's). This evaluation constitutes the second stage of the history of philosophy as an enterprise. Although my primary task concerns the first stage of analysis, i.e., establishing the coherence of Aristotle's position, I close by offering some observations concerning the second stage, evaluating these arguments (and this position) so that they (and it) may be understood not only on their (its) own terms but in a broader context of the history of ideas and the choices and decisions that this history represents.

2

Nature and Motion

Before turning to Aristotle's arguments on place and void, we must consider the structure of *Physics* IV as a *logos* and its relation to the other books of the *Physics,* especially *Physics* II and III.[1] In *Physics* II, 1, Aristotle defines nature as a source of motion and rest (192b14); *Physics* III, 1, opens with the claim that in order to understand nature, we must also understand motion (200b12–14). In effect, *Physics* III starts from the subject of physics established in *Physics* II. In short, motion and nature are coextensive, they are found together, and they, and those things required by them, form the primary subject matter of physics as a science. Hence, Aristotle concludes, it is clear that universal and common things must be examined first, namely, motion and those things without which motion seems to be impossible, including the infinite, place, void, time, and the continuous.[2]

The Structure of the Arguments

Aristotle proceeds accordingly. First, "motion" is defined (*Physics* III, 1–3), and then "the infinite" (*Physics* III, 3–8). "Place," "void," and "time" occupy *Physics* IV, while "the continuous" along with the related notions "in contact" and "in succession," "points," and "lines" occupy *Physics* V and VI.[3] These terms follow motion and nature not only rhetorically – they are next in the text – but also logically. That is, the examination of them presupposes the definition of motion that he has just established. Hence,

---

1 Ross comments that "III and IV form a continuous work," 534.
2 *Physics* III, 1, 200b20–25. Ross, 359, conjoins the topics of III and IV on the basis of this text; cf. also 534, where he describes them as a continuous work.
3 Waterlow, *Nature, Change, and Agency in Aristotle's Physics,* 129–131, suggests that *Physics* VI represents a departure from the arguments of III. More recently Brunschwig argues for the logical coherence of the entire *Physics* (with the possible exception of Book VII) in "Qu'est-ce que la *Physique* d'Aristote?" 11–40, cf. esp. 29–30 for an argument that *Physics* II–VI forms a continuous line of argument; Berti, "Les Méthodes d'argumentation et de démonstration dans la *Physique* (apories, phénomènes, principes)," argues, as do I, that *Physics* III, 1, announces a program that then occupies III-VI, 66.

place and void are considered, defined, and evaluated in terms of Aristotle's definitions of motion and ultimately nature.

Several consequences (which will be discussed later in detail) follow from this relation. Aristotle never considers "place" or "void" independently of motion in natural things; rather, his arguments examine place and void only as things without which motion, as he defines it, "seems to be impossible." For example, he proves that there is no void apart from things not by considering whether a void exists, but by arguing that a void cannot serve as a cause of motion (*Physics* IV, 8, 214b12–16). Finally, he confirms place and rejects void because he defines nature and motion as principles of order; place renders the cosmos determinate and, as determinate, orderly, whereas the void is a principle of indeterminacy. (As I shall later argue in examining inclination, the full account of how place serves as a cause of motion lies in the problem of the motion of the simple elements.) In this sense, Aristotle's definitions of nature, natural things, and motion establish the criteria for the examination – and consequent success and failure – of place and void. Consequently, in Aristotle's physics, place and void are neither independent of natural things nor are they examined in themselves; rather, they operate only within a cosmos of natural things and Aristotle's examination of them presupposes motion and nature as defined in *Physics* II and III. In this sense, place and void are, as concepts, "subordinated to" motion and nature.

Herein lies my first claim about Aristotle's cosmos and his physics as a science of that cosmos: Aristotle's definitions of nature and motion define his accounts of place and void by providing the criteria for his arguments about them as causes operating within the cosmos. He establishes and then presupposes his definition of nature; as I shall examine in more detail later, nature is everywhere a cause of order because it is an intrinsic source of being moved and being at rest. With this definition of nature, he establishes and then presupposes his definition of motion, i.e., the actualization of the potential as such. Universal and common terms, e.g., place and void, are examined only *after* motion is defined because they are proposed as those things without which motion *on this definition* seems to be impossible. When place succeeds and the void fails, they do so not as independent topics considered in their own right, but as required by Aristotle's definition of motion, which in its turn presupposes nature and things that are by nature.

Consequently, Aristotle's arguments about place and void (as well as the other common and universal terms) are not "neutral" explorations of independent topics as they would be if he were considering possible positions on the nature of place and void to determine which one is best. Furthermore, the definitions established in *Physics* III through VI are not serial – first motion, then the infinite, then place and void, time, and the continuous; treating them as serial fails to identify the way in which these topics presuppose the definition of motion. Rather, Aristotle's related definitions of nature and motion establish the domain of the subsequent arguments: terms such as place and void appear and are analyzed because and only insofar as they are required by the initial definitions of motion and, ultimately, nature.

I shall not consider other universal and common terms, except insofar as the arguments concerning place and void refer to them. However, I would suggest that Aristotle's analysis of each term, e.g., the infinite, also presupposes his definitions of motion and nature. Consequently, universal and common terms are parallel to one another in the sense that each presupposes these definitions, although they are not subordinated to one another (with the important exception of void, which, as I shall argue, does presuppose the definition of place). So, for example, whereas it is not necessary to examine the infinite in order to examine place, it is necessary to examine nature and motion in order to examine either the infinite or place.

Thus rather than presenting a choice between competing alternatives equal at the outset, Aristotle develops his arguments about place and void solely on the basis of his definitions of motion and ultimately nature. Indeed, the logical structure of *Physics* III through VI may itself be thought of as teleological in that these arguments presuppose an end, i.e., providing a full account of motion and nature, for which they are developed.[4] Each argument opens with the defining point relative to which subsequent arguments are developed and concepts evaluated. In short, the definition of motion serves as the end for arguments concerning place and void.

This structure raises a serious methodological issue. Contrasting "subordinated arguments," i.e., those that presuppose a definition that has been established on independent grounds, with "serial arguments," i.e., those that may be treated independently of one another, shows what is at stake in the relations among arguments and our analysis of them. When arguments form a series, each is independent of the others. Theoretically, each argument could be analyzed independently as self-contained. For example, place and void could be thought of as equally possible terms and a case made for each; its relation to other terms, such as motion and nature, would be established subsequently. Whenever arguments are structured serially, one must make the case for the true answer (and for the larger position) only after all the possibilities have been examined. If Aristotle's arguments were of this type, we would both expect and require that such a case be made because without it there would be no philo-

---

4 Although the term "teleology" is regularly applied to Aristotle, it "is a modern one, and is quite definitely fixed in meaning by contemporary use," Owens, "The Teleology of Nature in Aristotle," 136.

sophic position as such. But such a case is entirely absent from *Physics* IV (and the *Physics* generally). It is absent not because Aristotle fails to make his case, but because the structure of the arguments does not follow this model. Consequently, any analysis that assumes that these arguments are serial is doomed from the outset by virtue of this misunderstanding.[5]

These arguments are teleological in the sense of determined by an initial definition that subsequent arguments presuppose, develop and support. And any analysis of them, if it is to assess their logical structure (and hence success), must proceed with this relation in view. I shall first consider Aristotle's definition of nature and motion because, as I shall argue, Aristotle's arguments about place and void are "subordinated to," i.e., presuppose the definition established by, the account of nature and motion.

We must consider place, Aristotle says, because "everyone thinks that things that are are someplace [ποú]" (*Physics* IV, 1, 208a29). The expression "things that are" presumably includes both natural things and artifacts.[6] All of these, either immediately or mediately, contain an intrinsic principle of motion and rest. So the question at stake is *where* (ποú) are things that are? There are two possibilities for "where" things may be: place (τόπος) or void (κενóν). In Aristotle's list of categories, the category usually translated "place" is in fact "the where" (τó που).[7]

In the *Physics* that category is defined: it is place (*Physics* IV, 1–5) and not void (*Physics* IV, 6–9). Place is considered first and established as "where" things that are are; its position as "the where" of things is further supported by a rejection of its competitor, void. The order of these arguments is neither accidental nor arbitrary. Place, the constructive term, appears first and solves a variety of problems; void fails to account for motion, and its rejection leaves place in "first place." This order informs the logical structure of each argument in several ways; we shall see this structure at work as we progress through the arguments, but its characteristics may be anticipated here.

(1) The *first* term examined in *Physics* III, motion, is primary and "that

---

5 For an example of the account, which is thus rejected and which I shall criticize throughout, cf. Hussey, who claims that *Physics* III and IV "consist of five essays . . . composed fairly early in Aristotle's career as an independent thinker. Probably they were originally written soon after the *Topics* was completed, and not (except for the essay on time) much reworked later . . . most of the time the modern reader will not be far out if he takes Aristotle to be doing in these essays much what a modern physicist or philosopher does when discussing certain very general topics, such as the nature of time or space, in general terms" (ix). The essay is a Renaissance form which is quite different from Aristotle's *logoi*. Aristotle is not doing what a modern physicist or philosopher does, and neither these topics nor Aristotle's discussion of them should be characterized as "general." On the rhetoric of Aristotle's arguments and the difference between an essay and a *logos*, cf. Lang, "Topics and Investigations," 416–435.

6 Aristotle, *Physics* IV, 1, 208a29; compare *Physics* II, 1, 192b8–9.

7 For examples, cf. *Cat.* 4, 1b26, 2a1; 8, 10b23; 11b11; *De Gen. et Corr.* I, 3, 317b10; *Metaphysics* V, 7, 1017a26.

for the sake of which" later terms and the arguments concerning them
are developed. Subsequent arguments are neither independent, nor
self-contained, nor the development of an initial seminal meaning.
Rather, each argument presupposes the definition of motion in that this
definition establishes the criteria for the success or failure of subsequent
terms, e.g., place and void. This logical structure, i.e., a primary definition in relation to which subsequent arguments and terms are developed, is an important mark of Aristotle's teleological conception of
physics.

(2) Taking this argument at its primary moment, we must say that
physics is the science of things that are by nature and that nature is an intrinsic "source or cause of being moved and of being at rest."[8] This definition of nature necessitates an investigation of motion and the terms
without which it seems to be impossible, including place and void (*Physics*
III, 1, 200b20).

(3) Things that are by nature require an answer to the question "*where
are they?*" This question arises as a requirement of motion, and, as we
shall see, Aristotle's definition of motion establishes the criteria that the
answer to this question must meet.

(4) Within the requirements set by the definition of motion, the question "where are things?" receives two competing answers: place and void.
As we shall see, Aristotle considers a variety of issues concerning place;
he then defines it and ultimately concludes that place rather than void
must be that without which motion seems to be impossible.[9] The force of
the arguments about place is determined by the way in which they presuppose the definition of motion.

(5) Understanding the arguments as structured in this way not only reflects their logical and conceptual limits but also renders their order intelligible. Aristotle takes up void after place (*Physics* IV, 6–9). Void – κενόν
literally means "empty" – is defined as a special case of place, i.e., as place
in which there is nothing. As we shall see, Aristotle finds void incoherent
as a concept and so unable to serve as a term required by motion. Because
void is defined as a special instance of place and at the same time is an alternate answer to the question "where are things that are?," it presupposes first the concept of place and then motion. The rejection of void
turns explicitly on the fact that it would render motion (on Aristotle's definition) impossible. (Void is the only universal and common term to be
rejected outright.) Consequently, Aristotle never considers it independently of motion; rather, the logical limits of the arguments about the void

---

8 Aristotle, *Physics* II, 1, 192b23. Mansion, *Introduction à la physique aristotélicienne*, makes a
   similar point, although his larger interests are quite different, 80.
9 Although he is not considering the issue methodologically, Furley expresses the same
   point: "In fact, his [Aristotle's] own idea of what place is makes it impossible that there
   could be an empty one," in "Aristotle and the Atomists on Motion in a Void," 88.

are established first by the definition of place and ultimately by the definitions of motion and nature.

We begin with nature and motion, which, as I shall argue, are most important because they define the domain of the subsequent arguments. After considering the definitions of nature and motion, we shall turn to place and void, i.e., terms without which, Aristotle says, motion seems to be impossible. In the analysis that now follows, I shall argue for the structure of Aristotle's arguments as I have just outlined them and for the view that, given the definitions of motion and place, the cosmos is determinate and orderly. Here is the central thesis of this study: the cosmos is comprised of a principle of order, namely, place, and insofar as all things that are by nature or by art are composed of the elements, each is intrinsically ordered to its proper place. The cosmos is determinate and orderly by virtue of these two principles – order and being ordered – working together, and natural motion is nothing other than the expression of this order. Thus the cosmos and all natural things bear witness to Aristotle's principle: nature is everywhere a cause of order.

After making this case for place as a principle of order, I shall turn to one of the most specialized problems that Aristotle's physics must address: the problem of ῥοπή, inclination, and the particulars of how each of the four elements comprising all things in the world, both natural and artifacts, are moved. Here I shall argue that inclination constitutes the very nature of each element and is a principle of an element's being moved, which is oriented toward the respective proper place of each. Thus the cosmos is constituted by these two principles, which are intrinsically and causally related. Consequently, I shall conclude that Aristotle's view of the cosmos as determinate and orderly in this sense may constitute one of the deepest marks of his teleology.

Nature

"Since nature is a source of motion and change and our investigation concerns nature, what motion is must not remain hidden" (*Physics* III, 1, 200b12–14). This characterization of nature at *Physics* III, 1, refers to the earlier discussion and definition in *Physics* II, 1.[10] Things that are "by nature" both constitute the proper subject of physics as a science and provide the starting point and goal of the examination of motion.[11] Hence, we must return here to the definition of things that are by nature in *Physics* II, 1. I shall examine Aristotle's account of nature at some length

---

10 Aristotle, *Physics* II, 1, 192b14–15, 18–19, 21–23, 28–29.
11 Aristotle, *Physics* I, 1, 185a13–14. Solmsen, in *Aristotle's System of the Physical World*, 93, comments that Aristotle uses nature to replace Plato's doctrine of soul. For an account of the history and use of the word "nature," cf. Pellicer, *Natura*, 17–39.

because, as I shall argue, it is presupposed by his accounts of motion, place, void, and the elements.

The rhetorical and logical structure of *Physics* II is identical to that of *Physics* III through VI; that is, the primary term, nature, is set out first, i.e., in *Physics* II, 1. The remainder of the book articulates the meaning and force of this term, its implications and various topics (e.g., the four causes), insofar as they relate to nature and things that are by nature.

Nature As a Source of Motion and Rest

Things that are by nature include animals and plants, their parts, and the four elements, earth, air, fire and water, and such things are substances.[12] This list, if the heavens are animate, is complete.[13] Consequently, the definition and account of nature in *Physics* II applies to any and all things that are by nature, without exception.[14] Natural things, unlike artifacts that are not by nature, are unique because "each of them has within itself a principle of motion and rest" (*Physics* II, 1, 192b15). Furthermore, even artifacts contain such a principle insofar as they are made out of things that are by nature, e.g., stone or wood (192b20). Thus it seems:

that nature is a source or cause of being moved and of being at rest in that to which it belongs primarily, in virtue of itself and not accidentally. [192b22–23: ὡς οὔσης τῆς φύσεως ἀρχῆς τινὸς καὶ αἰτίας τοῦ κινεῖσθαι καὶ ἠρεμεῖν ἐν ᾧ ὑπάρχει πρώτως καθ' αὑτὸ καὶ μὴ κατὰ συμβεβηκός.][15]

At first glance, however, this definition of nature is problematic. The verb "κινεῖσθαι" may be either middle or passive – the forms are homographic. Because the middle voice implies that nature is a self-mover, whereas the passive voice clearly indicates being moved, i.e., moved by something, the choice between them is crucial.[16] Aristotle's conception of nature is at stake here: is a thing that is by nature a self-mover, or is it moved [by some-

---

12 Aristotle, *Physics* II, 1, 192b9–10; 192b33; cf. also *Metaphysics* VII, 2, 1028b10–11.
13 Aristotle calls the heavens animate at *De Caelo* II 2, 285a29, and includes a discussion of the heavens in the *De Motu Animalium*. Cf. also *Metaphysics* VII, 2, 1028b10–13, which seems to imply that the heavens would be included among animals.
14 So, e.g., Wildberg, *John Philoponus' Criticism of Aristotle's Theory of Aether*, turns to the definition of nature in *Physics* II, 1, as the background to the arguments of the *De Caelo*, 40–41.
15 Irwin, *Aristotle's First Principles*, 94, summarizes the entire discussion of 192b13–27 saying, "Things have nature in so far as they have an internal origin of change and stability, and in this way they differ from artifacts such as beds or tables"; he claims that "origin of change" is explained only in II, 3, with the introduction of the four causes. He does not seem to notice either the force of this line or the shift from a noun phrase to the fully spelled out verbs here.
16 Compare, e.g., the Hardie and Gaye translation, i.e., "that nature is a principle or cause of being moved and of being at rest," with that of Charlton, "that nature is a sort of source and cause of change and remaining unchanged." Charlton's translation is presumably reflected in Waterlow's translation, "nature is a principle and cause of change and stasis," 2, in *Nature, Change, and Agency*. Witt, *Substance and Essence in Aristotle*, discusses things that are by nature as possessing an "inner source, or cause, of change and rest," 68 ff., but never mentions the verb form κινεῖσθαι. The same comment applies to

thing]? If the former, then Aristotle must explain self-motion, and if the latter, then he must identify a mover and explain the mover/moved relation. Hence, the problem here goes beyond the definition of nature to embrace Aristotle's definition of motion and his account of the motion of the elements in *Physics* VIII, 4. As I shall argue, the verb here must be passive: what is by nature must be moved by something; motion is a mover/moved relation defined in terms of potency and actuality, and each element presents a unique potency moved by its appropriate actuality.

In a recent paper, Furley sums up one version of the problem:

To anyone who reads *Phys.* a little incautiously it might appear that since nature is declared to be an internal source of change and rest . . . all the things specified at the beginning of *Phys.* II.1 should be self-movers: living things and their parts, plants, and simple bodies, earth, water, air and fire.

But this turns out, of course, to be too generous. We are told explicitly in *Phys.* VIII.4, 255a5–10 that the bodies that move by nature up or down cannot be said to move themselves. . . .

The refinement, according to *Phys.* VIII is a difference in the voice of the verb: the natural bodies, as opposed to things with souls, have a source not of causing movement or of acting (κινεῖν, ποιεῖν) but of being acted on (πάσχειν).[17]

Such incautious readings abound and it is worth mentioning a few. Guthrie argues, "In this first chapter of *Phys.* B, Aristotle has already given a rough preliminary description of what he means by natural objects – those, namely, which seem able to initiate their own motions of growth, etc. From which it follows that φύσις itself is to be described as that within objects by virtue of which they move or grow."[18] Rist too takes nature as "apparently" a self-mover, i.e., animate, and because the elements are included among things that are by nature, the elements are included among living things; this view then, according to Rist, must be modified in *Physics* VIII.[19] In his commentary on *Physics* II, 1, Charlton contends that "despite his general protestations" (presumably *Physics* VIII, 4), Aristotle

Graham, *Aristotle's Two Systems*, where Graham "quotes" Aristotle's definition of nature as "a certain source and cause of motion and rest," 200; Graham, "The Metaphysics of Motion," repeats this translation, 172, and treats the sense as active, 173; cf. also the summary of *Physics* II, 1, given by Lettinck, *Aristotle's Physics and Its Reception in the Arabic World*, 117. Perhaps Charlton's translation also influences the recent translation of Couloubaritsis, which reads: "la nature est quelque principe et cause de se mouvoir et se tenir en repos" – a phrase that is simply repeated in the commentary in *Aristote: Sur la nature.* (*Physique II*), 41, 80.

17 Furley, "Self-Movers," 3–4; Lear, *Aristotle*, treats it simply as active without further consideration, 15. For further examples of confusion concerning nature in *Physics* II, cf. McGuire, "Philoponus on *Physics* II," 241–267; for Aristotle's argument that "everything moved must be moved by something" (*Physics* VII, 1, 241b34) and the specific problem of translating the verb forms here as middle or passive, cf. Weisheipl, "The Specter of *Motor Coniunctus* in Medieval Physics," 99–101.

18 Guthrie, "Notes on Some Passages in the Second Book of Aristotle's Physics," 70–76, esp. 70.

19 Rist, *The Mind of Aristotle*, 123–124, 130; cf. also 204. The same reasoning appears in Meyer, "Self-Movement and External Causation," 73, n. 12.

would argue that the elements are self-moved.[20] Likewise, for Waterlow nature is a self-mover, and she concludes that there is a major problem between *Physics* II, 1, and *Physics* VIII, 4.[21] More recently, Cohen asks how a motion caused externally can be natural (*Physics* VIII) if being natural implies "an internal principle for natural motions."[22]

The identification of nature as a self-mover in *Physics* II, 1, goes back at least as far as Philoponus; but he replaces the verb κινεῖσθαι with a phrase, ἀρχὴ κινήσεως, because the verb here is not ambiguous: κινεῖσθαι is always passive, not middle (or reflexive), in both Plato and Aristotle and, furthermore, is never used to express self-motion.

Because the middle voice in Greek can on occasion be mildly reflexive, it is natural to think of κινεῖσθαι as like the English intransitive, e.g., the stone moves downward. However, a reflexive sense is the least common use of the middle in Greek. Generally, the middle expresses a vested interest of the subject in an object. And the most common use of κινεῖσθαι is found in the obscene expression "the men 'move' the women," which warrants a separate entry in Liddell and Scott and gives us the word κινητήριον (brothel).[23]

The crucial evidence appears, however, if we ask how Plato and Aristotle express self-motion. Neither uses the middle voice; rather, they use either an active verb with an object and a reflexive pronoun or a passive accompanied by an agent. Denying that the elements move themselves, Aristotle argues that it is unreasonable to think that something self-moving would exhibit one motion only or that something continuous could "move itself by itself:" ἄλογον δὲ καὶ τὸ μίαν κίνησιν κινεῖσθαι μόνην ὑφ᾽ αὑτῶν, εἴγε αὐτὰ ἑαυτὰ κινοῦσιν. ἔτι πῶς ἐνδέχεται συνεχές τι καὶ συμφυὲς αὐτὸ ἑαυτὸ κινεῖν.[24] Plato's arguments that the soul must be a

---

20 Charlton, 92.
21 Waterlow, *Nature, Change, and Agency*, 193, 240, and chap. 5, 204–257 *passim*. Waterlow's argument is criticized (as is her conclusion) by Furley in his review of this book, 110. Witt, *Substance and Essence in Aristotle*, agrees with Waterlow "that natural beings are self-sufficient to determine the pattern of their typical changes," 69, n. 6.
22 Cohen, "Aristotle on Elemental Motion," 152–153. I shall answer this question in Part II: elemental motion is natural when it is toward its proper place.
23 Eupolis 8, *Fr.* i b(2). 5D. Although Liddell and Scott does not say that κινέω is always transitive, no examples of intransitive usage are given; indeed, in this respect, κινέω contrasts with φέρω, which is specified as sometimes intransitive.
24 Aristotle, *Physics* VIII, 4, 255a10–13. For the elements as continuous, cf. 255a15. This entire passage turns on such usage, as does an apparent reference to it as *De Caelo* IV, 3, 311a12. For additional examples in the *Physics*, cf. *Physics* VII, 1, 242a38: "τὸ κινούμενον κινήσεται ὑπό τινος." VIII, 5, 256a21: "ἀνάγκη αὐτὸ ὑφ᾽ αὑτοῦ κινεῖσθαι." 33: "τὸ αὐτὸ αὑτῷ κινοῦν . . ." 34–256b: "ἀνάγκη αὐτὸ αὑτὸ κινεῖν" b-2: "ὃ κινούμενον ὑπὸ τοῦ αὐτὸ κινοῦντος κινεῖται . . ." 22–23: "ὃ κινεῖται μέν, οὐχ ὑπ᾽ ἄλλου δὲ ἀλλ᾽ ὑφ᾽ αὑτοῦ . . ." 257a32: "εἴ τι κινεῖ αὐτὸ αὑτό . . ." 257b2: "τὸ αὐτὸ αὑτὸ κινοῦν . . ." 257b13–14: "ὅτι δ᾽ οὐκ ἔστιν αὐτὸ αὑτὸ κινοῦν . . ." 18–19: "τὸ μὲν τὸ ὑπ᾽ ἄλλου κινούμενον αὐτό, τὸ δ᾽ αὑτῷ . . ." 29: "κινεῖται αὐτὸ ὑφ᾽ αὑτοῦ . . ." and 258a7: "τὸ αὐτὸ αὑτὸ κινοῦν . . ." For a parallel usage when Aristotle refers to Plato's notion of a self-mover, cf. *Metaphysics* XII, 6, 1072a1–2.

self-mover reflect exactly the same usage; arguing that the soul is death-less, Plato uses a variety of phrases, all of which are either active with an object or passive with an agent: "τὸ δ᾽ ἄλλο κινοῦν καὶ ὑπ᾽ ἄλλου κινούμενον . . . μόνον δὴ τὸ αὐτὸ κινοῦν, ἅτε οὐκ ἀπολεῖπον ἑαυτό, οὔποτε λήγει κινούμενον, ἀλλὰ καὶ τοῖς ἄλλοις ὅσα κινεῖται, . . . τὸ αὐτὸ αὐτὸ κι-νοῦν . . . ἀθανάτου δὲ πεφασμένου τοῦ ὑφ᾽ ἑαυτοῦ κινουμένου, . . . πᾶν γὰρ σῶμα, ᾧ μὲν ἔξωθεν τὸ κινεῖσθαι, ἄψυχον, . . . μὴ ἄλλο τι εἶναι τὸ αὐτὸ ἑαυτὸ κινοῦν ἢ ψυχήν, . . ."[25] In short, both Plato and Aristotle express self-motion in either of only two ways: (1) actively, to move itself by itself, or (2) passively, to be moved by itself.[26]

Indirect evidence also exists. Aristotle regularly uses these forms to contrast with the active, and they can only be passive.[27] Furthermore, the middle and passive forms are distinct in the future tense, and here Aris-totle uses the unequivocally passive (and rare) form of φέρω, οἰσθήσε-ται.[28] When Philoponus, who wishes to make nature reflexive, quotes Aristotle's definition of nature, he eliminates the verb κινεῖσθαι alto-gether and substitutes the phrase "ἀρχὴ κινήσεως," which he interprets as self-motion.[29] We may speculate that were it possible to exploit κινεῖσθαι as a middle verb and thus reflexive in meaning, Philoponus would surely do so; therefore, his substitution indicates that it is not

25 Plato, *Phaedrus* 245C-246. This usage is also reflected at *Laws* X, 895B–896A2: τὴν αὐτὴν ἑαυτὴν δήπου κινοῦσαν . . . ὅταν αὐτὸ αὐτὸ κινῇ; . . . τὴν δυναμένην αὐτὴν αὑτὴν κινεῖν κίνησιν; τὸ ἑαυτὸ κινεῖν. . . . Cf. also *Timaeus* 57E. In *Simplicius: On Aristotle's Physics 7*, trans. Hagen, 102, n. 10, asserts without evidence that the middle of κινεῖν can be used as intransitive active; but no evidence for such usage can be found in either Plato or Aris-totle.

26 Waterlow, *Nature, Change, and Agency,* suggests that the active κινεῖν can be used in-transitively and that in this case the subject would be changed, 162. Not only is there no clear case of such usage in either Plato or Aristotle, but the texts suggest the opposite: κινεῖν is always transitive with an object, and the subject is moved when the verb is pas-sive, with or without the agent expressed. For an interesting example in *Physics* IV, cf. *Physics* IV, 1, 209a22.

27 Examples are so numerous, it is impossible to list them. For one very clear case, cf. *Physics* VIII, 4, 254b7–12.

28 Aristotle, *Physics* III, 5, 205a13; IV, 8, 214b21, 215a24, 216a5, 17. Hussey translates the first occurrence ambiguously as "will be in motion" but treats the latter two as active: "it will be found to traverse" and "what reason could there be for its moving faster." Hardie and Gaye treat the first as passive "being carried along," but they take the second two as active "to traverse" and "why should one move faster." But this verb must mean "it will be carried"; as we shall see, it also occurs in one of the arguments utilizing ῥοπή, *De Caelo* III, 2, 301b13.

29 Cf. esp. Philoponus, *In Phys.*, 196, 15–17: ἀποδίδωσιν οὖν ἔνθεν τὸν ὁρισμὸν τῆς φύσεως, ὅτι ἔστιν ἡ φύσις ἀρχὴ κινήσεως καὶ ἠρεμίας, ἐν ᾧ ὑπάρχει πρώτως καθ᾽ αὑτὸ καὶ μὴ κατὰ συμβεβηκός. Also 195, 24–26. Lacey, in his translation of Philoponus, *On Aristotle's Physics 2,* 149, refers to *Physics* 192b22 as "slightly misquoted"; I am suggesting that there is an important vested interest at stake in the "misquotation." That is, substituting the noun for the verb allows Philoponus to go ultimately to an active rather than a passive verb. On the noun κίνησις as the nominalization of both the active and passive verb forms, and as such ambiguous, cf. Gill, "Aristotle's Theory of Causal Action in *Physics* III, 3," 147, n. 18.

possible.[30] Finally, when mathematicians require a sense of intransitive motion, e.g., a sphere "rotates" about its center, they use the verb περι-φέρω – the verb περικινέω, meaning "turn [something] around," is apparently not used in classical Greek.

The same point may be made on substantive grounds. According to Aristotle, "everything moved must be moved by something."[31] The self-mover is a special case of motion by another, whereas a truly "first" mover is not self-moved but unmoved.[32] A first mover moves another but contains no intrinsic source of being moved and being at rest and so is unmoved; being separate from nature and the cosmos, the first unmoved mover of the heavens is *not* "by nature."[33]

Living things, plants and animals, are "by nature" and, Aristotle says, are clearly moved by something, e.g., body is moved by soul.[34] The elements too, earth, air, fire, and water, like all natural things, must be moved by something. The problem lies in identifying the mover. When fire goes up or earth down, each goes to its respective natural place, but there is no obvious external mover and, being undivided, they cannot be self-movers. Indeed, Aristotle's identification of a mover for the elements remains problematic – we shall return to it in the second half of this study – but the sense in which the elements, being by nature, contain a source of motion is (as Furley makes clear) unambiguous:

So it is clear that in all these cases [i.e., the elements] the thing does not move itself, but it contains within itself the source of motion – not of moving something or of causing motion, but of suffering it.[35]

Whereas what possesses no ability to be moved, e.g., god, is outside nature, things that are by nature, i.e., plants and animals and their parts and the elements, are always moved by another. Thus things that are by nature

---

30 On Philoponus's reading of *Physics* II, 1, cf. H. Lang, "Thomas Aquinas and the Problem of Nature in *Physics* II, 1," 411–432.

31 Aristotle, *Physics* VII, 1, 241b34; *Physics* VII, 1, is devoted to a proof of this proposition; *Physics* VIII, 4, also constitutes a proof of this proposition and is especially important for its treatment of the four elements, which are so expressly "by nature"; for the conclusion itself, cf. 256a2. For an analysis of both these arguments, cf. H. Lang, *Aristotle's* Physics *and Its Medieval Varieties*, 35–84. The entire discussion of potency and act in *Metaphysics* IX is characterized by the language of that which acts and that which is acted upon; cf. *Metaphysics* IX, 1, 1046a18–29. I shall discuss this argument in Chap. 7.

32 Aristotle, *Physics* VII, 1, 241b37–242a46; VIII, 5, esp. 256b28–258b8; also *Physics* VIII, 4, 254b24–33.

33 Aristotle, *Metaphysics* XII, 10, 1075a12–15; cf. also *Physics* VIII, 10, 267b5–27.

34 Aristotle, *Physics* VIII, 4, 254b15–31; *De Anima* I, 3, 406a30–406b; II, 4, 415b8–12, 15, 22–27.

35 Aristotle, *Physics* VIII, 4, 255b29–31: ὅτι μὲν τοίνυν οὐδὲν τούτων αὐτὸ κινεῖ ἑαυτό, δῆλον· ἀλλὰ κινήσεως ἀρχὴν ἔχει, οὐ τοῦ κινεῖν οὐδὲ τοῦ ποιεῖν, ἀλλὰ τοῦ πάσχειν. For the argument that the elements cannot be self-moved, cf. 255a5 ff.; for the classification of the elements as both natural and non-self-moving, cf. 255b34–256a; this classification in turn derives from the fourfold division with which the argument opens, 254b12–14, cf. H. Lang, *Aristotle's* Physics *and Its Medieval Varieties*, 63–84.

by definition possess a source or cause of being moved, and κινεῖσθαι must be passive in meaning.

The characterization of nature as a source or cause of being moved and being at rest essentially and not accidentally is followed immediately by the rejection of an apparent case of self-motion, a doctor who cures himself.[36] Aristotle argues that this case is *apparent* self-motion because the doctor cures himself only accidentally (*Physics* II, 1, 192b24–33). That is, the doctor does not cure himself *qua* doctor but because the man who happens to be sick also happens to be a doctor: the two are combined only by accident with the result that the doctor cures himself (192b32–33). The sick man *is cured* by the doctor and so, properly speaking, this motion too is produced by another, even though mover and moved are both contained within the same individual. Consequently, even apparent self-motion is nothing other than being moved by another.

The target of the argument about self-motion is undoubtedly Plato's doctrine that soul, defined as self-moving motion, serves as the origin of all motion in the cosmos. This view forms the backdrop to Aristotle's arguments about nature and so is worth considering.[37] Plato argues that because what is moved by another may cease to move and cease to live, it cannot serve as the source of unceasing motion for the cosmos; therefore, if motion (and the cosmos that is in motion) is to be eternal, and not collapse into immobility, it requires a first principle that is imperishable and that cannot be destroyed (*Phaedrus* 245C5-E). This first principle must be a self-mover, and to be moved by itself rather than by another is "the very essence and definition" of soul, which is immortal, does not die, and is not born.[38]

---

36 Cohen, "Aristotle on Elemental Motion," wishes to solve "the problem" between *Physics* II and VIII "by denying that Aristotle holds that fire, *by its nature*, has a natural motion . . . The potentiality for natural elemental motion, and thus the motion that is the actualization of that potentiality, is accidental to the elements," 156 (his italics). This claim violates the definition of nature in *Physics* II, 1, and leads to serious confusions. Daniel Graham, "The Metaphysics of Motion," says of the example of the doctor, "His [Aristotle's] real objection is not that healing is a transitive action, but that there is no necessity in the subject's being identical to the object. On the other hand, he gives no clear example of a nature that involves a transitive action. The examples he does give . . . seem to suggest an intransitive model of a subject changing without affecting some object even itself," 174. This entire account assumes the definition of nature to involve intransitive motion, i.e., to move intransitively, which is false. As I shall argue in a moment, for Aristotle intransitive self-moving motion breaks the law of noncontradiction because it implies that the same thing is both agent and patient at the same time and in the same respect. The example of the doctor healing himself is an example of a transitive motion, i.e., divided between a mover and a moved within a single individual.

37 For an explicit rejection of Plato's self-moving soul as the origin of the motion of the heavens, cf. *Metaphysics* XII, 6, 1071b37–1072a2.

38 Plato, *Phaedrus* 245D4–246. Cf. also *Laws* X, 896-E. Wardy, *The Chain of Change*, argues that in *Physics* VII, 1, Aristotle intends to deny that there are any real self-movers, 94. The same view underlies the argument of Wedin, "Aristotle on the Mind's Self-Motion": "Everyone knows that Aristotle's animals are self-movers, and almost everyone knows

Aristotle's account of nature systematically rejects every feature of
Plato's account of motion and nature. Plato and Aristotle agree only
about art: art is that which has its source of motion from without. Their
disagreement about nature begins immediately: Plato describes nature as
a divine work of art produced by a master craftsman using a perfect
model, whereas Aristotle contrasts nature with art – indeed, for Aristotle,
nature is the original or model and art is derivative.[39] Because, for Plato,
nature is a work of art produced by a maker extrinsic to it, nature must
have a beginning, whether chronological or causal, whereas Aristotle unequivocally calls nature, i.e., things in motion, eternal.[40] And nature, for
Aristotle, is not, as in Plato, constituted by an *external* source of motion,
such as self-moving soul descending into body, but by an *intrinsic* source
of motion. Again, *contra* Plato, nature is not a principle of self-motion, but
a principle of being moved and being at rest.

As we shall see in a moment, on Aristotle's view, Plato's self-moving
soul, i.e., an identity of mover and moved, breaks the law of noncontradiction: the same thing is both mover and moved (agent and patient) in
the same respect at the same time. Criticizing Plato, Aristotle complains
that he fails to provide a principle of motion for nature (*Metaphysics* I, 9,
992a24–b9); presupposing his own view, Aristotle requires an *intrinsic*
source, and of course, Plato, given his view, provides only extrinsic sources
of motion.

Central to any account of nature is the relation of body to soul or form.
For Plato, body resists soul, which descends from without, whereas for
Aristotle, potency yearns for form immediately and intrinsically because
soul is the first entelechy of body (*De Anima* II, 1, 412a28–29). Plato's
demiurge sends soul as his divine messenger into resistent recalcitrant
body, but Aristotle needs no messenger, no intermediary, to connect form
and matter because, on his view potency desires form, e.g., the heaven,

that Aristotle has a problem explaining this. If for example, everything that is moved is
moved by something else, then nothing will move itself," 81. I have argued that the point
of this argument is to deny Plato's notion of a self-mover, cf. H. Lang, *Aristotle's* Physics
and Its Medieval Varieties, 35–44. Furley, *Cosmic Problems*, makes exactly the same point in
respect to the argument of *Physics* VIII, 4: "Aristotle does not *reject* the concept of self-
movers in *Physics* VIII. . . . It is evidently quite legitimate, in Aristotle's view in these chapters, to call the whole a self-mover, provided that the moving part is itself unmoved except accidentally," 124 (his italics); Furley repeats this point in "Self-Movers," 7.

39 Aristotle, *Physics* II, 2, 194a21–22. In "Nature and Craft in Aristotelian Teleology," Broadie
claims that Aristotle "likens" nature to craft "for the crafts are principles of activity vested
in particular individuals who live, move and have their being within the physical world.
Craft in its active exercise is evidently end-directed, and to Aristotle the same is true of nature, although less evidently so. Thus it is craft that provides the model for nature, not the
reverse," 390. Whereas it is true that both nature and art are "end-directed," it is not clear
that for Aristotle it is less evidently so in nature; furthermore, this "end-directedness" need
not imply the priority of "craft." Indeed, it is universally recognized that for Aristotle art
imitates nature, not the other way around.

40 Plato, *Timaeus* 28A2, 28C5; Aristotle, *Physics* VIII, 1, 252b5–6.

being moved by the unmoved mover acting as an object of love, "runs always."[41] Because potency, for Aristotle, is immediately oriented toward its natural form, no "third cause," either a demiurge or soul, is needed to unite them; Plato's notion of "chaos" or "the receptacle" as that which is not intrinsically related to form becomes, on Aristotle's view, not only unnecessary, but meaningless.

When Philoponus comments on *Physics* II, 1, his account of nature – nature is a mover, form descended into matter – derives from Plato's account, not Aristotle's (*In Phys.* 197.33–198.4). For animals, the intrinsic mover is form or soul and for the elements inclination (ῥοπή) (*In Phys.* 198.6–8; 195.29). I shall consider inclination in the second part of this study, but we see here that it does not appear at all in *Physics* II and *a fortiori* is not identified as a mover. Philoponus introduces it here because having identified nature as an intrinsic mover he requires such a mover for the elements.[42] I shall argue that for Aristotle, just as nature is an intrinsic source of being moved and being at rest, so too is inclination the principle of being moved and being at rest in each element.

The notion of an "intrinsic source" is itself problematic. Irwin argues that Aristotle's claim that nature is an "origin of change" is not explained until he introduces the four causes in *Physics* II, 3, and that, at least *prima facie*, nature seems to be most properly identified with the "efficient cause."[43] This view is also expressed by Meyer, who claims that "[a] self-mover is a kind of efficient cause" and that "Only a self-mover can properly be called the origin (ἀρχή) of an outcome."[44] Not surprisingly, for both Irwin and Meyer, "Aristotle's position" turns out to be confused and in need of modification.[45] But *Physics* II, 1, does not raise the problem of the four causes; rather, it defines nature as a source or cause of being moved and being at rest. Hence, we require an account not of the four causes, but of motion and rest. And this we find in *Physics* III, 1, with its reference to *Physics* II, 1.

For Aristotle, the active orientation of potency toward actuality is crucial to the account of "things that are by nature." Indeed, it expresses the teleology of nature and so is central to his notion of nature as "everywhere a cause of order," particularly, as I shall argue, his account of elemental motion. But "active orientation" is a problematic notion. Charlton, for example, refers to "passive powers" in Aristotle and claims that "passive

---

41 Aristotle, *Physics* I, 9, 192a22–24; *De Caelo* I, 3, 270b23; *Metaphysics* XII, 7, 1072b3.
42 On this topic, cf. Macierowski and Hassing, "John Philoponus on Aristotle's Definition of Nature," 82, who argue that Philoponus presents "a major deviation from Aristotle" on this point.
43 Irwin, *Aristotle's First Principles*, 94.
44 Meyer, "Self-Movement and External Causation," 65–66.
45 Irwin, *Aristotle's First Principles*, 94–98. Meyer, "Self-Movement and External Causation," 67 ff.

power is defined in terms of active: an object has the passive power to become f if something else has the active power to make it f."[46] But Charlton's "passive" refers to Aristotle's notion of "to be affected" or "to be moved," and for Aristotle these phrases do not mean "to be passive."[47] The ability of a natural thing to be moved is always potential *for* something, i.e., is never neutral to its mover.[48] Hence, Aristotle identifies nature with ὁρμή, an intrinsic active striving that contrasts with external force.[49] The active orientation of potency to actuality rests on the intimate relation between them: the moved, or potential, has its very definition in that which is actual, i.e., the mover.[50] Consequently, when nothing intervenes between them, the potential is moved by its proper actuality because that actuality constitutes the definition of the moved. In effect, the moved is actively oriented toward its own definition, its own being as actual. As I shall argue, Aristotle's teleology should be identified with this active orientation of a thing toward its own being. Indeed, for natural things, the relation of the moved to its proper mover is so strong that if nothing intervenes between them, what is potential *cannot fail* to be moved by its actuality.[51] The force of Aristotle's definition of nature as an intrinsic source of being moved lies here: nature is uniquely defined by an intrinsic active orientation of the moved, potency, toward its mover, actuality.

Because Aristotle defines nature as a source of being moved *and* of being at rest, we must consider "being at rest" (ἠρεμεῖν). First, it is unequivocally an active infinitive. "To be at rest" is not a passive state such as absence of motion and, indeed, Aristotle calls it "contrary to motion" (*Cat.* 14, 15b). It is especially important for the elements. When what is potential is moved, or actualized, it becomes actual and so identical with its mover. For example, fire is by nature "up"; so when it is out of its nat-

---

46 Charlton, "Aristotelian Powers," 278.
47 What is often referred to as "passive potency" in Aristotle is just this capacity to be affected or acted upon. The text most frequently cited concerning passive potency, *Metaphysics* IX, 1, 1046a11–13, reads: ἡ μὲν γὰρ τοῦ παθεῖν ἐστι δύναμις, ἡ ἐν αὐτῷ τῷ πάσχοντι ἀρχὴ μεταβολῆς παθητικῆς ὑπ' ἄλλου ἢ ᾗ ἄλλο. It is misleading to think of it as "passive" in the sense of neutral to the mover. For an example of confusions that follow from such an assumption, cf. Gill, "Aristotle on Self-Motion," 246–254. Graham, "The Metaphysics of Motion," speaks of "passive power" in explicating *Physics* II, 1 (175) and *Physics* VIII, 4 (177, n. 12).
48 Cf. also *Metaphysics* IX, 5, 1048a-2, where Aristotle emphasizes that the potential is specified toward its actuality, i.e., it is potential for something at some time and in some way. At *Metaphysics* IX, 7, 1049a34–1049b, what is potential is always relational and must be of something or for something.
49 Aristotle, *Physics* II, 1, 192b18; *Metaphysics* V, 5, 1015a27, b2; 23, 1023a9, 18, 23; *Post. An.* II, 95a1; *Eudemian Ethics* II, 8, 1224a18-b9. On the strong sense of active striving as the force of ὁρμή, cf. Charlton, 92.
50 Aristotle, *Physics* III, 3, 202a15–16; cf. Edel, "'Action' and 'Passion,'" 60.
51 Aristotle, *Metaphysics* IX, 5, 1048a5–7: τὰς μὲν τοιαύτας δυνάμεις ἀνάγκη, ὅταν ὡς δύνανται τὸ ποιητικὸν καὶ τὸ παθητικὸν πλησιάζωσι, τὸ μὲν ποιεῖν τὸ δὲ πάσχειν . . .

ural place, it is actively oriented upward. Fire's potential is actualized by being moved upward, and when it is completely "up," it rests there.[52] Consequently, fire's nature is to be up, and this nature expresses itself either as potential, by an active orientation upward and by being moved upward if nothing hinders, or as actual, by resting in and resisting being moved out of its natural place.[53]

Like motion, "to be at rest" can be either constrained or natural; it is constrained when a body is held in a place to which its movement was constrained (*De Caelo* III, 2, 300a20–30; 300b5–6). So, for example, heavy things by nature are moved downward and are moved upward only by force; when a heavy thing, such as a roof, is moved upward by force and then held in place, e.g., by pillars, that "resting in place" is also by constraint. Thus if we remove the pillars, the roof is immediately moved downward (*Physics* VIII, 4, 255b25).

But "to be at rest" is "by nature" when a thing is in a place to which its movement was natural (255b25). Thus, what is heavy goes downward by nature, and, when it is in a place that is down, it rests there and can be dislodged only by force applied from the outside, as when a lever lifts a stone upward (255a22). In the absence of external force, it rests in a downward place because that place is natural to it. Its *intrinsic* principle in this place is "to be at rest" because insofar as is possible it has actualized its potential to be moved.[54] In this sense, "to be at rest" is associated with completion, with what is by nature and with form. In short, being at rest is for Aristotle the activity associated with being immovable in that which *may* be moved (*Physics* III, 2, 202a3–4).

Many commentators omit mention of this part of Aristotle's definition.[55] And when they do mention it, they treat it as passive. Charlton, for

---

52 Aristotle gives this example at *Physics* II, 1, 192b35–193a. We shall see the force of this point more fully in the second half of this study.

53 Cohen, "Aristotle on Elemental Motion," initially ignores the second part, i.e., to be at rest, of Aristotle's definition of nature and so produces a very confused account: "The problem with this is that if the rock has its natural motion as its proper activity, then once the rock has arrived at its natural place it can no longer perform its proper activity," 155; later he treats being at rest as an alternative principle to being moved: "Being inanimate, earth does not have an internal principle of motion – all its motions have an external principle. But it will nonetheless be 'a thing that exists by nature' if it has an internal principle of rest or stasis," 156. "Being moved" and "being at rest" in *Physics* II, 1, are not disjunctive: nature is a source or cause of being moved and being at rest.

54 I say "insofar as is possible" to indicate external constraints that may prevent further actualization. Obviously, if such constraints were removed, there would again be potential to be moved, which if actualized would lead again to a thing's being at rest. The only time there is no further intrinsic potential is when a thing is in its natural place, absolute down, i.e., the center of the cosmos, for what is absolutely heavy, i.e., earth; or up, i.e., the periphery, for what is light, i.e., fire.

55 For a classic example, cf. Ross, 499–500: ". . . Aristotle does establish a distinction between two classes of things, one consisting of things which as such have an internal principle of movement (i.e., animals and their parts, plants, and the four simple bodies earth, water, air, fire), the other of things such as beds and clothes which as such have

example, translates ἠρεμεῖν as "to remain unchanged" and refers to it in his commentary as "staying put."[56] Waterlow translates it as "stasis."[57] But "to be at rest" is neither "stasis" nor "to remain unchanged"; rather, it is an activity because it implies that potential is fully actualized and activity ensues.[58]

As part of the definition of things that are by nature, "to be at rest" takes us to being unmoved in things that are movable. Nature is both an ability to be moved and to be at rest: the former when a thing is out of its proper place and the latter when it is in its proper place. Clearly, "where" a thing is bears an intrinsic relation to its very nature; thus the account of "where things that are are" will have an intrinsic relation to the definition of nature.

Nature as Form or Matter

The remainder of *Physics* II, 1, reflects the importance of this issue as Aristotle asks whether nature is more properly identified with matter or with form. Some say nature is the matter, i.e., the underlying constituent taken without arrangement (193a9–12). On this view, nature resembles art, persisting throughout change as the underlying reality of a thing while various arrangements or shapes come and go and so are only accidental to it (193a12–29).

The identification of nature with matter – a view often attributed to materialists – makes matter prior to form, its temporary accidental arrangement.[59] But Plato too argues that the foundation of the physical world, the receptacle, has no formal character of its own but is something underlying that is able to receive temporarily whatever form is imposed on it (*Tim.* 50D-51B5). On his view, the cosmos is a work of art with form imposed on recalcitrant matter by a master craftsman. Just as Aristotle rejects Plato's view of form as separate from nature and nature as moved by self-moving soul, so too he rejects this construal of the form/matter relation. In so doing, he concludes that nature cannot be primarily identified with matter.

Nature may also be identified with the shape or form specified by the definition (*Physics* II, 1, 193a30–31). On this view, nature is "the shape, namely the form, not separable except in definition, of things having in

---

no internal principles of movement." Charlton first refers to an internal source of "a thing's changing and staying put" but begins his more extended commentary "What does Aristotle mean by an internal source of change, and is he right in thinking that it is what differentiates natural objects?" 88.

56 Charlton, 88.
57 Waterlow, *Nature, Change, and Agency*, 2.
58 Aristotle, *Physics* VIII, 4, 255b18–24; also *De Caelo* IV, 3, 311a-4.
59 Cf. Furley, *The Greek Cosmologists*, 178–179; cf. H. Lang, *Aristotle's* Physics *and Its Medieval Varieties*, 29–33.

themselves a principle of motion."[60] And Aristotle unambiguously confirms that: "this [the shape and the form] rather than matter is nature" (193b6–7). But as soon as he has confirmed that nature is form, Aristotle includes matter as part of physics and as part of nature, albeit in a weaker sense than form. Thus, he does not reject matter as nature absolutely so much as confirm the primacy of form and so prepare the way for his own construal of the form/matter relation – a construal that differs radically both from that of Plato with his separate forms and from that of the materialists with the primacy of matter.[61] This construal is central to his account of things that are by nature and so is central to any argument that presupposes this account.

We must first consider Aristotle's identification of nature with form. It asserts the independence and primacy of nature as a subject and a substance (*Physics* II, 1, 192b33–34). There are two reasons why nature is more properly identified with form than with matter: (1) a thing is what it is more properly when it is actual than when it is potential – in this respect nature and art are alike: there is nothing artistic about a potential bed or natural about flesh, blood, and bones that are not yet specified by form (193a32–193b2); and (2) form is that toward which a thing tends or grows (193b7–19). In short, form is a thing not as derivative or accidental, but as complete and as specified by the definition. And nature is just that.

There are two serious problems here: (1) form, according to Aristotle, is a mover, not a principle of being moved, whereas nature was initially characterized as an intrinsic principle of being moved and of being at rest; and (2) Aristotle began by contrasting things that are by art with those that are by nature – natural things alone possess an intrinsic source of being moved whereas artifacts must have their form imposed from without; but here art resembles nature in being primarily identified with form. Is the identification of nature with form consistent with the definition of nature, in contrast to art, as an intrinsic principle of being moved and being at rest?

Clues in the arguments of *Physics* II, 1, indicate that Aristotle is using two distinct but closely related notions, namely "nature" and "by nature." The distinction between them is crucial to physics: "nature" is primarily identified with form, whereas what is "by nature," e.g., a man, is a combination of form and matter.[62] "Nature" is what a thing is, and a *thing* is "by nature."

---

60 Aristotle, *Physics* II, 1, 193b3–5: ὥστε ἄλλον τρόπον ἡ φύσις ἂν εἴη τῶν ἐχόντων ἐν αὑτοῖς κινήσεως ἀρχὴν ἡ μορφὴ καὶ τὸ εἶδος, οὐ χωριστὸν ὂν ἀλλ᾽ ἢ κατὰ τὸν λόγον. On the separability of the subject matter of theoretical science, which includes physics, cf. Grene, "About the Division of the Sciences," 10; Groot, "Philoponus on Separating the Three-Dimensional in Optics," 161, claims that Aristotle ruled out separability even in thought of physical forms from matter.

61 Cf. Cherniss, *Aristotle's Criticism of Plato and The Academy*, 175–176.

62 Aristotle, *Physics* II, 1, 193b5–6: τὸ δ᾽ ἐκ τούτων φύσις μὲν οὐκ ἔστιν, φύσει δέ, οἷον ἄνθρωπος.

Nature is primarily form (although the physicist must include a reference to matter). Form is what we know when we know a thing properly speaking, and form is the primary constitutive principle of anything whether art, nature, or even god.[63] Nature, as we have seen, is a substance and subject, and in a primary sense substance and subject must always be identified with form because form alone is able to be apart.[64] Because it can be apart and is primary in definition, in knowledge, and in time, substance (and hence form) is the primary subject matter of first philosophy (*Metaphysics* VII, 1).

When we speak about anything that is "by nature," however, even though we may be speaking primarily of form, we do not speak solely of form. "Things that are by nature" are a combination of form and matter, and this fact is crucial to our understanding of them, first because they constitute the proper subject matter of physics as a science and second because "common and universal terms," such as place and void, are analyzed within the science of physics. For this reason, Aristotle goes on in *Physics* II, 2, to argue that the physicist must know nature in both its senses, form and matter, even though form is primary and matter is relative to it.[65] Thus although he unambiguously identifies nature with form, this identification does not altogether exclude matter from things that are by nature; rather, he grants matter a second place in the characterization of natural things. And such a view accords with the account of matter in the *Metaphysics:* matter is substance but as potential rather than actual.[66]

Since things that are "by nature" include both form and matter, the relation between them is central to nature and to things that are "by nature." This relation constitutes the crucial difference between things that are by nature and those that are by art. And Aristotle could hardly be

---

63 God, properly speaking, is identified with actuality (*Metaphysics* XII, 7, 1072b8, 27) and actuality with form. Cf. *Metaphysics* VIII, 2, 1043a20; 3, 1043a33, b1; IX, 8, 1050a16, b2; *De Anima* II, 1, 412a10; 2, 414a17; *Physics* II, 7, 198b2–4.

64 Aristotle, *Metaphysics* VII, 3, 1028b33–1029a33. Graham, "The Metaphysics of Motion," takes the distinction between "nature" and "by nature" to imply that "we cannot identify nature either with the substance that is its subject nor with activities or properties. It is something else that *inheres in* the substance and *causes* certain motions or changes without being the substance or the changes in question," 173 (his italics). This confusion results from taking nature as an intrinsic source of motion, i.e., a mover rather than a source of being moved.

65 Aristotle, *Physics* II, 2, 194a26–27, 28–33; 194b8–9. Frede in "The Definition of Sensible Substance in *Met.* Z," 129 concludes that "Aristotle does think that in natural science sensible substances ought to be defined in terms of both their form and their matter. But, unless one sees that it is only in a qualified sense that he subscribes to this view, many details of the discussion in *Met.* Z remain unintelligible. . . . For in *Met.* Z he clearly takes the position that there is no place for a reference to matter in the definition of a substance, strictly speaking." Replying to Frede, cf. Morrison, "Some Remarks on Definition in *Metaphysics* Z," 131–144.

66 Aristotle, *Metaphysics* VIII, 1, 1042a25–33; XII, 5, 1071a5–10.

more explicit about it: matter too must be grasped by the physicist because "matter is among things which are in relation to something; for there is different matter for different form" (*Physics* II, 2, 194b8–9). So, for example, the matter of an oak differs from that of a human. Furthermore, when a natural thing grows, it does not produce matter first or in a primary sense, but only in relation to form – any natural thing grows toward its form and toward nature in the sense of form (*Physics* II, 1, 193b13). So an acorn does not produce "wood," but becomes an oak; and it does so because it possesses an intrinsic ability to be moved.

The intrinsic ability to be moved requires a moved, which is identified with matter, and a mover, form, without which *a fortiori* there can be neither motion nor matter that is moved. Form, as we have seen, is nature in the primary sense, that toward which a thing grows and the proper object of the definition. But matter is the "impulse to change," the source of being moved intrinsic to things that are by nature.[67] As Aristotle says in another context, matter runs after form, it desires and yearns for form (*Physics* I, 9, 192a19–24). Here is the central point of Aristotle's notion of an intrinsic source of being moved: in natural things, matter is never neutral to form, and form never needs to impress itself or be impressed (by another) upon matter.[68] Here is Aristotle's teleology of nature: form acts as an object of desire – indeed, form is a final cause when it acts as a principle of motion – and matter immediately desires form as its nature and definition.[69] Matter is potential and is moved by form because it is actively oriented toward its proper form. For this reason, plants, animals, their parts, and the elements, once generated (and in the absence of hindrance), are immediately moved toward their nature, i.e., form or completion. In Aristotle's example, "to be carried upward" belongs to fire by virtue of what it is (*Physics* II, 1, 192b35–193a).

Nature may be constructively contrasted with art. Matter has no innate impulse toward artistic form and so an artist *must impose* such form on it; if the process is interrupted, the work remains incomplete. But in nature,

---

67 Aristotle, *Physics* II, 1, 192b18: Aristotle says of things that are by art that each οὐδεμίαν ὁρμὴν ἔχει μεταβολῆς ἔμφυτον. Charlton notes how active the word ὁρμήν is, 92.

68 This view, of course, contrasts sharply with that of Plato. Lear speaks as if form according to Aristotle were imposed on matter in nature, *Aristotle: The Desire to Understand*, 18–19. Meyer, "Self-Movement and Causation", 72–73, refers to *Physics* II, 1, 193b6–7, to make an odd claim: "The central claim of Aristotle's physics and biological works is that the nature of an organism is its form, more than its matter" (*Phys.* II.1 193b6–7). She defends this thesis in a variety of ways, sometimes by claiming that the formal cause of the organism is its unmoved mover, whereas its material causes are only its moved movers." But Aristotle explicitly identifies moved movers with moving causes in contrast to matter (*Physics* II, 7, 198a27–28); matter is identified with what is potential, i.e., contains a source of being moved. One text cited by Meyer (*De Gen. et Corr.* II, 9, 335b29–31) could not be more explicit: "For, to suffer action, namely to be moved belongs to matter; but to move, namely to act, belongs to a different power."

69 Aristotle, *Physics* I, 9, 192a23–24; II, 7, 198a25; 198b2–3; *Metaphysics* V, 4, 1015a10.

form, once in contact with matter, immediately causes the matter to be moved, and the matter is moved immediately because it is oriented toward form. So the father brings the seed *into contact* with matter that yearns for it as its own fulfillment.[70] A new individual, one that is "by nature," is immediately generated, and if material conditions such as nourishment are met, the individual develops according to its own now intrinsically possessed principles: form acting as a final cause and matter yearning for fulfillment in that end. No imposition of form is ever required for natural generation or growth because natural things possess an intrinsic principle of being moved and being at rest. This principle is matter aimed at and presupposing form, i.e., nature in the primary sense: in natural things, form and matter go together "by nature."

This difference between art and nature can be misconstrued, and the issue at stake is important. Lear, for example, claims that Aristotle "relies on the analogy between art and nature to give one some idea of the form of a material object. A craftsman can impose a form on various bits of matter. . . ."[71] But this is the exact point of disanalogy between nature and art: in natural things matter is dynamically aimed at form, and so there are no such "bits of matter." Furley makes the point when he argues that the quarrel between the Atomists and Aristotle lay primarily "in an epistemological preferences for the bits and pieces of things on the one hand, and for whole forms on the other."[72]

As Furley implies, there is no conflict between identifying nature with form and specifying things that are by nature as possessing an "intrinsic source and cause of being moved and being at rest." Aristotle identifies nature primarily with form but includes a reference to matter in things that are by nature and constitute the objects of physics. As identified with form, nature resembles both art and god – neither of which is nature or "by nature." But the full specification of things that are by nature – they contain an intrinsic principle of being moved and being at rest – distinguishes natural things from both god and artifacts and at the same time, establishes Aristotle's distinctive conception of form, matter and their relation. That natural things contain an intrinsic principle and cause of being moved and being at rest means that they possess *both* matter – actively, dynamically, aimed at form – *and* form – the object of the definition, which is presupposed by matter, i.e., by being moved and being at rest.[73] Indeed, just this conception of nature requires an account of motion and ultimately of the terms without which it seems to be impossible.

---

70 For this example, cf. *Metaphysics* IX, 7.
71 Lear, *Aristotle: The Desire to Understand*, 17.
72 Furley, "The Cosmological Crisis in Classical Antiquity," 19.
73 Cf. Aristotle, *Metaphysics* VII, 17, 1041b8. For an account of the two senses of nature, cf. Weisheipl, "The Specter of *Motor Coniunctus* in Medieval Physics," 105.

Motion

In *Physics* II, the subject matter of physics is established as "things that are by nature," including both form and matter, and "nature," is identified primarily with form. Because natural things contain an intrinsic source of being moved and being at rest, if we are to understand nature, we must understand motion; because motion is entailed by the definition of nature, the present account of motion both presupposes the definition of "nature" and is directed towards clarifying it. After defining motion, Aristotle takes up terms, including place and void, without which motion seems to be impossible and follows the same procedure; by establishing their implications and requirements, his examination of place and void further explains motion and nature in things that are by nature (*Physics* III, 1, 200b21–25). In this sense, motion and nature provide the purpose, or end, for the examination of place and void.

Aristotle's examination of motion appears first (*Physics* III, 1–3) because motion is immediately implicated in the definition of nature; other terms follow because without them motion seems to be impossible. He claims that there is no motion apart from things that are by nature. This claim both implicitly criticizes Plato, who argues that the first motion, that of soul, is by definition apart from physical things, and reaffirms his own definition of nature because motion is inseparable from things that are by nature.[74]

Hence, an account of motion is in some sense an account of things in motion, and Aristotle now distinguishes among movable things. His distinctions, which turn on the notions of potency and actuality, take us to the definition of motion as an actualization of the potential. And again the structure of the argument is one of subordination. The definition of motion appears on the first page of *Physics* III, 1; the remainder of the argument serves to support it further by (1) explaining problems traditionally associated with motion, (2) demonstrating the strengths of this definition by solving these problems, and (3) giving a pathology of alternate views.

Aristotle begins his account by referring motion to the categories of being. Throughout the categories, some things are only actually, whereas others are both actually and potentially (*Physics* III, 1, 200b26–28). What is relative is spoken of generally about both what produces motion and what is able to be moved. For what produces motion does so in the movable and the movable, is moved by that which produces motion.[75]

---

74 Cf. Aristotle, *Physics* VIII, 1, 250b14–15, where motion is characterized as like some life to all things constituted by nature. In Plato, cf. *Phaedrus* 245C5 and 246C-D, where soul looses its wings and so descends into body, which thereby becomes animate, i.e., moved.

75 Aristotle, *Physics* III, 1, 200b28–31. Cf. *Metaphysics* IX, 1, 1046a10–13; 6, 1048a30–1048b9. We may note parenthetically that Aristotle's language and distinctions here and in subsequent arguments further support the view that "κινεῖσθαι" in *Physics* II, 1, must be passive rather than middle.

Aristotle drives home his point, motion must be the motion of something, by identifying the four categories in which motion or change may occur:

There is no motion apart from things. For what changes always changes either in respect to substance, or quality, or quantity, or place and it is not possible, as we say, to find anything common to these which is neither this nor quantity nor quality nor any of the other categories; therefore, there will certainly be neither motion nor change apart from the things mentioned, because nothing is apart from the things mentioned.[76]

Natural things possess an intrinsic principle of motion, and motion is of these things: motion is a relation (yet to be defined) between mover and moved.

Being falls immediately into the categories (*Metaphysics* IV, 2, 1004a5). That is, there is no being apart from or prior to the categories that somehow comes to be present in them; rather, being is in the categories immediately and nonderivatively. Because (a) motion and change are found within the categories of substance, quality, quantity, and place, and (b) there is nothing apart from these categories, motion can be neither defined nor explained apart from these categories of being. Hence, an account of motion can be only an account of things that are by nature.

Within the categories, motion and change belong to things in two ways, i.e., in actuality (or completion) and in potency (or as incomplete) (*Physics* III, 1, 201a3–10). In the category of substance, these two ways are form and privation, in quality, white and black; and in quantity, complete and incomplete (*Physics* III, 1, 201a4–6). For locomotion, which is crucial for place and void, the ways in which motion is found are upward and downward, or light and heavy (201a7–8). For example, when the light (or upward) is held downward, it is only potentially in its proper place, and conversely, when the heavy (or downward) is held upward; but when the light is upward (or the heavy downward) each is in its respective proper place, their respective motions are complete, and they are actually. Potency and actuality are two kinds of being in each of these four categories, and so it may be said that "there are as many kinds of motion and change as there are of being" (201a8–9).

Using the terms potency and actuality, Aristotle now defines motion: ". . . having distinguished in respect to each genus what [is] on the one hand actually and on the other what [is] potentially, the actualization of what is potentially insofar as it is such is motion."[77] Granting that motion

---

76 Aristotle, *Physics* III, 1, 200b32–201a3; cf. also *Metaphysics* IX, 3, 1047a30–1047b. This may be an implicit criticism of Plato for whom the forms and soul are apart from the things mentioned here. Self-moving soul most certainly moves according to Plato.

77 Aristotle, *Physics* III, 1, 201a9–11: διῃρημένου δὲ καθ' ἕκαστον γένος τοῦ μὲν ἐντελεχείᾳ τοῦ δὲ δυνάμει, ἡ τοῦ δυνάμει ὄντος ἐντελέχεια, ᾗ τοιοῦτον, κίνησίς ἐστιν, . . .

can occur only in things, the crucial point rests on the relation between the potential and the actual such that the potential is actualized, i.e., becomes actual, within each of the four categories in which motion occurs. Indeed, for Aristotle, an account of motion is nothing other than an account of the relation between what is potential and what is actual.

We shall see later, in the account of elemental motion, that this relation fully expresses the teleology of nature. An important feature of it may be anticipated here. Although he does so in rather different terms than those given here, Lear emphasizes that Aristotle's teleology is incompatible with any form of mechanical explanation.[78] And, it seems to me, Lear is correct. In "mechanistic accounts," motion is transferred from one body to another and so *a fortiori* must be separable from these bodies; furthermore, the two bodies are ontologically indistinguishable from one another as motion is transferred from one to the other. Aristotle in effect denies both these points: (1) Motion is never separable or transferable but must be in the moved; and (2) the two bodies, far from being equal, are distinguished as mover and moved, agent and patient.

The order of Aristotle's arguments is important methodologically. He first clarifies the definition of motion and thereby reveals what he takes to be the most serious problems entailed by it. Repeating his definition, he emphasizes the strength of his account, namely, that it solves these problems. Against this strength, he provides, in *Physics* III, 2 and 3, a pathology of his opponents' views and then solutions to additional problems associated with motion. Consequently, his clarification provides his account of the problems entailed by "motion" as well as the superiority, from his view, of his definition.

In his account, Aristotle first explains what he means by "insofar as it [the potential] is such [i.e., potential]." So, for example, insofar as a thing may be altered, it is alterable, and its fulfillment (or activity) is called alteration; when a thing is able to be moved, it is movable, and locomotion is the actuality of the movable (*Physics* III, 1, 201a12–15). So too, "when the buildable insofar as we say it is such, is in actuality, it is being built and this is building" and likewise with several other examples, such as learning and doctoring.[79]

Here Aristotle specifies the relation between a mover and what is able to be moved. The potential both is and is able to be moved only in relation to its respective actuality, which moves the potential and is its fulfillment or activity. So, what is potential is always and only moved in exactly that respect, i.e., insofar as being potential it is relative to its proper actuality. Consequently, the same thing cannot be both potential and actual

78 Lear, *Aristotle: The Desire to Understand*, 36; cf. also Cooper, "Aristotle on Natural Teleology," 197–222.
79 Aristotle, *Physics* III, 1, 201a6–19; the example of learning is used to illustrate the motion of the elements at *Physics* VIII, 4, 255b-5.

in the same respect at the same time. Insofar as a thing is potential, it is able to be acted upon, i.e., to be moved, because it is relative to what is actual. But insofar as a thing is actual, it is fulfilled or complete rather than merely *able* to be such and so is that which produces motion in something potential. If a thing were both potential and actual in the same respect at the same time, it would be both incomplete and complete, able to be fulfilled and actually fulfilled – which would clearly violate the law of noncontradiction.

The target of this argument may again be Plato. According to Plato, self-moving soul, which is the origin of all motion in the cosmos, is an identity of mover and moved; indeed, soul resembles form in the sense that the self-identity of soul allows it to enter into a relation with body, which in itself is lifeless, and, as a consequence of this relation, soul moves body. Again, the pattern of agreement and disagreement between Plato and Aristotle is striking. They agree that "everything moved is moved by another";[80] but for Plato this proposition implies a first mover that must be *self-moved* as an identity of mover and moved, whereas for Aristotle it implies an *unmoved* first mover. The immovability of the first mover follows directly from Aristotle's division between potency, i.e., ability to be moved, and actuality, i.e., an actual mover, along with "the fact" that the moved is actively oriented toward its respective mover.[81]

This point is explained in *Physics* III, 1. While the same thing cannot be both actual and potential at the same time and in the same respect, it can be actual and potential in different respects (*Physics* III, 1, 201a20–21). Consequently, two things will act and be acted upon by one another in various ways – and what is an agent, i.e., actual, in one respect will be a patient, i.e., potential, in another (201a21–25). Although this complex relation, Aristotle claims, has led some to think that every mover is moved, later arguments will show that there is something that produces motion while being itself unmoved.[82]

---

80 Plato, *Phaedrus* 245C; *Laws* X 896A-C; Aristotle, *Physics* VII, 1 241b34, and VIII, 4, 256a2. On the argument of *Physics* VII, 1, as specifically directed against Plato's account of self-motion, cf. H. Lang, *Aristotle's* Physics *and Its Medieval Varieties*, 35–62.

81 The sequence of arguments in *Physics* VIII is telling on this point. Aristotle first argues that "everything moved is moved by another" (*Physics* VIII, 4) and only after establishing this proposition takes up the question of whether the first mover is unmoved or self-moved (*Physics* VIII, 5). And he rejects the self-mover for the unmoved mover (*Physics* VIII, 6).

82 *Physics* III, 1, 201a25–27. The reference may be to *Physics* VIII. The thinkers may be Plato and his students; Plato apparently never considers a case in which the cause of motion in the cosmos is unmoved. In both the *Phaedrus* and *Laws* X, motion by another immediately implies self-motion so that every mover is moved either by another or by itself. Furthermore, we may note again that Aristotle's use of the active voice and the verb κινεῖσθαι requires that it be passive, not middle. Wardy, *The Chain of Change*, gives an odd gloss on this argument: "In *Physics* VIII, 5, Aristotle attempts to establish the thesis that whenever something is changed, but not changed by itself, the ultimate source of the κίνησις must be an agent which *does* change itself," 89 (italics in original). But *Physics*

Having distinguished between mover and moved, Aristotle expands and clarifies his definition of motion. His emphasis again lies with the potential as movable. "The actuality of something potential (whenever being actual it operates not as itself but as movable) is motion" (201a27–29). For example, bronze is potentially a statue; but motion is not the fulfillment of bronze *qua* bronze because if to be something bronze and to be something potential were the same, they would be the same "absolutely and according to definition" [ἁπλῶς καὶ κατὰ τὸν λόγον], in which case the actuality of the bronze *qua* bronze would be motion (201a29–34). That they are not the same is clear from contraries such as sickness and health. For, on the one hand, to be capable of being sick and to be capable of being healthy are different, since if they were not, there would be no difference between being healthy and being sick; on the other hand, the subject of both health and sickness is one and the same (201a34–201b2).

The point concerns the ways in which a thing may be potential. Bronze, Aristotle's first example, may be potential either as bronze, i.e., according to its definition, *or* as a work of art. Motion need not be the actuality of a thing's definition, i.e., the bronze as bronze, but must be the actuality of some potential in the thing, e.g., potential to be a statue or a shield.

Several conclusions now follow. Clearly the actuality of the potential *qua* potential is motion.[83] The relation between potency and actuality is further defined: something potential is moved when its actuality occurs, and neither before nor after (*Physics* III, 1, 201b7). Because motion is a relation between actuality as mover and the potential as moved, there cannot be motion in the absence of actuality, i.e., the mover. The importance of this point appears in a moment, when Aristotle specifies that mover and moved must be together, and again, in *Physics* VII and *Physics* VIII, when he argues that "everything moved must be moved by something." Here

VIII, 4, concludes: "all moved things must be moved by something"; 5 begins with the assertion that this must come about in one of two ways: the mover either is moved by another or is moved by itself, and in the latter case, *if* everything moved is moved by something and *if* the first mover is moved, it must be moved by itself (256a19–21). And Aristotle provides several arguments for this conclusion. But the larger argument continues as Aristotle "makes a fresh start and asks: if a thing moves itself how and in what way does it do so" (257a31–33)? He reaches two unambiguous conclusions: (1) The self-mover does not impart motion as a whole nor is it moved as a whole; rather one part moves and another is moved (and the self-mover must always be divided into at least two parts) (258a25–27). (2) Clearly, that which first imparts motion is unmoved; whether we reach a first unmoved mover or consider a self-mover: "On both assumptions, the first mover for all moved things is unmoved" (258b5–9). This conclusion completes *Physics* VIII, 5. Furley, "Self-Movers," sums up the point: "It is evidently quite legitimate, in Aristotle's view in these chapters [*Physics* VIII, 4 and 5] to call the whole a self-mover, provided that the moving part is itself unmoved, except accidentally," 7.

83 On using "actuality" in this sense, rather than "actualization," cf. Kosman, "Aristotle's Definition of Motion," 43 (and *passim*). In this article, Kosman offers very telling criticisms of Ross's construal of this definition; cf. Ross, 361–362, 536–538.

the requirement of actuality follows from the definition of motion as the actualization of the potential *qua* potential, i.e., of motion as a relation between what is actual and what is potential.

Furthermore, the actuality is the actuality of the potential not after the motion is complete, e.g., when what is buildable has become a house, but while the potential is being actualized, e.g., the house is being built.[84] When motion is complete, there is only actuality, e.g., a house, and no potential as such to be actualized.[85] So motion entails that there must be an actuality, and it is the actuality of the potential insofar as it is potential. And motion is nothing other than this relation.

The force of this account, Aristotle claims in *Physics* III, 2, becomes clear if we consider what others have said and because motion is not easily described in any other way.[86] Motion cannot be explained in any other way because it cannot be placed in any other genus; the ways in which other thinkers place motion in their categories, e.g., inequality or nonbeing, and try (unsuccessfully) to explain it show how difficult the problem is. The problem is this: motion seems to be something indefinite that cannot be placed without qualification under either the potential or the actual because neither, taken by itself, is necessarily moved (*Physics* III, 2, 201b25–33). Motion may be thought to be actuality of a sort, but whereas actuality is complete, motion is incomplete, and the cause of this incompleteness is the potential (201b33–34).

The solution to the problem lies in seeing that motion is an actuality of sorts: an actuality of what is potential insofar as it is potential. This relation always obtains for things that are by nature because nature is a principle of motion and rest. Several points now follow that spell out the features of this relation. Because, as I shall argue, Aristotle's accounts of place and void are subordinated to his account of motion, these features provide the "purpose" for the later accounts. That is, to serve as terms "without which motion seems to be impossible," place and void must explain (or provide the conditions for) these features.

The first point concerns natural movers, whose immobility is rest: each produces motion by being together with the moved so that the mover both produces motion and is acted upon (202a7–9). Furthermore, the mover will always carry some form, and this form will always be the source and cause of the motion produced in the moved, as Aristotle's favorite example emphasizes: an actual man makes a man from what is potentially a man.[87] The requirement of being together and the identification of

---

84 Cf. Aristotle, *Metaphysics* IX, 8, 1049b30–1050a34.
85 On this point, cf. Hussey, 62.
86 Hussey schematizes this section and rightly points out that it serves "as indirect confirmation of Aristotle's own definition," 62–63.
87 Aristotle, *Physics* III, 2, 202a9–12. He uses this example in affirming that nature is primarily identified with form, *Physics* II, 1, 193b8; cf. *Metaphysics* IX, 8, 1049b24–25.

form as the source of motion are both crucial to Aristotle's account of motion as a relation of the potential to the actual.[88] I shall consider them in order.

Natural movers, because they are together with the moved and possess rest as their immobility, differ from the first unmoved mover, which is outside nature. The first mover contains no intrinsic ability to be moved; because it is unmoved both essentially and accidentally, it does not experience "rest," i.e., actuality or activity following upon the completion of motion. For this reason, the first mover is outside the cosmos and is not "by nature." Furthermore (although an account lies beyond this study), the first mover is not together – at least not in the same sense – with that which it moves; thus in this case alone, the mover is *not* also acted upon when it moves another (cf. *Metaphysics* XII, 7).

Natural things by definition contain an intrinsic source of being moved and being at rest. Thus when unmoved, a natural mover is characterized not as immovable, but as "at rest," i.e., movable but at present in the contrary state. In this sense, to be movable and to be at rest are closely related for natural things: they are contraries within nature and exclude reference to what is outside nature. Again, this account of motion supports the definition of nature as it contains two parts, i.e., being moved and being at rest.

Although Aristotle does not pursue the point here, he adds that natural movers move by being together with the moved, so that, when it produces motion, the mover is in its turn acted upon by the moved.[89] This point too may be anti-Platonic. According to Plato, soul as self-moving motion produces physical motion as its natural by-product, but soul is not itself affected by this by-product; for Aristotle, soul moves body but in so doing is at the same time accidentally moved by body (*De An.* I, 3, 406bff; 408a30–33).

In the mover, form is the source or cause of motion; form in turn is the highest expression of being, the first candidate for substance and actuality;[90] actuality is always identified with a thing as a fully developed complete reality, whether it be something natural or god outside of nature.[91] The identity between form and actuality unites being and motion in Aristotle.[92] Because actuality is identified with form, an account of

---

88 Cf. Weisheipl, "The Spector of *Motor Coniunctus* in Medieval Physics," for a variety of problems associated with this requirement, 99–120.

89 Aristotle, *Physics* VII, 2. In this argument, the first mover is not specified as unmoved, a point that puzzles some of Aristotle's readers. But in fact it may not be unmoved. Cf. H. Lang, *Aristotle's Physics and Its Medieval Varieties*, 35–62.

90 Aristotle, *Metaphysics* VII, 8, 1033b5–1034a7; IX, 6, 1048a34–1049b.

91 Aristotle, *Metaphysics* IX, 3, 1047a30–1047b; 6, 1048a31–35; XII, 7, 1072a25–26; 1072b5–8; 1072b23–29.

92 Being and becoming, of course, remain always separate in Plato; cf. *Republic* VI, 509D5–510C, the divided line, and *Timaeus* 27D5.

motion is always implicitly an account of being. In the fullest sense possible, these terms specify motion by specifying the relation between movers and moved things.

The actuality of the mover is not other than that of the moved, but is the actuality of both (*Physics* III, 3, 202a14–16). A thing is a possible mover [κινητικόν] because it is capable of producing motion in another (or in itself *qua* other), but it is a mover [κινοῦν] because (and when) it acts; because a thing is able to act, or actually acts, on the moved, the actuality of both mover and moved is one and the same, namely that of the mover (202a16–18). This issue is crucial for Aristotle's definition of motion as the actualization of the potential *qua* potential (by what is actual). Why does he insist that the actuality of the mover is the actuality of the moved, and what follows from it?

Most importantly, the form of the mover is the form of the moved. As Aristotle argues elsewhere, natural things possess only one substantial form (mules possess more than one form but cannot reproduce their own kind for precisely this reason); thus, to share the same actuality is to be identical in form. The form constitutes the being of a natural thing in a primary sense and, along with a reference to matter, is the proper object of the definition.

Consequently, the definition of the actual *is* the definition of the potential, and the potential has no definition apart from its actuality.[93] For example, the definition of acorn is potential oak, child is potential human, etc. What is actual, however, can be defined independently of the potential because the actual is more properly identified as being in the sense of form, i.e., being in the primary sense. Indeed, *because* the actual is prior in being to the potential, the potential is defined in terms of the actual while the actual is independent of the potential. Thus, the acorn is defined in terms of the oak, but the definition of the oak need not refer to the acorn.

Because the actual is identified with form and being in the primary sense, it serves as the mover of the potential: if nothing intervenes, something potential cannot fail to be moved by its proper actuality.[94] The importance of this point appears if we consider Aristotle's two causes of motion, moving cause and final cause. Although they differ in the sense that moving causes produce motion from the outside whereas final causes move as objects of desire, nevertheless both produce motion by being actual relative to what is potential. Hence, to identity a mover as such we need not know whether it is a moving cause or a final cause; rather, we need know only that the mover is actual relative to a potency and that nothing intervenes between them.[95] And in nature, Aristotle's teleology

---

93 Cf. Kosman, "Aristotle's Definition of Motion," 56.
94 This point is emphasized throughout *Metaphysics* IX; cf. esp. chap. 7.
95 Hence Aristotle's argument for the eternity of motion and time in *Physics* VIII, 1, operates exclusively in terms of actuality and potency and never mentions moving or final causes.

is not restricted to final causes, but includes moving causes as well because they too produce motion by being actual.

This point returns us to the relation between act and potency: potency never fails to be completely oriented toward its proper actuality; thus, if nothing intervenes between them, what is potential cannot fail to be actualized by its proper actuality.[96] What is potential is related to what is actual in two ways, (1) in virtue of its definition and (2) in virtue of an orientation toward the actual as its mover. On the one hand, the potential can neither be nor be moved apart from the actual; on the other hand, in the absence of hindrance the potential cannot fail to be moved by its proper actuality.

Again, the contrast between Aristotle and Plato is telling. For Plato, body (or matter) resists soul and the imposition of order that soul represents; for example, necessity, the errant cause of the *Timaeus*, requires persuasion, and even the prisoner freed from the cave in *Republic* VII must be dragged along by the scruff of the neck.[97] At the same time, form can be found in the physical world, the world of becoming, because and only because it has been imposed upon the recalcitrant receptacle by the demiurge using soul as his messenger.

For Aristotle, however, matter is neither resistant to form nor passive on contact with it; rather, matter yearns for form.[98] Form need never be imposed on matter because they go together "naturally." Thus, unlike artistic form and matter, natural form and matter go together without reference to any third cause: matter is moved immediately by the form that is its actuality and definition. This is the very definition – the specific difference – of things that are by nature as opposed to things that are by art.

Final causes present the clearest example of this relation because they move as objects of desire. But what of moving causes? The father is a moving cause of the child and the artist of the statue (*Physics* II, 3, 194b30–32). But the father moves "by nature" while the artist moves "by art." If both are moving causes, what differentiates the one as natural from the other as artistic?

The father and the artist are both moving causes because both move from the outside and both bring form to matter. But the artist *imposes* form on matter that is not naturally oriented toward it, and the imposition must continue until the work is complete; if the artist's activity is interrupted, then the work remains incomplete because matter has no orientation toward artistic form. The father, unlike the artist, brings form

96 Aristotle, *Physics* I, 9, 192a22–24; cf. *Metaphysics* IX, 5, 1047b35–1048a24. On the argument of *Metaphysics* IX, cf. H. Lang, "Why the Elements Imitate the Heavens," 335–354, and Kosman, "Substance, Being and Energeia," 121–49.
97 Plato, *Timaeus* 47E–48B2; *Republic* VII, 515E–516A3.
98 Aristotle, *Physics* I, 9, 192a20–24; cf. *De Caelo* I, 3, 270b23–25; *N. Ethics* I, 1, 1094a–2.

into contact with matter that naturally desires it. On contact, a new individual is immediately generated because in nature matter is never neutral to form, and the ensuing development is independent of the father (*Metaphysics* IX, 7, 1049a-18). The father causes not by imposition, but by nature, by bringing form (i.e., actuality) together with matter oriented toward that form. Consequently, in nature (unlike art), even though what is potential is moved (and in this sense is caused) by its proper actuality, what is potential is not thereby passive: in natural things what is potential is caused by its proper actuality because it is actively oriented toward it.[99] And this active orientation of the potential for the actuality that completes it lies at the heart of the order and teleology of nature.

This position stands in sharp contrast not only to Plato but also to later philosophy, including the Stoics and Philoponus. Frede argues that according to the Stoics, "if an object acts on another object so as to make it react in some way it does so by imparting a force or power to it: there is a transfer of force. . . . It is difficult not to suspect that this may be the ultimate source of Philoponus' theory of imparted forces."[100] And Frede is right: the source of Philoponus' notion of imparted force is not Aristotle, according to whom motion is not transferred from one thing to another at all, but is in the moved, itself aimed at its mover.[101]

A serious epistemic consequence – one that I shall now consider at length – follows from Aristotle's definition of the relation of potency to actuality. For natural things, because (1) the definition of the actual is the definition of the potential, *and* (2) the potential is by definition oriented toward what is actual, an identification of what is actual in relation to the potential completely explains motion. The condition required between mover and moved for natural things is mentioned here and further specified by Aristotle later: they must be together, or touching.[102] Given the orientation of potency to actuality, and the identification of actuality as the definition and mover of what is potential, actuality is always a cause of motion: [when together] the potential cannot fail to be actualized by what is actual.

Here we reach the full force of Aristotle's account of motion in *Physics* III. Motion is coextensive with nature because nature is a source of being moved and being at rest. Motion is always of something, always a relation between mover and moved. By definition, motion is the actualization of the potential *qua* potential by what is actual. In natural things, the

---

99 Although he makes it in somewhat different language, I believe this is, at least in part, Kosman's point, "Aristotle's Definition of Motion," 55–56.

100 Frede, "The Original Notion of Cause," 149–150.

101 On the development of matter as a passive principle by medieval Aristotelians, cf. Weisheipl, "The Principle *Omne Quod Movetur Ab Alio Movetur* in Medieval Physics," 89–90.

102 Aristotle, *Physics* VII, 1, 242b59; [see p. 37, n. 7], cf. also IV, 5, 212b30–33; *De Gen. et Corr.* I, 6, 323a25–30.

mover must be together with the moved and produces motion by virtue of form. The moved is actively oriented toward its actuality because there is but one actuality for mover and moved and because that actuality is the primary being and definition of both.

Aristotle's definition of motion as an actualization is crucial to his arguments concerning place and void – "terms without which motion [conceived as an actualization] seems to be impossible" (*Physics* III, 1, 200b21). As we shall see, place is a limit that differentiates the cosmos into actually "up" and actually "down"; thus, it serves as the actuality of what is potentially up, i.e., fire, of what is potentially down, i.e., earth, and of the middle elements, water and air. Furthermore, Aristotle rejects the void because it is undifferentiated; here he introduces the natural locomotion of the elements up and down because as potencies they require an actually up and an actually down to serve as their respective movers. He concludes that the void cannot cause elemental motion and asks "of what will it be a cause?" (*Physics* IV, 8, 214b13–17).

At the outset, the arguments concerning place and void are defined as accounts of that without which motion [in things] seems to be impossible. And they are in this sense subordinated to it: place causes motion whereas the void fails to do so; hence within the framework of this argument, place must be affirmed and the void rejected. The definition of motion as an actualization of what is potential *qua* potential establishes the conceptual framework for arguments about place and void.

3

Place

When, at the opening of *Physics* III, 1, Aristotle lists those things without which motion seems to be impossible, the infinite appears first, followed by place, void, and time. He examines the infinite after completing his account of motion (*Physics* III, 4–8). I omit examination of it here for two reasons. (1) As noted earlier, the infinite, place, void, and time are subordinated to the definitions of motion and nature, but they are not subordinated to one another. Thus it is not necessary to examine the infinite in order to examine place and void. (2) Furthermore, the infinite plays a largely negative role in the order of nature. There is no such thing as an infinite magnitude, and indeed the being of the infinite is matter and even privation (*Physics* III, 7, 207a35, 208a). However, when Aristotle identifies the infinite with matter and privation, he anticipates his account of place in an important way.

As matter or privation, the infinite is identified with what is surrounded rather than with what surrounds, which is form (207a35–36; 208a3–4). Place, as we shall see, is not itself form, but strongly resembles form: both are limits (*Physics* IV, 4, 211b13). Ultimately, Aristotle defines place as the first unmoved limit of that which surrounds (212a20–21). As I shall argue, place, as a first limit, serves as a cause of order: it renders the cosmos determinate in respect to "where things are and are moved." Hence, place is at once a term without which motion seems to be impossible and a cause of motion insofar as it is a source of order. We shall return to place in Part II because this issue reappears in *De Caelo* IV, and I shall conclude that as a source of order, place may be associated with the nature of the cosmos itself. And so I now turn to place.

Aristotle first considers place in *Physics* IV, 1–5, and then a special case of place, namely, "empty place" or void, in *Physics* IV, 6–9. His account of place (1) raises a number of problems associated with place, including a preliminary definition of what it means to be "in" place; (2) establishes four characteristics rightly thought to belong to place; and (3) rejects three false accounts of place. After these preliminaries, Aristotle (4) gives a true account of place and (5) concludes his argument. I shall consider these five moments of the account of place in order.

As I shall argue, the problems associated with place are entailed by natural things changing place and are, on Aristotle's view, the crucial problems that a true account of place must resolve. In this sense, his account of place presupposes – is "subordinated to" – his definitions of motion and nature, and it solves the problems posed by these definitions. Insofar as his account of place is designed to solve these problems, "place" is never considered independently of natural, i.e., moved, things. Rather, motion in things seems to be impossible without place precisely because it solves these problems and thereby accounts for where natural things both are and are moved. For this reason, place appears within the science of physics as a common and universal term without which motion [in things] seems to be impossible.

Problems Involved in Place

Physics IV, 1, opens by explaining why the physicist must understand place:

And likewise, just as concerning the infinite, the physicist must also know [things] concerning place, namely, if it is or not, and how it is, and what it is. For everyone thinks that things that are are "somewhere" (for what is not is nowhere; for where is a goat-stag or sphinx?) and of motion the most common and noble is according to place, which we call "locomotion." [208a26–32: Ὁμοίως δ' ἀνάγκη καὶ περὶ τόπου τὸν φυσικὸν ὥσπερ καὶ περὶ ἀπείρου γνωρίζειν, εἰ ἔστιν ἢ μή, καὶ πῶς ἔστι, καὶ τί ἐστιν. τά τε γὰρ ὄντα πάντες ὑπολαμβάνουσιν εἶναί που (τὸ γὰρ μὴ ὂν οὐδαμοῦ εἶναι· ποῦ γάρ ἐστι τραγέλαφος ἢ σφίγξ;) καὶ τῆς κινήσεως ἡ κοινὴ μάλιστα καὶ κυριωτάτη κατὰ τόπον ἐστίν, ἣν καλοῦμεν φοράν. Cf. also *Physics* VIII, 7, 260a28–29.]

Aristotle intends to explain "where" things are and "where" they change place – locomotion being the primary motion.[1] The only things that fail to be "somewhere" are things that are not; all [natural] things must be "somewhere." (God, of course, "is" but is unmoved and so is neither "by nature" nor "somewhere." Soul too "is" but is unmoved essentially; it is moved accidentally, because of its location in a body, i.e., "where" it is, and so is accidentally in place.)

"Where" [ποῦ] appears regularly when Aristotle lists the categories of being.[2] Although this word is sometimes [wrongly] translated "place," Aristotle's analysis of the problem of "where" things are here in *Physics* IV, reveals the important distinction between "where" [ποῦ] and "place"

---

1 On the priority of locomotion, cf. Aristotle, *Physics* VIII, 7, 260a27–261a13.
2 Aristotle, *Categories* 4, 1b26; 4, 2a1; 9, 11b11; 10, 10b23; *De Gen. et Corr.* I, 3, 317b10; *Metaphysics* V, 7, 1017a26. On the orthography of ποῦ, cf. Hoffmann, "Les catégories *où* et *quand* chez Aristote et Simplicius," 218–220.

[τόπος].[3] Everyone agrees, Aristotle argues, that all things that are must be "somewhere"; hence the inclusion of πού in the categories. And neither here nor elsewhere does Aristotle argue for this category; rather, he asserts it as both necessary and universally agreed upon. But "where" things are raises the question "what is the where of things?" – is it place, as Aristotle thinks, or is it a void (the empty), as his opponents believe?

The problem at stake for Aristotle's analysis of place and void appears here. In answering the question "where," "place" and "void" are not independent topics considered sequentially, and the question is not "does place exist?" or "does void exist?" Rather, because things that are must be "somewhere" [πού], the problem is whether place or void is the requisite "where" [πού]. The immediacy with which Aristotle first treats place [τόπος] and then void [κενόν] indicates that on his view they are the only possible answers to the problem of "where things are and are moved."[4] Consequently, they are among the common and universal terms without which motion in things seems to be impossible, and Aristotle considers them only vis-à-vis things that are by nature. "Void," as we shall see, is examined after place because it is a special case of place, i.e., place with nothing in it.[5] Whether place or void solves the problems posed by "'where' things are and are moved" is the sole issue in Physics IV, 1–9. Hence, as I shall argue, we learn nothing about place or void independently of this problem because the analysis that answers the question "where are natural things?" takes its starting point in the definitions of both motion and nature.

Next Aristotle provides evidence for the generally accepted view that place is. First, place seems to be different from all bodies that occupy place, e.g., where there once was water there is now air.[6] Second, the re-

3 For the translation of πού as "place" in the Categories, cf. Creed and Wardman, The Philosophy of Aristotle, and the Oxford translation of Edgehill; in the De Gen. et Corr., see the Clarendon Aristotle Series translated by Williams (the Oxford translation of Joachim translates πού as "position"). Here in the Physics, Hussey translates it as "somewhere" and Apostle as "where," although in his translation of the Metaphysics Apostle uses "whereness" and in that of the Categories "somewhere"; Ross in the Oxford translation of the Metaphysics renders it "where." William of Moerbeke translating the Metaphysics renders πού as ubi, not locus. Ackrill translates πού in the Categories as "where" and adds the following note, which is very much to the point: "'Where', 'when': the Greek words serve either as interrogatives or as indefinite adverbs ('somewhere,' 'at some time'). 'Place' and 'time' are best kept to translate the appropriate Greek nouns," Ackrill, "Aristotle's Categories, Chapters I–V," 108.
4 Aristotle's procedure here, going directly from a problem to specific possible answers, is found in other arguments. Cf. Metaphysics XII, 1, where he begins by saying that the subject of the investigation is οὐσία and then immediately subdivides οὐσία into three kinds. On the force of this procedure, cf. H. Lang, "The Structure and Subject of Metaphysics Λ," 257–280.
5 Cf. Aristotle, Physics IV, 1, 208b25–27; 6, 213a14–19.
6 Aristotle, Physics IV, 1, 208b–8. Referring to this point (and to the next concerning locomotion), Furley comments, "We need the concept of place, he says, for two reasons: (1) we want to talk about displacement . . . (2) we want to talk about natural place, i.e. the destination, the resting place, of bodies with respect to their natural motion," "Aristotle and

spective locomotions of each element show not only that place is but also that "it has some power."[7] Each of the four elements, unless hindered, is carried to its own place, e.g., fire is always carried up and earth down.[8] The distinctions among places, i.e., up, down, left, right, front, and back, are not just relative to us, but are given in nature.[9] Because these distinctions are given in nature and because each of the elements always goes to the same place when nothing hinders or interferes, i.e., when their motion is natural, it seems that places differ not only by position, e.g., up or down, but also by possessing distinct powers (*Physics* IV, 1, 208b19–22).

Aristotle's evidence for place provides evidence about Aristotle's place. First, this evidence derives directly from things that are – the elements and things made of them, i.e., all natural things (and artifacts insofar as they are made of natural things). Hence all things that are and are moved, whether by nature or by art, bear a direct relation to place; consequently, a correct account of place will in this respect explain all things that are either by nature or by art. Second, the cosmos is determinate – in and of itself the cosmos (and place within it) are determined as "up," "down," etc. These determinations are not just relative to us, but are given in nature. Aristotle connects the determinateness of place to the regularity of elemental motion and cites it as evidence that place "has some power." His reliance on the notions of regularity and determinateness indicates that he operates within his basic view of nature as a principle of order.

The notions of nature as orderly and place as determinate clearly contrast with the modern concept of "space," which is by definition indeterminate. As indeterminate, space is internally undifferentiated – two spaces are identical, if they are of equal dimensions – in contrast to Aristotle's concept of place, which is immediately and intrinsically differentiated in six directions.[10] Therefore Aristotle's concept of τόπος is not the

---

the Atomists on Motion in a Void," 88; but Aristotle here is providing evidence that place is.

7 Aristotle, *Physics* IV, 1, 208b10–11: ἀλλ' ὅτι καὶ ἔχει τινὰ δύναμιν. . . . Hardie and Gaye translate this phrase as "exerts a certain influence." Machamer, "Aristotle on Natural Place and Natural Motion," 378, rightly objects to their translation.

8 I translate the verb here [φέρεται γὰρ ἕκαστον . . .] as passive, in part because we have seen in *Physics* II, 1, that the elements are by nature and nature is a principle of being moved and in part by reference to an argument that will not be considered here, namely, *Physics* VIII, 4. There Aristotle argues that "everything moved is moved by another" and explicitly argues the case of the elements. It is interesting to note that the Oxford translation of Hardie and Gaye also takes this verb as passive, whereas the translation of Hussey makes it active (cf. also his commentary, 101). Ross connects this reference to the power of place to *De Caelo* IV, 3, 310a22, where place is apparently identified as form. The topic lies beyond the bounds of this study, but Philoponus uses the notion of the power of place to develop the notion of ῥοπή as an intrinsic mover in the elements. Cf. *In Phys.* 499.2–13, 18–23.

9 Aristotle, *Physics* IV, 1, 208b14–19; cf. *De Caelo* I, 4, 271a26–35.

10 On a moment in the history of philosophy in which Aristotle's account of place is redefined into space in such a way that the difference between them becomes apparent, cf. H. Lang, "The Concept of Place," 245–266.

same as the modern notion of "space," and it is misleading to attribute a notion of space to Aristotle.[11]

The claim, which Aristotle does not explain, that place has "some power," presumably in respect to the elements, presents two serious problems. (1) The word here, δύναμις, is his usual word for potency. Thus, Machamer, for example, claims that a "more literal translation would read: 'it (place) has some potency.'"[12] But there is no indication that Aristotle intends it in a technical sense here. It is more natural (and no less literal) to take the word in the ordinary sense of "power," as do Hussey and Ross.[13] Indeed, δύναμις appears two more times in less than a page and in each instance is best translated "power."[14] (2) Machamer and Hussey both find this claim difficult on substantive grounds. In little more than a page, Aristotle will deny that place can be any of the four causes. Given this denial, Hussey concludes that place cannot be "an agent of change" and thus it is hard to see how it can have power.[15] Machamer too notes that place cannot be among the four causes, but draws the opposite conclusion: "What is needed is an account of the power of place which is somehow not causal and also an account of the causes of the natural motion of the elements."[16]

Aristotle's remarks here serve to provide evidence that place is. This evidence leads him to emphasize the relation between the elements and place, i.e., place has power. But this relation, and hence the force of this phrase, remain unspecified beyond the claim itself. He next argues that although place and body are in the same genus, place cannot be a body, and then he raises the question of how place may be a cause. Again, he gives virtually no constructive specification of his position, and the reason why is not difficult: the purpose here is to identify issues requiring solution. The solution itself will appear later, in Aristotle's constructive ac-

---

11 Hussey, 99: "'About place': Aristotle's use of the noun τόπος ('place') is variable in the same ways as ordinary English use of 'space.'" Cf. also Sambursky, *The Concept of Place in Late Neoplatonism*, 11, who speaks of "space or place" as if these terms were interchangeable. He also claims that place for Aristotle is passive, which, I shall argue, is false. Grant, "The Principle of the Impenetrability of Bodies in the History of Concepts of Separate Space from the Middle Ages to the Seventeenth Century," 551–571, uses "place or space" throughout. Sorabji makes no distinction between "place" and "space," using the term "space" throughout *Matter, Space, and Motion;* for one clear example, cf. 142 for "Aristotle's rejection of extracosmic space . . . " These examples can be multiplied many times over.

12 Machamer, "Aristotle on Natural Place and Natural Motion," 378.

13 Hussey translates at 208b8: "Again, the locomotions of the natural simple bodies (such as fire and earth and the like) not only show that place is something but also that it has some power, . . ." Cf. Ross, 563.

14 So Hussey translates it. Cf. Aristotle, *Physics* IV, 1, 208b22, 34.

15 Hussey, 101.

16 Machamer, "Aristotle on Natural Place and Natural Motion," 378. For the same view, along with a very interesting account of this problem among Aristotle's immediate successors, cf. Algra, "'Place' in Context," 150 ff.

count: as the first limit of what surrounds, place is a cause of order without being one of the four causes. And as a cause of order, it has power and is a term "without which motion seems to be impossible."

The evidence that place is concludes with a reference to Hesiod – even the poets grant priority to place.[17] "And if this is such, the power of place is something marvelous and prior to all things" (208b34–35). Because all things are composed of the elements, nothing will be without place; but place can be without the others and therefore must be first (208b35–209a2). What this power is and how it operates remains to be seen, but we may anticipate Aristotle's solution to this problem: respective proper place is the actuality and hence the mover of each element. This account, according to Aristotle, is the best because it solves these problems by explaining "the facts" presented by natural things: the regular natural motion of the elements out of which all things, natural and artistic, are composed (cf. *Physics* II, 1, 192b19–20). In short, the order of nature and the fact of that order are explained.

Now Aristotle specifies the the first step in his investigation of place: "... there is the problem if place is, what it is, whether some bulk of body or some other nature. First, let us investigate its genus. For it has three dimensions, length, breadth, and depth by which all body is bounded" (209a2–6). Many commentators, we shall see, claim that for Aristotle place is a surface. Because a surface must be two-dimensional, place too must be two-dimensional. But here Aristotle unequivocally asserts that place has three dimensions. As three-dimensional, it is in the same genus as body, which raises a serious problem: place might be thought to be body – which is impossible (209a6).

Aristotle argues that even though both place and body are in the genus of three-dimensional entities, place cannot be a body. In his summary of this argument, Lettinck gives a very different sense of what is at stake. The issue he raises is important because it anticipates the position (and supporting arguments) that lie ahead.

But if we start to investigate what place is, we are confronted with several problems which make us even doubt whether it exists: 1) A body has three dimensions, so the place of a body must be three-dimensional too. Therefore, place is also a body (everything that has three dimensions is a body), and consequently two bodies would coincide, which is impossible.[18]

Several problems here reappear in various guises later. (1) Aristotle declares it settled that place is before this argument begins (*Physics* IV, 1, 209a3). Hence, the point of this argument cannot be to throw doubt on

---

17 Aristotle, *Physics* IV, 1, 208b27–33. I have argued elsewhere that references to the poets can serve as "punctuation marks" that close a given section of argument; cf. H. Lang, "The Structure and Subject of *Metaphysics* Λ," 257–280.

18 Lettinck, *Aristotle's* Physics *and Its Reception in the Arabic World*, 268.

the existence of place. (2) Here he says that he will investigate the genus of place, namely, that it is in the genus of three-dimensional things. If we anticipate his position, we see that place cannot be a body, but must be three-dimensional because it is the first unmoved limit of the containing body. In short, it is false that "everything that has three-dimensions is a body." (3) A puzzle results that must be solved by Aristotle's account, if it is to be successful: place must be in some sense three-dimensional without being a body.

The entire point of this section of the argument is to identify the most important problems that any true account of place must be able to solve. That an account *can* solve them bears witness to its truth as an account of place. Here I shall simply identify the problems. We shall see them more fully later – when Aristotle clearly thinks that his account meets the test.

If place were a body, there would be two bodies in the same place, namely, place and the body in place, which is impossible (209a6–7). Again, place might be thought to be a limit of body. But this view too is problematic. The limits of a body, i.e., its surfaces, must, like the body itself, be in place and so place must be distinct from them too (209a7–9). If we think of a point, which is without dimension, we reach the most difficult case because it and its place will not be distinct (209a9–13). In short, however place is defined, the requirement that these things too, i.e., body, surface, and points, have a place distinct from themselves must be met.

This problem is formidable. Place must have magnitude because it belongs to the genus of things having three dimensions. But it can neither be an element nor be composed of elements, whether the supposed elements be thought of as corporeal or incorporeal. Consequently, although it has magnitude, place can neither be body nor come from "intelligible things," because such things do not produce magnitude (209a13–18). I shall argue that place is unique: it is three-dimensional but not a body, a limit but not a surface. These characteristics and an account of place based on them allow Aristotle (I believe) to solve these problems.

Without pause, Aristotle raises a second equally serious question about place: Of what in things is place a cause (209a18–19)? For it cannot be one of the four causes, i.e., neither the matter, nor the form, nor the end, nor the moving cause (209a18–22). The puzzle here lies in an expectation that place will be a cause of something, but what that is remains entirely unclear. Later, the same question will be asked about the void, and Aristotle will argue that its proponents think the void causes motion when in fact it fails to do so (*Physics* IV, 7, 214a24–25). This failure, i.e., the failure to cause motion (or anything else), lies at the heart of his rejection of the void. Conversely, his conclusion that place succeeds as "the where" of things that are and are moved (and hence is a term without which motion seems to be impossible) seems finally to assert that somehow place must be a cause. As I shall conclude, place, resembling form, is a primary cause of the order of nature.

The problem of how place can serve as a cause raises a broader issue that I shall address here only briefly, as we must return to it with a fuller account of place. Hussey – and he is not alone in this – takes Aristotle's denial that place can be one of the four causes as a rejection of the claim that place is a cause at all.[19] Sorabji comments that "the denial . . . that place can serve as any of the four causes, or modes of explanation, is merely part of a puzzle or *aporia*."[20] And Sorabji is clearly right – Aristotle is defining the "puzzles" involved in the notion of place, not setting out his account.

However, beyond this methodological point lies substantive issues, first about the status of the four causes and second about the very possibility of a cause in addition to these. Hussey's argument implicitly claims that if something cannot be identified as one of the four causes, it cannot be a cause at all. That is, the four causes provide a systematic and exhaustive set of all possible causes. But Aristotle clearly does not intend them in this sense and, indeed, does not himself use them in this way.[21] For example, his accounts of motion, of the eternity of the world, and of god all turn on the notion of actuality as a cause – and it is not one of the four causes.[22]

This point may be made more specifically for the issues at stake in *Physics* III and IV: the definition of motion as a relation between moved and mover does not require that we identify one of the four causes; rather, it requires only potency relative to proper actuality. Furthermore, the four causes explain things that are by nature, and these things appear within the cosmos and are composed of both matter and form.[23] Given that place cannot be another body within the cosmos, it cannot be another thing that is "by nature" – at least not in the same sense that contained bodies are. Because the four causes operate in respect to all natural things

19 Hussey, 103: "For the four Aristotelian 'causes' (types of explanatory factor), see *Physics* II.3. *Prima facie* place will not figure as an explanatory factor of any type, and so will be superfluous in science . . ."; for the same view, cf. Machamer, "Aristotle on Natural Place and Natural Motion," 378. Arguing in a rather different context that the four causes are not "results of an analysis of linguistic usage," but are "grounded in more ambitious if hardly recondite theoretical preoccupations: the identification of explanatory projects," Schofield seems to agree with Hussey that the four causes exhaust the possibility of causal explanation. "Explanatory Projects in *Physics*, 2.3 and 7," 38.
20 Sorabji, "Theophrastus on Place," 160, n. 6.
21 For a very sensible note on this point, cf. Sprague, "The Four Causes," 298–300.
22 Aristotle, *Physics* III, 1, 200b26–201a4, 202b27; VIII, 1, 251a9–251b10, 251b30–252a6; *Metaphysics* XII, 6, 1071b22; 7, 1072b7, 20–30.
23 The best-known account of the four causes is *Physics* II, 3, 194b17–195a2; 7, 198a15–198b8. On the necessity of including matter within nature cf. *Physics* II, 2. One may object that god in *Metaphysics* XII, 7, appears as one of the four causes, a final cause, outside nature. But this objection is not valid. That god acts as a final cause need not mean that god is one of the four causes in nature, and indeed, in *Physics* II, 7, a final cause outside nature is mentioned but not discussed precisely because it is outside nature.

and place is not another natural thing, we should not expect the account of place to involve the four causes. Aristotle confirms this view and, saying that place cannot cause as if it were one of the four causes, he could hardly be more explicit: place does not cause as does matter, as does form, as does an end, and it does not move things that are (*Physics* IV, 1, 209a20–22). Nevertheless, place does seem to be a cause, i.e., to be responsible in some sense, and it appears on the list of things without which motion seems to be impossible. Again, an aporetic conclusion follows. If Aristotle's account of place is to be successful, it must solve this puzzle: in what sense is place a cause without being one of the four causes?

Another problem concerning place is raised by Zeno: if everything that is has a place and place is, place too requires a place, and it a place, etc. (209a23–25). This problem, as we shall see, is less difficult. Place, as Aristotle defines it, constitutes place for all things within the cosmos, but is not itself in place; hence it need not have a place, and there is no infinite regress.

Finally, just as every body is in place, so every place has a body in it.[24] And place is neither less than nor greater than things in place. If a thing grows, does its place grow with it? A closely related problem appears later: if the cosmos is already full, how can place grow? Would the whole bulge? Aristotle will argue that there is no need for place to change to accommodate "local" changes within the cosmos (*Physics* IV, 5, 212b24).

This introduction to place shows that the questions "what place is" and even "if it is" present problems that must be resolved by a successful answer to the more general question of "'where' things are and are moved" (*Physics* IV, 1, 209a29–30). And after defining and giving his account of place in *Physics* IV, 4, Aristotle explicitly returns to and (on his view) resolves these problems. We may conclude that he takes them to be the most serious and difficult problems that can be posed for any account of place. Hence from his viewpoint, his ability to resolve them firmly establishes the success, the superiority, of his account of "'where' things are and are moved." But before turning to his own account, Aristotle raises two additional sets of difficulties beyond the question of if or what place is. These will be treated at length later, and they too establish what an account of place must account *for*.

Something may be said of a subject either in virtue of itself or in virtue of another; for place there is the common (κοινός) place in which all bodies are and there is the proper (ἴδιος), i.e., "local," place in which a body first is.[25] So, for example, individuals are in their local places, which are

---

24 Aristotle, *Physics* IV, 1, 209a26–27; cf. *Physics* III, 5, 206a2.
25 Aristotle, *Physics* IV, 2, 209a31–33. We may note that the Hardie and Gaye translation of the second part of the sentence is confused: "and there is place which is common and in which all bodies are, and which is the proper and primary location of each body." Hussey has it right: "and place may be either (a) the 'common' place in which all bodies are; or (b) the special place which is the first in which a body is."

"in" the earth, the earth is in the air, and the air in the heaven; therefore individuals may be said to be in the heaven.[26] So individuals are in "common" place by virtue of being in another (i.e., earth, air, and heaven), and they are in proper, or local, place immediately and in virtue of themselves. If we look at what primarily contains, i.e., local place, it would be some kind of limit and would appear to be the form or shape of each body; but if we view place as "the interval of the magnitude" [τὸ διάστημα τοῦ μεγέθους], it appears to be matter.[27] Here these relations are raised as problematic, but Aristotle offers few clues to his position. Ultimately, he will reject any identification of place with matter and form; he will define place as a limit, but as the limit of the container (unlike form, which is the limit of the contained).

The distinction between common (κοινός) and proper (ἴδιος) place requires a moment's reflection. As we have seen, at *Physics* III, 1, Aristotle announces that the upcoming investigation, including that of place, will consider what is "common and universal for all things" [πάντων εἶναι κοινὰ καὶ καθόλου ταῦτα] while the investigation of proper things will come later [ὑστέρα γὰρ ἡ περὶ τῶν ἰδίων θεωρία τῆς περὶ τῶν κοινῶν ἐστιν]. After he defines place as a limit, he turns immediately to the heaven: because place is the first motionless boundary of what contains, the middle of the heaven and the extremity are thought to be up and down in the strict sense.[28] In short, this account of place is immediately identified with common rather than local (or proper) place, the heaven in which all things that are are. However, a successful account of place must explain *both* senses of being in place. And although the account of *Physics* IV is primarily an account of place not in the proper but in the common sense, Aristotle believes – so I shall argue – that his account does explain both senses. Place turns out to be "common and universal" in just this sense: a single "place," acting as a determinative principle, accounts for "the where" of all things that are and are moved.

Aristotle next comments on Plato's account in the *Timaeus* and proceeds with problems consequent on identifying place either as form or as matter, although he will ultimately agree that it is a limit. If place were either matter or form, we would expect difficulties, but it is not hard to show that it is neither (*Physics* IV, 2, 209b17–22). In natural things, form and matter are not separable in fact from the individual, but place is (209b22–31). That is, place is separable from what is in place, and a true account of place must explain its separability – as Aristotle believes his account does.

26 Aristotle, *Physics* IV, 2, 209a33–209b; this example is repeated at *Physics* IV, 4, 211a23–29. For an interesting account of a skeptical treatment (that of Sextus Empiricus) of this distinction, cf. Burnyeat, "The Sceptic in His Place and Time," 232–240.

27 Aristotle, *Physics* IV, 2, 209b1–10. The translation of διάστημα will be considered shortly.

28 Aristotle, *Physics* IV, 4, 212a20–23. Ross comments on 209a32 that κοινὸς τόπος "is co-extensive with the οὐρανός," 565.

Furthermore, if place were either matter or form, how could a body be carried to its own place (210a2–3)? Again, the question concerns the relation between the elements and place: "it is impossible for that which has no relation to [either] motion or up or down to be place; therefore, place must be sought among such things" (210a3–5). Because motion must occur in what is movable and all things that are movable are composed of the four elements, whatever place is, it must somehow be causally related to the natural motion of the elements (although without being one of the four causes).

Finally, when a thing changes place, the place may be said to have been destroyed in the sense of no longer being occupied by that thing. But what can such "destruction" mean (210a10–11)? Aristotle does not answer the question here; but given that on his view place must be separable from what is in place, we may predict that place will be destroyed only in the weakest and most accidental sense.

Aristotle now raises the problem of the different senses in which one thing is said to be "in" another, distinguishing eight senses of "in" and briefly indicating how the last will solve Zeno's problem that place will require a place (*Physics* IV, 3, 210a14–210b22). A full account of the meaning of "in" when we say "a thing is 'in' place" awaits the constructure account of *Physics* IV, 4. Lewis suggests that even here Aristotle lists ordinary uses of the phrase "one thing is *in* another" not so as to select the most appropriate common usage "but to dissociate his technical use of 'in' from *any* of the ordinary uses."[29] When Aristotle develops and defends his own definition and account of place, Lewis's conclusion, it seems to me, is quite right; but the technical sense of "in" required by an account of place does not become apparent until that point in the argument.

(Although Aristotle does not raise the closely related problem here, his first concern after defining place will be to explain how the heaven is [or is not] in place. Is the heaven in place in the same sense in which any body is in place? The answer will turn on the meaning of "in" and is central to the success of any account of place. Perhaps Aristotle omits this question here because it is very technical and this "preliminary" account of the meanings of "in" will not sustain even the question.)

Here the list of meanings seems to be the ordinary uses forming part of the backdrop to Zeno's problem and to the problem generally of how a thing is *in* place. Of the eight senses given, only the last is immediately relevant. (1) A part is "in" the whole while in another sense (2) the whole

29 F. Lewis, *Substance and Predication in Aristotle*, 58–59, n. 17. Lewis relates this problem to comments about the meaning of "in" in the *Categories*, cf. 2, 1a24–25; 5, 3a29–32. We may note that his view specifically criticizes that of Ackrill, 74, and agrees with Dancy, "On Some of Aristotle's First Thoughts about Substance," 344–347. For a more general discussion of the origins of technical terms and their relation to ordinary language, cf. Edel, "'Actions' and 'Passions,'" 59–64.

is "in" the parts; (3) a species is "in" a genus while in another sense (4) the genus is "in" the species; (5) form is "in" matter; (6) in a sense, things are "in" their first mover; (7) in a sense, things are "in" the end, that for the sake of which. Most importantly, for the problem of "where" (8) a thing is "in" a vessel and generally "in" place (*Physics* IV, 3, 210a14–24). A number of problems follow for place, and here I would suggest we find the first clues to Aristotle's own position.

The first problem concerns whether a thing can be in itself. The short answer is that a thing cannot be in itself directly, but a whole can be in itself in the sense that all the parts are in the whole. For example, given an amphora of wine, neither the wine nor the amphora are in themselves (respectively) but both are parts of the same whole (210a30–33). Furthermore, although a thing cannot be in itself primarily, it can be in itself in the sense that white is in body or science in mind (210a33–210b).

Aristotle extends the example of the wine and the amphora, in part in anticipation of his own view that place is a container of sorts while the cosmos is contained. Within the whole, the two parts remain distinct in virtue of what each is: the amphora is the container and the wine contained (210b-17). As we shall see with place and the cosmos, or the heaven, in order to say that a thing is in place the distinction between container and contained must be maintained. This distinction will be central to Aristotle's account.

Aristotle returns to his first point to conclude his preliminary analysis of "in": a thing can be "in" itself neither accidentally nor essentially (210b13–22). This distinction yields the solution to Zeno's problem of place being in place and the consequent infinite regress: nothing prevents place from being in place – not as another thing contained and so requiring a container, but as the container; thus there will be no infinite regress (210b24–27). Furthermore, because form and matter are always identified with what is contained, Aristotle again concludes that place can be neither form nor matter (210b27–31).

Because an account of place must resolve these problems, they set the program for such an account. First and foremost, place must have "some power" and in some sense serve as a cause for the natural motion of the elements. But it can be neither one of the four causes, nor an element, nor composed of the elements. Being three-dimensional, it must have size while in some sense being separable from and prior to natural things that are "somewhere." Finally, place cannot be "in" another as a thing contained is "in" its container and so can be neither matter nor form. Rather, place and what is in place will form a whole as container and contained.

These requirements for a successful account of place are formidable. But then so too, as I shall argue, is Aristotle's account. The power of place lies in its operation as a cause – a cause without which motion seems to be impossible. Now he is prepared to begin his account.

The Four Characteristics of Place

Having completed the preliminaries to a successful account of place, Aristotle begins his constructive account by listing the characteristics rightly thought to belong to place in virtue of itself.[30] Because they belong to place, these characteristics provide a foundation for a true account of place. But Aristotle's list – and hence the characteristics themselves – is problematic. The divisions among these characteristics and hence their number – are there four or six? – as well as their substance and the relations between them is unclear. I shall give my own divisions, but indicate important alternatives in parenthesis. (1) Place is what first surrounds that which is in place and is no part of the thing contained. (Ross and Hussey find two characteristics here: [1] place surrounds and [2] is not part of the thing.) (2) Again, place is neither less than nor greater than the thing contained; (3) again, place is separable from what is in place. (4) "In addition to these, every place has 'up' and 'down,' and each of the bodies [i.e., the four elements] is carried by nature and [they] remain in their proper places, and this makes either up or down." (Ross and Hussey divide this characteristic: [1] place has "up" and "down" and [2] each of the bodies . . .)[31]

The divisions (and hence the number) of the characteristics of place carry important substantive implications for Aristotle's arguments. First, there is linguistic evidence. Aristotle, as we shall see again later, divides his lists internally with the disjunctive "ἔτι" (again);[32] this word appears before points (2) and (3) (as I show them); the final point is demarcated

---

30 Aristotle, *Physics* IV, 4, 210b33: Again, the verb δοκεῖ may simply mean "seem," but, unlike the argument of *Physics* III, 1, we are clearly in the realm of accounts of place here.

31 Aristotle, *Physics* IV, 4, 210b34–211a6: ἀξιοῦμεν δὴ τὸν τόπον εἶναι πρῶτον μὲν περιέχον ἐκεῖνο οὗ τόπος ἐστί, καὶ μηδὲν τοῦ πράγματος, ἔτι τὸν πρῶτον μήτ' ἐλάττω μήτε μείζω, ἔτι ἀπολείπεσθαι ἑκάστου καὶ χωριστόν εἶναι, πρὸς δὲ τούτοις πάντα τόπον ἔχειν τὸ ἄνω καὶ κάτω, καὶ φέρεσθαι φύσει καὶ μένειν ἐν τοῖς οἰκείοις τόποις ἕκαστον τῶν σωμάτων, τοῦτο δὲ ποιεῖν ἢ ἄνω ἢ κάτω. Cf. Ross, *Physics*, 374. Hussey's translation reads: "We require, then, (1) that place should be the first thing surrounding that of which it is the place; and (2) not anything pertaining to the object; (3) that the primary [place] should be neither less nor greater (than the object); (4) that it should be left behind by each object [when the object moves] and be separable [from it]; further, (5) that every place should have 'above' and 'below'; and (6) that each body should naturally move to and remain in its proper places, and this must do either above or below." Hussey's division is repeated with his numbers and without comment by Lettinck, pp. 278–279. Hardie and Gaye, as published in 1930, subdivide the first characteristic, as do Ross and Hussey, but do not subdivide the last. For a more natural translation, cf. Barnes's correction of Hardie and Gaye: "We assume first that place is what contains that of which it is the place, and is no part of the thing; again, that the primary place of a thing is neither less nor greater than the thing; again, that place can be left behind by the thing and is separable; and in addition that all place admits of the distinction of up and down, and each of the bodies is naturally carried to its appropriate place and rest there, and this makes the place either up or down." Apostle lists the characteristics as I do, but fails to translate the πρὸς δὲ τούτοις.

32 For a parallel use of this word, cf. *Physics* IV, 8, 215a14, 19, 22, 24.

by the stronger disjunctive phrase "πρὸς δὲ τούτοις." Ross and Hussey further subdivide Aristotle's list of characteristics at the appearance of the conjunction "and" [καί].[33] However, "and" is never used as a disjunction – disjunction of individual points within the list is the explicit function of ἔτι and πρὸς δὲ τούτοις and is the very point of difference between these words and καί.

The issue here is not only grammatical or stylistic. Conjoined terms work together substantively and disjoined terms do not. The conjunction "καί" can be used in an explicative sense for precisely this reason. So, for example, in the first characteristic, the connective "καί" indicates that "to surround" is either closely related to, or even means, to be no part of the thing contained. Separating these points, as if "and" were a disjunction rather than a conjunction suppresses this relation. In either case, i.e., whether we take the "and" as an explicative or in a weaker sense, those things that it conjoins must be kept together.

Hence there are not six but four characteristics of place, and two of them are complex. More importantly, each characteristic asserts a specific relation between place and what is in place. Looked at methodologically, this relation implies that there is no identification or specification of place apart from things that are in place. Finally, because these four characteristics are "rightly" thought to belong to place, they establish the foundation for a true account of place: it must account for these characteristics.

The first three points are not difficult. (1) That place is what first surrounds without being part of the contained reflects the discussion of "in," which distinguishes between the amphora and the wine as container and contained. As we shall see, it reappears in Aristotle's definition of place.[34] (2) The claim that place is neither less than nor greater than what is contained in place establishes a primary condition for a thing's being "in" place. (3) Place and what is in place must always be separate; Aristotle will argue in a moment that they are never continuous, but must be divided.

The fourth characteristic, that place is differentiated and that this differentiation, in conjunction with the elements being carried by nature to their natural place, constitutes "up" and "down," is problematic. Its meaning certainly seems obscure, although it seems to reflect the earlier claims that the regions of place are not just relative to us and that place seems to have "some power" relative to the elements. I shall argue later that the differentiation of "up," "down," etc., is the most important feature of place both for Aristotle's account of place as in some sense required by

33 In Hussey's translation, this part of the sentence divides into two characteristics marked off by the "καί"; but this conjunction never indicates a division of this sort.

34 This was apparently a common assumption in ancient physics. Cf. Plato, *Parmenides* 138b2–3. On its appearance in Plotinus as a common assumption, cf. Strange, "Plotinus on the Nature of Eternity and Time," 41.

things in motion and for his account of elemental motion in the *De Caelo*.
For now, several conclusions follow about this characteristic, but they are
quite limited. First, it is no accident that this fourth and last characteristic is set off by the strong phrase "in addition to these"; second, its two
parts, conjoined by "and," must somehow be related; and, finally, it completes the specification of the characteristics of place: thus, it is the most
important characteristic of place.

To be κάλλιστα, an account of place, Aristotle now claims, should
explain what place is in such a way that problems associated with it are
solved, characteristics thought to belong to it do in fact belong, and, finally, the source of the difficulties concerning place should become clear
(and be resolved) (*Physics* IV, 4, 211a7–11). That is, the account must explain (1) "where" things are, (2) motion according to place, (3) why the
characteristics just given attach to place, and finally, (4) why other
philosophers are confused about place. And Aristotle intends to provide
just such an account (cf. *De An.* I, 1, 402b22–25).

These remarks may be thought of as methodological insofar as they
specify the job to be done "most effectively." But they are not methodological in a post-Cartesian sense. (1) They neither open the argument
nor precede the determination of any of its content; they follow a long introduction to the problem of place and an initial listing of the characteristics rightly thought to belong to place. (2) Rather than being separate
from or prior to the substantive problems of place, these remarks refer to
these problems and "attributes" and the need to consider them. Thus they
are best thought of as a "marker," summing up the criteria for the upcoming account and indicating that the introduction is over. The account
now begins – or begins again.

The next sentence begins: "First, therefore, it is necessary to understand that place would not be considered if there were not motion according to place [i.e., locomotion]" (211a12–13). That is, Aristotle provides evidence "that place is" – a problem explicitly raised earlier. And the
heaven most of all is thought to be in place because it is always in motion,
i.e., it always exhibits circular locomotion (211a13). In fact, two kinds of
motion involve place and so lead us to think about it: (1) locomotion and
(2) increase and decrease.[35] The latter involves place because what was
in one place has changed to another that is larger or smaller (211a15–
17). In this sense, increase and decrease presuppose locomotion; hence,
locomotion is prior and an account of it will also account for change of
place during increase and decrease.[36]

Locomotion, in its turn, raises further distinctions because, as we saw

---

35 Aristotle, *Physics* IV, 4, 211a14–15; cf. *Physics* VIII, 7, 260a27ff, where alteration too is
    shown to involve and in some sense depend on locomotion.
36 For a full account of the priority of locomotion, cf. Aristotle, *Physics* VIII, 7, 260a29ff.

in *Physics* III, motion must occur in the moved, and a thing may be moved either (a) by virtue of itself or (b) accidentally.[37] Again, before proceeding with the argument, there is a serious issue of translation and hence interpretation. Hussey translates "καθ᾽ αὐτό" as "in itself" and expands it in his commentary to mean "in (respect of) itself."[38] On this basis, he claims, the distinction here contrasts "in respect of itself" with "in respect of something else" and concludes that this distinction shows "that for Aristotle locomotion is, in the primary sense, change of position *relative to immediate surroundings*, not relative to any absolute landmarks."[39] But Hussey's proposed topics bear virtually no relation to Aristotle's argument. Furthermore, "καθ᾽ αὐτό" and "κατὰ συμβεβηκός" are, respectively, Aristotle's regular terms for "by virtue of itself" and "accidentally." (They occur in his full definition of nature, for example, at *Physics* II, 1, 192b22.) And the contrast clearly lies between these terms.

(a) When a thing is moved by virtue of itself, it is moved as a result of a direct relation to a mover by which it is actualized, as when soul moves body.[40] (b) But here Aristotle is primarily interested in accidental motion. What is moved accidentally may be subdivided into two kinds: (1) that which is capable of being moved by virtue of itself but happens at the moment to be moved accidentally, such as a nail in a ship or the parts of the body, e.g., when the hand is moved the fingers are moved because they are part of the hand, and (2) that which cannot be moved except accidentally, such as whiteness or science (*Physics* IV, 4, 211a19–22). So, for example, (1) a nail is moved by virtue of itself when a hammer pounds it into the ship, but is moved accidentally when the wind moves the ship and everything connected to it, including the nail.[41] But (2) things such as whiteness or science can never be moved directly but are moved only when that to which they belong, e.g., a person, changes place; thus they are capable only of accidental motion.

Straightforward as the distinction between essential and accidental motion seems, it returns us to an earlier problem. All accidental motion rests on the fact that one thing is "in" another. But what does it mean to be "in" another, e.g., the nail is "in" the ship or a thing is in place in the heaven? The earlier distinctions concerning the meaning of "in" raised two issues: (1) how can things be both "in" their "local" or immediate place and "in" place as a whole, and (2) whether place is "in" place, which results in an infinite regress. Now Aristotle gives a "technical account" of

37 Aristotle, *Physics* IV, 4, 211a17–19. This distinction appears throughout Aristotle's works. In the *Physics*, cf. for example, *Physics* II, 3, 195a6, 30–195b2; VII, 1, 241b37–39; VIII, 4, 254b8–12.
38 Hussey, 112.
39 Hussey, 113 (italics in original).
40 Cf. Aristotle, *Physics* VIII, 4, 254b12–17, 20–32.
41 Aristotle, *Physics* IV, 4, 211a22–23; for an exact parallel, cf. *Physics* VIII, 4, 254b8–12.

what it means to say that one thing is in another. This account is central
to the problem of place and establishes the distinctions that will solve it.

Aristotle requires three more terms, "touching," "contiguous," and
"continuous," which he uses to distinguish the ways in which one thing
can be "in" another, and announces that it will now be clear what place
is (211b5–6). Hussey characterizes the arguments concerning how one
thing can be in another as "based upon a dichotomy," i.e., continuous
and divided from; he comments that it is not "clear exactly how this distinction is related to that of the previous section, though the relationship
must be close."[42] Although "continuous" and "divided from" are certainly
important here, the problem of what "in" means, especially "in place," is
the central issue.

I would suggest that casting back over the argument reveals an important feature of its structure, which I would call "subordinated." The
argument begins with the primary evidence for place – evidence without
which we would not think about it at all. Locomotion provides universal
evidence that place is, and locomotion divides into two types, essential
and accidental. Accidental motion in its turn raises the problem of what
"in" means. Thus the analysis of "in" here in *Physics* IV bears directly only
on accidental motion and is at one remove from the problem of place.
Again, "in" requires the further terms "touching," "contiguous," and
"continuous." These terms relate directly only to the problem of what "in"
means and are at one remove from the problem of locomotion and two
removes from the problem of place. In effect, Aristotle traces the terms
required for an account of place through progressively more specialized
problems that allow him to specify just those meanings ultimately required for the solution of the initial problem: what is place.

An account of what it means to say that one thing is "in" another – the
account that will be put to work in solving the problem of place – now follows. We say, Aristotle explains, that a thing is in place in the heaven because it is in the air and the air is in the heaven (*Physics* IV, 4, 211a24–25).
This example appeared earlier in the distinction between common place
and proper place. Here it draws out an ambiguity with the notion of "in"
when we say a thing is "in" place because it is "in" the air.[43] That is, a given
object is not in "all" the air – if it were, its place would be greater rather
than equal to the object contained, which violates the second primary

42 Hussey, 113.
43 Ross's comment on this passage is confused. He seems to think that Aristotle is talking
    about proper place, "the place of a thing is the nearest unmoved boundary of a container,
    the first you would come to in working outwards from the thing" (575), and he concludes,
    the requirement that place be unmoved is ultimately in conflict with the requirement
    that it be "the limit of the containing body at which it is in contact with the contained
    body" (575–576). But Aristotle is not raising this distinction here, and we shall see that
    he is not in the sort of trouble that Ross claims. For a somewhat different criticism of
    Ross's comment here, cf. Burnyeat, "The Sceptic in His Place and Time," 232, n. 15.

characteristic of place; rather, we say something is in the air because of the air immediately surrounding it (211a25–29).

(At the end of *Physics* IV, 4, Aristotle comments that "place is thought to be some kind of surface" [212a28], and this view is often attributed to Aristotle himself. We shall consider it after completing his account of place. Here though, we may note Hardie and Gaye as well as Wicksteed and Cornford translate "τὸ ἔσχατον" as "surface" rather than as "the last," or what immediately surrounds.[44] "τὸ ἔσχατον" cannot mean surface, and Ross later notes that Aristotle uses two words – neither of which appears here – for the technical sense of surface [ἐπίπεδον and ἐπιφάνεια]. Aristotle's word "heaven" here, οὐρανός, carries a strong connotation of the circular vault of the heaven; so, thinking of place in the "common" sense, he undoubtedly thinks of the heaven as enclosing the air and hence in an indirect way the objects in the air.)

The meaning of "in" remains ambiguous, however, there are two ways in which one thing may be "in" another that surrounds it: either (1) the contained is undivided from and continuous with the container, or (2) the contained and the container are divided and touching. When the container is not separate from but is continuous with the contained, the thing is said to be in the container not as in place but as a part in a whole (211a29–31). A thing is "in" the air in the second sense, i.e., it is divided from the immediately surrounding air but touched by it. This distinction turns out to be crucial to the problem of place:

> But when it [i.e., the contained] is divided and touching, it is immediately in the extremity of that which surrounds it, which is neither a part of what is in it, nor greater than its dimension, but equal to it; for the extremities of the things that are touching are in the same [place].[45]

Indeed, this relation defines place as a container and the cosmos as contained: they are not continuous but are divided and touching. And, of course, it recalls the first two characteristics rightly thought to belong to place: place surrounds but is in no way part of the thing contained, and place is neither less than nor greater than the thing contained.

This distinction between the contiguous (divided and touching) and the continuous (undivided) explains the difference between being moved "in" and being moved "with" another. Because a crucial part of any account of place lies in explaining how a thing is *in* place, this difference is important. When one body is continuous with another, it is not moved

---

44 Hussey translates it as "the limit," which, I would agree, is the sense here; but "limit" should be reserved to translate τὸ πέρας.

45 Aristotle, *Physics* IV, 4, 211a31–34: ὅταν δὲ διῃρημένον ᾖ καὶ ἁπτόμενον, ἐν πρώτῳ ἐστὶ τῷ ἐσχάτῳ τοῦ περιέχοντας, ὃ οὔτε ἐστὶ μέρος τοῦ ἐν αὐτῷ οὔτε μεῖζον τοῦ διαστήματος ἀλλ' ἴσον· ἐν γὰρ τῷ αὐτῷ τὰ ἔσχατα τῶν ἁπτομένων. Again, both Hardie and Gaye and Wicksteed and Cornford translate τὸ ἔσχατον as "surface" rather than as "extremity." This issue will be discussed shortly.

"in" it, but "with" it; conversely, when they are divided, the one is moved "in" the other (211a34–36). (The question of whether the container is moved or not is at this point irrelevant [211a39–211b].) For example, the pupil in the eye or the hand in the body are described as parts of a whole and so are moved "with" the body; water in a cask or wine in a jar are described as separate from the cask or jar and so are moved "in" their respective containers (211b-4). The second example provides the model for place and what is in place. It recalls the eighth meaning of "in" (where Aristotle used the example of an amphora and wine) and accounts for the third characteristic of place, namely, place is separable from what is in place because what is "in" place is divided from but touching the place which it is "in."

This distinction introduces "contiguous" and "continuous," which, along with "touching," are crucial for Aristotle's account of place. But they are not explained here. Because of their importance for the upcoming account of place, we must briefly anticipate Aristotle's account of them in *Physics* V.

"Touching." Things touch when their extremities are "together," i.e., in one place (*Physics* V, 3, 226b23). Aristotle insists that container and contained are in one place; thus they are "together" and do indeed touch (226b21–23). But what place means here is unclear, and if Aristotle's account of place is to avoid the charge of circularity, he cannot presuppose place as a condition of touching and then use "touching" to define place.[46] I shall return to this issue shortly.

"Contiguous." One thing is in succession to another, if the one follows the other immediately with nothing of the same kind in between – and a thing that is in succession and touches is contiguous (*Physics* V, 3, 226b34–227a9). Again, we shall see that like the container and the contained, place and what is in place are in succession and touch. That is, they are contiguous.

"Continuous." Finally, the continuous is a subdivision of the contiguous; things are continuous when the touching limits of two things become one and the same – continuity is always impossible if the limits remain two.[47] Place is never continuous with what is in place: their extremities touch, but are always divided. Thus they are not continuous but contiguous, and this relation will reappear in a moment when Aristotle gives his account of place as a limit.

Methodologically, the most striking feature of Aristotle's distinctions in both *Physics* IV and V is the immediacy with which he turns to moved things and the relations, e.g., touching and contiguity, between movers

---

46 For a statement of this charge, cf. Hussey, 114.
47 Aristotle, *Physics* V, 3, 227a10–17. Cf. *Metaphysics* V, 6, 1015b36–1016a17; X, 1, 1052a19–27.

and moved things. These distinctions drive home the sense in which place functions within a physics defined as the study of things that are by nature, i.e., things containing an intrinsic source of being moved and being at rest. Aristotle paves the way for his account of place by providing those distinctions between movers and moved things that solve the problem of place within physics as he defines it. Place is "the where" for such things insofar as they undergo the most common and noble type of motion, locomotion. Any account of place established apart from motion and things in motion has no possible role in Aristotle's physics and consequently does not appear here.[48]

Within these distinctions, Aristotle is concerned primarily with locomotion. (Increase and decrease appear because they presuppose locomotion.) There are two types of locomotion, essential and accidental; all accidental motion requires that one thing be "in" another, and so accidental motion raises the problem of how one thing is in another. One thing is in another either as a part is in a whole or as a thing is in place. In the first case, contained and container are continuous, i.e., their touching limits are one and the same. In the second case, container and contained are always divided and touching. This relation, divided and touching, must obtain between place as the container and what is moved as contained in place. These preliminaries conclude the preparation for a direct assault on the problem of place.

Three False Accounts of Place

"It will now be clear what place is" (*Physics* IV, 4, 211b5–6). Aristotle lists four candidates for place and rejects three of them: it can be neither shape, nor some interval between limits, nor matter (211b6–9). One remains and he argues for it directly: place is "the first motionless limit of what contains."[49] Although it might appear that Aristotle has already rejected form and matter in *Physics* IV, 2, there is a methodological difference between that argument and this: the earlier rejection was in the context of raising the problem of place whereas here Aristotle presents his constructive account of place.[50] Indeed, the rejection of the three false candidates for place further specifies the requirements for an adequate

48 Furley says of this argument in *Physics* IV, 4, "we do not need to think of place in abstraction from all body," *The Greek Cosmologists*, 190.
49 Aristotle, *Physics* IV, 4, 212a20–21: ὥστε τὸ τοῦ περιέχοντος πέρας ἀκίνητον πρῶτον, τοῦτ' ἔστιν ὁ τόπος. Two notes on this expression and its translation: (1) Hardie and Gaye translate πρῶτον as "innermost" rather than first, perhaps because they think of place as a surface, as indicated by their translation of τὸ ἔσχατον earlier. (2) Wicksteed and Cornford reflect the same understanding in their translation, "Thus whatever fixed environing surface we take our reckoning from will be the place."
50 Hussey, 115, comments that form and matter have already been "disposed of," presumably in *Physics* IV, 2. Cf. also Ross, 572, for the same comment.

account of place and so further prepares the way for Aristotle's own definition. Therefore, before turning to place as a limit, we must consider the rejected candidates: shape, interval, and matter.

(1) Place cannot be "shape" (μορφή) because shape (or form) fails to account for where things are and are moved. Nevertheless, shape resembles place in a serious and striking way: both form (εἶδος) and place surround and are limits. For this reason, form seems to be place:

> But because it is something that surrounds, the shape seems to be [place]; for the extremities of what surrounds and of what is surrounded [are] in the same [thing]. Both [place and shape] are limits but not of the same thing; rather, the form [is a limit] of the thing [contained] and place [is a limit] of the containing body.[51]

Because form and place are both "limits," the notion of "limit" that emerges in respect to form also applies to place. Hence, even though the argument rejects form, the comparison between place and form is constructive.[52]

Most importantly, form cannot be a limit in the sense of another material part, e.g., the last or outermost material part. Rather, it must limit the contained as a formal constitutive principle; material parts, on their side, yearn for form as their limit because it makes them to be actually what they already are potentially.[53] The point is not spelled out here – we shall see it when we consider the definition of place as a limit – but because place, like form, is a limit, it too cannot be another material part but must "resemble" form. What is the force of Aristotle's comparison of place and form?

Even though both are limits, form and place differ because form is the limit of the thing contained whereas place is the limit of the containing body. Again, shape or form is not material by virtue of the fact that it serves as the limit of the thing contained, which presumably does include matter; likewise, place will not be material by virtue of the fact that it serves as the limit of the containing body.[54] Both form and place are limits and as such are immaterial.

Any limit must be the limit of something. And although place and the form of natural things are both separable from things – separable in thought and definition – neither can actually be apart from that which they limit. Consequently, what is limited reveals why form fails to serve as

---

51 Aristotle, *Physics* IV, 4, 211b10–14: ἀλλὰ διὰ μὲν τὸ περιέχειν δοκεῖ ἡ μορφὴ εἶναι· ἐν ταὐτῷ γὰρ τὰ ἔσχατα τοῦ περιέχοντος καὶ τοῦ περιεχομένου. ἔστι μὲν οὖν ἄμφω πέρατα, ἀλλ' οὐ τοῦ αὐτοῦ, ἀλλὰ τὸ μὲν εἶδος τοῦ πράγματος, ὁ δὲ τόπος τοῦ περιέχοντος σώματος. Cf. *Physics* IV, 2, 209b-2.

52 Ross, 572, comments that this view "comes very near to the truth" because place and form are both boundaries.

53 This theme runs throughout Aristotle; for one example, cf. *Physics* I, 9, 192a20–24.

54 For another comparison between place and form, the contained and matter, cf. *De Caelo* IV, 4, 312a12.

a candidate for place. Form, or shape, fails to be place because form is a limit of the contained whereas place and what is in place must always be separate: place is the limit of the container.

As we shall see, the identification of place as a limit is central to Aristotle's definition and account of place. Place as a limit, an immaterial constitutive principle, must be explained and the problems raised by it resolved, if Aristotle's account of place is to be successful in terms of his own definitions of motion and nature. And, I shall argue, it is and they are.

(2) The interval between the extremities cannot be place. The interval is closely related to the notion of the void. (For this reason, later arguments identifying void as place often draw on this argument – void would seem to be a kind of interval, although an interval need not be void, i.e., empty.[55]) Aristotle gives two closely related arguments rejecting the interval as a candidate for place. The first claims that the interval cannot be separate, and the second claims that the interval entails an infinite regress of places. Both arguments resemble his later rejection of the void insofar as the interval depends upon the body that falls into it; because it depends on body, the interval (or void) looks indeterminate in and of itself.

And because of the frequent change of the body contained and separate while the containing body remains, such as water from a vessel, what is between [τὸ μεταξύ] [i.e., between the sides of the vessel] seems to be some interval, as being something in addition to the changed body. But this is not the case; rather, some chance body from those changing place and being by nature in contact falls in.

And if there were some interval being something natural and remaining, there would be infinite places in the same thing (for when water and air change places, all the parts will do the same thing in the whole as all the water [does] in the vessel). And at the same time [place] will be changing; therefore, there will be another place of place and many places will be together at the same time. But the place of the part, in which it is moved, is not different when the whole vessel changes place, but the same. For it is within place, the air and the water or the parts of the water change place, but not in which place they come to be, which place is a part of place which is place of the whole heaven. (211b14–29)

Before considering the substance of the argument, we must briefly consider Aristotle's term here, διάστημα, which I translate as "interval."[56] In this passage, διάστημα is often translated as "extension," with the result that Aristotle may be read as rejecting the notion that place is a "stretched out" or "spread out" ability to receive body.[57] This translation represents

---

55 Philoponus makes this association quite explicit. For one example, cf. *In Phys.*, 567.30–33.
56 Apostle translates διάστημα as "interval" here, as does Wolfson throughout *Crescas' Critique of Aristotle.*
57 Hussey renders it "extension," as do Hardie and Gaye. A full bibliography on this point would constitute a volume in itself. For two recent and interesting examples of the problem, cf. Sedley, "Philoponus' Conception of Space," 140–153 and Inwood, "Chrysipus

an understanding going back at least as far as Philoponus, who explicitly identifies "διάστημα" as a "stretched out power" to receive body; indeed, for the post-Cartesian reader the word "extension" immediately carries these overtones.[58]

But "διάστημα" does not mean "extension" etymologically, and Aristotle never uses it with that meaning. It means "interval," and he associates it with τὸ μεταξύ, what is between.[59] Furthermore, when he does wish to say "extended," he uses a more obvious word, which appears later in the arguments against the void, "ἐπεκτείνεται."[60] In Euclid, διάστημα is the technical term for "radius," i.e., any interval [straight line] between the center and circumference of a circle. Far from being associated with extension, i.e., stretched out ability to receive body, διάστημα is merely "the interval between two limits." At least in part, the conceptual difference between extension and interval is this: extension can, at least on some conceptions, be independent of anything else, whereas an interval must by definition lie between two limits and so cannot be independent.[61]

As I shall now argue, the proposed candidate for place here is the interval between extremities, considered independently of the body displaced; for example, if we imagine a jug first filled with water and then air, we might think of the interval between the sides of the jug as independent of the bodies that fill it and so identify it as place. But, Aristotle argues, there is a problem with this view because place must be separable from that which is in place.[62] Thus, to be a viable candidate for place, interval too must be separable. But it cannot be: there is no such thing as an interval that is independent of the bodies changing place in the interval.[63] In short, an interval is always an interval of something. There-

on Extension and the Void." Inwood cites the Greek terms he will use, translates διάστημα as "interval," and discusses "extension" without giving a Greek equivalent (246); later he seems to conflate them (247–249).

58 Philoponus, *In Phys.* 499.6–13, 18–23. I argue that Philoponus is anti-Aristotelian on just this point; cf. H. Lang, *Aristotle's Physics and Its Medieval Varieties*, 106–124.

59 Aristotle also uses it to mean the "interval between one and two," or the distance or interval between two lines; cf. *Metaphysics* X, 4, 1055a9; 5, 1056a36; XI, 9, 1066a32; XIII, 9, 1085b30; it also appears at *Physics* III, 3, 202a18, where Hardie and Gaye as well as Hussey translate it as "interval," and at IV, 1, 209a4, where they render it "dimension." It also appears at *De Caelo* I, 5, 271b30–32 (cf. also *De Caelo* II, 6, 288b12). In both these cases, Stocks translates it as "space," which has, again, connotations all its own even if they are different from those of "extension." Guthrie translates the first case as "space" but the second as "interval."

60 Cf. *Physics* IV, 9, 217b8–10; cf. also *De Caelo* III, 5, 303b28; 7, 305b16.

61 Cf. Aristotle, *Physics* II, 2, 193b32–35. Hussey gives a longish note on this argument, treating the interval as "extension" and generally finding the argument unsatisfactory. But what his note really shows is how "extension" as a concept implies something above and beyond body whereas "interval" is of or between bodies.

62 Aristotle, *Physics* IV, 4, 211a3; being separable from the contained is the third criterion for place.

63 Cf. Aristotle, *Physics* IV, 4, 211b8; 2, 209b10.

fore, there is no "interval" between extremities separable from the bodies that change place in the interval.

The force of the argument becomes clear if we recall the fourth characteristic of place: place within the cosmos is differentiated into up, down, etc., while each of the elements is carried to its respective natural place. An interval between extremities is presumably undifferentiated; but if (presupposing Aristotle's view) the various elements are by nature up or down, then whatever element happens to fall into the interval defines it as up or down. So, if earth happens to fall into the interval, it will be down; but if fire, then up – and so on. As the elements in the interval change, the interval (here identified as place) changes; hence, all place, i.e., internal place differentiated into up, down, etc., will be in the same interval. If the interval were place, many places would be in the same place – which is impossible.

A second argument rejecting the interval now follows and assumes (contrary to what has just been shown) an interval, separate, in virtue of itself, and remaining as things in it change (*Physics* IV, 4, 211b19–29). This argument is dialectical in the sense of resting on a false assumption, and its presence here may indicate the importance of the interval as a possible concept of place.[64] Furley suggests that the idea "appears to depend on certain difficulties that arise . . . when you consider either the parts of the contained substance taken as divisible *ad infinitum* or the parts of something moving in a complex of moved containers," whereas Hussey asserts that "[t]he situation envisaged seems to be that the surrounding body is moving and that, inside it, either water or both water and air are circulating."[65] But Aristotle mentions neither divisibility of parts nor "circulation," although the argument rests on a view of many contained parts, all of which are in motion in an unspecified way while the container too moves.

The argument is best understood as presupposing the immediately preceding argument, i.e., as the element in the interval changes, the interval too changes and in this sense depends upon the elements. The argument shifts the point from the elements as a whole to each part of whatever element falls into the interval. For example, if water and air are mixed, wherever the parts are, each will define its interval. Thus not only will place be redefined with each element that falls into it, but there will be as many places as there are parts of the element.

---

64 Cf. Apostle, 244, n. 21. Hussey's commentary, which takes διάστημα as "extension," fails to recognize the dialectical status of the argument and finds that the entire passage "does not make much sense," 115–116. Furley, *The Greek Cosmologists*, 190, notes that the text here is uncertain and argues that the Atomists (if they are the target of this criticism) would be unmoved by it. Drummond, "A Note on *Physics* 211b14–25," 221–222, argues "the reason he can so confidently assert that there is no such interval-extension is that his own theory of prime matter entails that such *physical* extension is an *inseparable* aspect of matter or physical bulk."

65 Furley, *The Greek Cosmologists*, 190; Hussey, 116.

Introducing parts raises the related issue of the parts of the element as they are moved within place and place as a whole. Again, the key assumption is that place is differentiated into up, down, etc., for everything that is or is moved within place. So, if a jug is moved as a whole, the interval within the jug should nevertheless remain the same in respect to "up," "down," etc. But such would be the case if and only if the directions up, down, etc., are constituted by place and not by what is contained, i.e., the elements. If all the parts moving within place define it, then place will be redefined every time the parts move within it; since the parts would be moved if place as a whole were moved, place will be redefined, if the whole is moved. Hence, if interval were place, not only would there be many places in the same place, but the relation between the place of the parts and place as a whole would be inexplicable.

Aristotle returns to the argument that the interval is defined by whatever body falls into it in *Physics* IV, 5 (212b26–27). But there the point is somewhat different: the cosmos as a whole is not a body, and so place cannot be an interval of body. Body is always what is contained within the cosmos. Hence, as a candidate for place, the interval fails on two counts: it is not separable, and it confuses the relation between body as contained and place as a limit resembling form. Thus, the interval fails to meet the criteria of place and cannot solve the problems associated with place. Furthermore, identifying the interval as place entails mistakes not only about place, but about the cosmos as a whole.

As with the rejection of form, the rejection of the interval establishes criteria for Aristotle's own (successful) account of place. Most importantly, he must explain the relation between place and the elements. It clearly rests on the view that while each element is moved to its proper place, place determines the directions "up," "down," etc. In short, Aristotle must explain place as a determinative principle. I shall argue later that although his definition and account of place do explain it in this sense, a full account of the relation between place and the elements remains unclear throughout *Physics* IV. In fact, a full account requires the arguments of the *De Caelo*, where Aristotle explains both elemental motion and the causal relation between place and the elements.

(3) Matter is the weakest candidate for place. It may be thought to be place because, like place, it can "hold" change; for example, something formerly black is now white, or something once soft is now hard (*Physics* IV, 4, 211b31). But matter holds only qualitative change; unlike the interval, it cannot hold first one object, e.g., water, and then another, e.g., air (211b35). Finally, matter is rejected because it can neither be apart from nor surround the object in place, whereas place is both apart from and surrounds the contained.[66]

66 Aristotle, *Physics* IV, 4, 212a; cf. *Metaphysics* VII, 3, 1029a27–31.

Although matter is the least interesting candidate for place, neverthe-less it raises a serious problem for Aristotle's own account – another prob-lem that will not be fully resolved until the *De Caelo*. Aristotle criticizes matter because it fails to "hold" change from one object to another, e.g., there is now water where earlier there was air; thus he implies that a true account of place must explain just such change. The problem of how place contains first one object and then another is not only the most dif-ficult problem for Aristotle himself, it is one of the most difficult prob-lems historically with his physics. I shall consider it explicitly when taking up Aristotle's constructive account of place as a limit and return to it again in examining inclination in the *De Caelo*.

These three candidates for place must each be rejected. Place can be neither form, nor interval, nor matter. These rejections, however, specify the criteria of place as well as problems entailed by it. Place must sur-round and be a limit of the container analogous to form as the limit of the contained. As a limit, place must (in a sense yet to be explained) ren-der the cosmos determinate. Place must be separable not in the weak sense of separable from this or that body but in the stronger sense of sep-arable in principle from all contained body. And, finally, although place cannot be matter, nevertheless it must "hold" first one object and then another. Aristotle clearly believes that his account of place will do the job, i.e., account for locomotion in natural things, including the heaven, and solve the problems associated with place.

The True Account of Place: Place Is a Limit

If, then, place is none of the three, neither the form, nor the matter, nor some interval always underlying and different from the [interval] of the thing which changes place, place must be the remaining of the four, the limit of the contain-ing body [by virtue of which it is conjoined to the contained]. But I call the con-tained body what is movable by locomotion. [*Physics* IV, 4, 212a2–6: εἰ τοίνυν μηδὲν τῶν τριῶν ὁ τόπος ἐστίν, μήτε τὸ εἶδος μήτε ἡ ὕλη μήτε διάστημά τι ἀεὶ ὑπάρχον ἕτερον παρὰ τὸ τοῦ πράγματος τοῦ μεθισταμένου, ἀνάγκη τὸν τόπον εἶναι τὸ λοιπὸν τῶν τεττάρων, τὸ πέρας τοῦ περιέχοντος σώματος ⟨καθ᾽ ὃ συνάπτει τῷ περιεχομένῳ⟩. λέγω δὲ τὸ περιεχόμενον σῶμα τὸ κινητὸν κατὰ φοράν.]

First, a note about the text and translation. Ross retains the bracketed words "by virtue of which it is conjoined to the contained" even though the textual evidence for them is problematic, because, he claims, they are required by what follows.[67] But the δέ here is probably adversative and contrasts the contained body directly with the containing body. This con-trast may reflect a similar contrast made less than a page earlier at 211b11–14 where Aristotle says that place and form are both limits, the

---

67 Ross, 575. Hussey makes no mention of this difficulty.

one of what contains and the other of what is contained but of different things. Place is a limit of the containing body and form of the object.

Carteron includes the bracketed words in a note, but omits them from the text, while Prantl not only excludes them from the text but fails to mention them even in the *apparatus criticus*. The problem is that these words appear only in the Arabo-Latin translation.[68] Hence different editions reflect different evaluations of the Arabo-Latin translation. This translation is often (and notoriously) interpretive, and Ross's assertion that these words are required by the meaning proves nothing more than that he shares a particular interpretation. Given their complete absence from the Greek tradition, they are probably spurious and should not be retained, as reflected by both Carteron and Prantl. Furthermore, not only are they unnecessary for the meaning of this text, they are not confirmed by any other passage in the account of place.[69]

In fact, the bracketed phrase constitutes the only appearance of the verb συνάπτει in *Physics* IV. Both Hardie and Gaye and Hussey translate the verb as "in contact with" – place is "the limit of the surrounding body, at which it is in contact with that which is surrounded."[70] But "in contact with" implies that place and what is surrounded are two bodies, i.e., two bodies in contact with one another. There are two problems here. (1) For "in contact with" in the sense of one body in contact with another, Aristotle usually uses ἅπτεσθαι.[71] συνάπτει is clearly stronger; I translate it as "conjoin," but it often seems to mean "unite directly" or "coincide exactly."[72] (2) As a limit, place resembles form and cannot be another body. Indeed, Aristotle compares place and form as limits and contrasts place as a limit with body.

A limit contrasts in several ways with what is limited: it is a formal con-

---

68 Ross, 575.
69 I believe these words are spurious. As I shall discuss, Aristotle's account of place was read in a materialistic way, i.e., place is somehow like matter, as early as Theophrastus and certainly in the Arabic tradition. The same understanding (or misunderstanding) underlies a number of Ross's comments on Aristotle's account of place, his decision to include these words, Hussey's acceptance of them without comment, and Hussey's objection here.
70 Hussey's translation; he gives no indication either in the text or in his notes (cf. 117) of the textual problem. As noted earlier, Hussey's translation is strongly suggested by the summaries of "Aristotle's Position" given by Lettinck, *Aristotle's* Physics *and Its Reception in the Arabic World,* who deals specifically with Philoponus and the Arabic tradition of commentary on the *Physics;* Lettinck, again apparently following Hussey, summarizes, "So place must be what remains of the four: the boundary of the surrounding body where it is in contact with the surrounded body," 280.
71 Ross too clearly takes it in this sense, 375. For some examples of ἅπτεσθαι meaning "in contact with," cf. *Physics* V, 3, 226b23; 227a18, 24; VI, 1, 231a22; here in *Physics* IV, 4, 211a34, and 5, 212b23; cf. also *Gen. et Corr.* I, 6, 323a3, 10; *Metaphysics* V, 4, 1014b16–26; XI, 12, 1068b27; *De Caelo* II, 4, 287a34; *De Anima* II, 11, 423a24.
72 For some examples, cf. *Physics* VIII, 8, 264b27; *Hist. of Anim.* II, 11, 503a16; 17, 507a28; IV, 5, 530b14; *Gen. Anim.* II, 4, 740a34; 6, 744a2; *On Youth, Old Age* 17, 478b34; here in *Physics* IV, cf. 11, 218b25. Bonitz does not seem to mention this occurrence of συνάπτει.

stitutive part, indivisible, more closely identified with substance and more honorable.[73] What is limited, in contrast to a limit, is a body capable of motion. Hussey asks why Aristotle specifies body as movable here: because the specification of place as a containing limit and the contained as a movable body reveals the relation between them – and so the meaning of συνάπτει, conjoin (if we accept this phrase).[74] Place and movable body are not conjoined as two bodies in contact, but as a limit (place), and what is limited (movable body). However, Aristotle has already asserted that the conjunction between container and contained most nearly parallels that of form and matter: the limit and the limited are conjoined as constitutive principle and that which is constituted. In this sense, place is not "next to," "just beyond," or "in contact with" the first contained body any more than a surface or a form is merely next to or in contact with what is bounded by it.[75] Rather, the limit and the limited together comprise one being, the first heaven, as boundary and bounded.

This issue returns us to the earlier discussion of what it means to say one thing is "in" another. Is Aristotle's argument circular, as Hussey charges, because the limits of things that are touching are in the same place with the result that touching both defines and presupposes place? No. A limit and what is limited have an asymmetrical relation, i.e., they are not equal partners within a whole. "Conjoined" (again, if it is Aristotle's word) should be read in terms of his definition of "in," i.e., the contained is immediately "in" the limit of what surrounds it (211a33–34). By definition a limit must limit *something*, and place is the limit of the surrounding body. But the contained should not be thought of as contained by another body; rather, it is immediately in the limit, i.e., place. Hence, although the limit (of the containing body) and the contained are obviously "touching," they are not two bodies that are divided but touching, e.g., water and a jug when the water is "in" the jug. Rather, the relation of the limit to the limited resembles the relation of form and matter.

This relation answers the question of circularity. Aristotle's definition of place would be circular only if it presupposed place, and it would do so only if place were one body touching another, the contained, in such a way that their extremities are "in the same place." But place is not another body. It is a limit and so touches not in the sense of being in the same place as the contained (and so presupposing place), but in the sense of constituting place and so touching the contained, which is in place by virtue of this limit. As the limit of the first containing body, place, like form, is a constitutive principle that defines what it is for the contained to be in place. (Hence, place is not in place.)

73 Cf. Aristotle, *Physics* I, 2, 185b18; VI, 5, 236a13; *De Caelo* II, 13, 293a32 and 293b13–14. The last text will be discussed later.
74 Hussey, 117.
75 A line is compared to a surface at *Physics* IV, 4, 212a28.

This point may be constructively compared to Aristotle's comments about place in *Categories* 6, where he discusses quantities (literally "the so much") that are either discrete or continuous.[76] His examples of discrete quantities are number and language which are discrete precisely because they have no common boundary at which they "conjoin" (συνάπτει) (4b25–29). Numbers are obviously discrete, and language too is always discrete because there is "no common boundary" where "the parts conjoin" (4b35–36). Hence there is just a series, first one number and then another, or one syllable, then another; this relation most obviously resembles contact, nor conjoining, and so here in the *Categories* they are described as discrete.

In contrast to the discrete, continuous quantities always possess a "common boundary in relation to which its points conjoin" [*Cat* 6, 5a-2: ἔστι γὰρ λαβεῖν κοινὸν ὅρον πρὸς ὃν τὰ μόρια αὐτῆς συνάπτει]. For example, a line is a continuous quantity because its parts are conjoined at a common boundary, namely a point; the same holds true for a surface – its common boundary being a line – as well as for body – its common boundary being formed by a line or a surface (5a-6). In each example, the common boundary is formally different from that to which it is conjoined. A point possesses no dimension and serves as a common boundary for a line, which possesses one dimension. Likewise a line, possessing one dimension, can serve as a common boundary for a surface that possesses two dimensions, and a line or a surface can serve as a common boundary for body that possesses three dimensions. The partners in each case – point and line, line and surface, line or surface and body – act as limit and limited precisely because the limit is formally distinct from what is limited.

Place and time too, the argument continues, are of such a kind:

Again, place is among continuous things. For the parts of a body occupy some place and these parts conjoin in relation to some common boundary. Surely then, the parts of place, which each of the parts of the body occupy, conjoin to the same boundary in relation to which also the parts of the body [conjoin]. Therefore, place too would be continuous, for its parts conjoin in relation to some common boundary. (5a8–14)

Place serves as a common boundary for body precisely because it is not another body, but acts as a limit. Now Aristotle makes a point that is not made for the other examples of continuous quantities: as the limited is limited by a limit that forms a common boundary, so too is the limit a limit of what is limited. So, for example, there are no points apart from lines because a point is by definition the limit of a line.[77] Likewise, there

---

76 Aristotle, *Cat* 6, 4b25–29. Ackrill translates συνάπτει as "join together" throughout this argument.
77 On points and lines, cf. Aristotle, *Physics* VI, 1, 231b9; *Metaphysics* XI, 2, 1060b15, 18; *De Caelo* II, 13, 296a17; III, 1, 299a27–299b8. A line has magnitude in one dimension only,

is no place apart from the first containing body of which place is a limit because a limit is by definition the limit of what is limited.

It has been suggested that this argument contradicts *Physics* IV because here in the *Categories* place must be conceived of as volume and extension; thus Aristotle must have changed his mind between the *Categories* and *Physics* IV.[78] But the notions of volume and extension are entirely extraneous to the argument of *Categories* 6. Aristotle argues here that place is in a sense continuous; but being continuous in this sense need not imply being a volume or an extension. Indeed, virtually all Aristotle's examples of both discrete and continuous quantities are "immaterial," e.g., numbers, lines, etc.

A much more serious objection can be raised. Calling place a quantity seems to contradict both the comparison of place and form and the rejection of matter as a candidate for place – quantity is most obviously associated with matter (or body). Further, as we have just seen in *Physics* IV, Aristotle denies that place can be continuous with what is in place, whereas here in the *Categories,* place and in some sense the relation of place and body do seem to be continuous. Hence we do not need a notion of volume or extension to find an apparent contradiction between *Physics* IV and *Categories* 6.

The problem of including place among continuous quantities strikes at the heart of Aristotle's account of place as a limit distinct from what is limited. However, if we recognize that Aristotle's *logoi* are defined by their topics and that their arguments are subordinated to the topic at hand, the problem disappears. Two points that we have already seen concerning place should be kept in mind. (1) Although Aristotle denies that place is a body, it is nonetheless in the same genus as body because it is three-dimensional. In respect to its genus, place in *Physics* IV may be rightly defined as a quantity. Place, as we shall see, is by definition separate from all things that are in place; but at the same time, as a limit it is "in" the first containing body. We have seen this sense of "in" in *Physics* IV, 3, and, indeed, this sense defines the genus of place as three-dimensional and as quantity.

(2) When considering the meaning of "in," we saw that container and

---

and as a limit of the line, a point cannot share in this magnitude; cf. *Physics* VI, 1; for an analysis, cf. Konstan, "Points, Lines, and Infinity," 2–5; for a response to Konstan's analysis of Aristotle's argument, cf. H. Lang "Points, Lines, and Infinity: Response to Konstan," 33–43.

78 Mendell, "Topoi on Topos," 206–231, cf. esp. 209. Also Inwood, "Chrysippus on Extension and the Void," 252–253. The origin of these issues may be *Aristotle's Categories and De Interpretatione* translated with notes by Ackrill, 93, where, commenting on *Categories* 6, Ackrill says, "Place is defined in the *Physics* IV, 4 as the limit of the containing body. The proof given here that place is continuous treats it similarly as filled by (or perhaps only fillable by) a body. This raises the question whether place has a right to count as [a]n independent primary quantity in addition to body."

contained are both "in" the whole, but in different ways. We may for the moment ignore the different senses in which the parts are "in" the whole and consider only the whole as comprised of place and the first containing body. If we consider place as three-dimensional, i.e., as quantity, we would consider the whole in this sense, i.e., as comprised of place and the first body. According to its genus, place is in the first containing body as any limit is in what is limited, as length, breadth, and depth are in any three-dimensional body. Conversely, a body may be said as body to be in its limits. This is just the sense in which both parts (limit and limited) are "in" the whole. And in this sense, these parts are continuous and not divided. Indeed, on this issue *Physics* IV, 3, and *Categories* 6 agree.

But the relation between the limit and the limited emerges only if we ignore the different senses in which they are "in" the whole and consider the whole as such. The whole is a quantity and hence place within it appears as quantity – and quantity is the topic of *Categories* 6. But if we wish to define place, the task of *Physics* IV, 1–5, then the relation of the three-dimensional whole composed of two parts must be ignored in favor of the specific and different senses played by each part, the limit (place) and the limited (what is contained in place) within the whole – and place resembles form.

Thus there is no contradiction between *Categories* 6 and *Physics* IV. Rather, these arguments are discussing different problems, "quantity" and "the where." Place is in the same genus as is body – both are three-dimensional – and answers the question "what is 'the where' of things that are?"; place is the limit of the containing body. Therefore, place appears in both accounts. It does so, however, in very different respects.

In *Physics* IV, 4, place is defined as the limit of the containing body. This body is "movable by locomotion." Several problems remain, and Aristotle turns to them directly, concluding that place must be the first unmoved limit of what contains. With this definition in hand, he considers why place causes the upward or downward motion of each element. *Physics* IV, 5, completes the account of place by answering various objections and showing how this account of place meets all the requirements established at the outset.

Aristotle claims that defining place as a limit explains (1) the characteristics that belong to place essentially, (2) how the heavens are and are not in place, and (3) the locomotion of all things within the cosmos. Consequently, we must understand place as the first unmoved limit of what contains, and why, on this definition, place causes elemental motion. Herein lies the reason why place serves as something without which motion seems to be impossible.

Place is thought to be something important and hard to grasp, both because the matter and the shape present themselves along with it, and because the displacement of the body that is moved takes place in a container that is at rest. For it appears to be possible that there be an interval between, being something other than the moved magnitudes. And also air, seeming to be incorporeal, contributes something [to this belief]. For it seems not only that place is the limits of the vessel, but also what is in between [the limits] seems empty. But just as the vessel is a movable place, thus also place is an immovable vessel. Therefore, whenever what is within something moved is moved and changed, such as a boat in a river, it uses [place] as a vessel rather than place in respect of the surrounding. But place is meant to be unmoved; therefore, the whole river, rather, is place because the whole is unmoved. Therefore, the first unmoved limit of what contains, this is place.[79]

In his conclusion, Aristotle defines place as a limit and specifies it as both first and unmoved. But the account leading to this conclusion has been the focus of a number of problems. I shall begin with this passage, including its conclusion, before turning to the problems commonly associated with it.

Aristotle first emphasizes the sources of difficulties in understanding place. Place is hard to grasp because (1) it appears alongside matter and form, rejected candidates for place, and (2) moved body changes place within an unmoved (literally "at rest") container. These difficulties, which beset anyone who proposes form, matter, or interval as place, are exacerbated (3) by the appearance of the third rejected candidate for place, an interval, together with (4) the incorporeality of air because together they give the appearance of an empty vessel between limits.

The constructive account now follows. Place may be compared to a vessel.[80] A vessel is a movable place, and place is an immovable vessel. Both "hold" change – a vessel holds bodies that are moved, and place holds all movable body. (The sense in which place "holds" change remains unclear – the fact is sufficient for now.) But they differ because a vessel is movable and place immovable. Therefore, when a thing, A, is moved within a vessel that is itself moved, e.g., a boat on a river, the vessel is a movable place for A, although this movable vessel cannot be place because it is movable and place must be unmoved. Thus for A, the river, which as a whole is unmoved, must be place. And Aristotle reaches his own definition: place is the first unmoved limit of that which contains.

---

79 Aristotle, *Physics* IV, 4, 212a8–21; δοκεῖ δὲ μέγα τι εἶναι καὶ χαλεπὸν ληφθῆναι ὁ τόπος διά τε τὸ παρεμφαίνεσθαι τὴν ὕλην καὶ τὴν μορφήν, καὶ διὰ τὸ ἐν ἠρεμοῦντι τῷ περιέχοντι γίγνεσθαι τὴν μετάστασιν τοῦ φερομένου· ἐνδέχεσθαι γὰρ φαίνεται εἶναι διάστημα μεταξὺ ἄλλο τι τῶν κινουμένων μεγεθῶν. συμβάλλεται δέ τι καὶ ὁ ἀὴρ δοκῶν ἀσώματος εἶναι· φαίνεται γὰρ οὐ μόνον τὰ πέρατα τοῦ ἀγγείου εἶναι ὁ τόπος, ἀλλὰ καὶ τὸ μεταξὺ ὡς κενὸν ⟨ὄν⟩. ἔστι δ' ὥσπερ τὸ ἀγγεῖον τόπος μεταφορητός, οὕτως καὶ ὁ τόπος ἀγγεῖον ἀμετακίνητον. διὸ ὅταν μὲν ἐν κινουμένῳ κινῆται καὶ μεταβάλλῃ τὸ ἐντός, οἷον ἐν ποταμῷ πλοῖον, ὡς ἀγγείῳ χρῆται μᾶλλον ἢ τόπῳ τῷ περιέχοντι. βούλεται δ' ἀκίνητος εἶναι ὁ τόπος· διὸ ὁ πᾶς μᾶλλον ποταμὸς τόπος, ὅτι ἀκίνητος ὁ πᾶς. ὥστε τὸ τοῦ περιέχοντος πέρας ἀκίνητον πρῶτον, τοῦτ' ἔστιν ὁ τόπος. The definition of the last line clearly echoes 212a5–6: ἀνάγκη τὸν τόπον εἶναι . . . τὸ πέρας τοῦ περιέχοντος σώματος . . .
80 Cf. Aristotle, *Physics* IV, 3, 210a24.

The image of the vessel and the river has caused considerable confusion. Ross suggests that Aristotle means "the place of a thing is the nearest unmoved boundary of a container, the first you would come to in working outwards from the thing."[81] That is, Ross identifies place with the vessel, which is "first" in the sense of nearest the moved. Consequently, he finds it impossible that place be unmoved: if A is in a moving body B, then there is "no body which both is unmoved and immediately surrounds A. From Aristotle's silence on the point, it seems as if he had not noticed the difficulty."[82]

Hussey raises another objection, which rests on the problem of what a first limit can be. Like Ross, he seems to assume that "first" means "nearest" and finds it problematic that only a "remote" limit can be found. He even suggests that this definition of place may be an interpolation:

If the river banks constitute the place, on this definition, of the boat, then the boat is not in locomotion at all, whether it is drifting downstream or being rowed upstream. Moreover, there is no guarantee that we shall find any such limit that is not very remote. For these reasons the definition is deeply suspect.[83]

He then designs a series of suppositions to "rescue" Aristotle's definition.[84]

Aristotle is not, I think, in as much (or the kind of) trouble that Ross and Hussey suggest. First, Aristotle rejects the boat as representing place: the river as a whole is place; therefore, "first" cannot mean "nearest" and that place is "remote" from what is contained in the vessel is not immediately problematic. Furthermore, there is no evidence that this text is suspect.

Aristotle often approaches his own definitions through what seems or is alleged to be thought by his opponents and/or the ancients.[85] He likes to show that his view is in accord with all that is true and best in popular opinion, the poets, and earlier philosophers. Owen points out, citing *Physics* I-V, that the phenomena Aristotle wishes to save "are the common convictions and common linguistic usage of his contemporaries, supplemented by the views of other thinkers."[86] So, for example, arguing in *Physics* IV, 1, that place must be distinct from things in place, he approvingly quotes Hesiod (208b29–31).

The ancients speak of a river, Okeanos, the source of all things; it sur-

---

81 Ross, 575. This interpretation is reflected in Hardie and Gaye's translation of 212a20–21: "Hence the place of a thing is the innermost motionless boundary of what contains it." I shall discuss this translation.
82 Ross, 576. For a more recent version of Ross's confusion, cf. Summers, "Aristotle's Concept of Time," 60.
83 Hussey, 117.
84 Hussey, 117–118.
85 For a clear example of this procedure, cf. *Metaphysics* XII, 8, 1074b1–14; 10, 1075a25–1076a3; cf. H. Lang, "The Structure and Subject of *Metaphysics* Λ," 257–280.
86 Owen, "Aristotle: Method, Physics, and Cosmology," 155.

rounds the earth, and the sun moves on it as a vessel on a river.[87] Aristotle first dismisses his opponents' view; then by rejecting the containing vessel and identifying the river with place, he emphasizes that his account agrees with the ancients for whom "the whole river" represents place.

Place, "the whole river," is unmoved. But what does unmoved mean within Aristotle's account of place? Apart from the particulars of this passage, there are several reasons why place must be unmoved. Just as a point, being the limit of a line, cannot have magnitude, place as the first surrounding limit must be motionless. That is, place as a limit cannot have the characteristics of what is limited or contained – if it did, it would be another part of the limited rather than the limit. And, as we have seen, Aristotle specifies that what is contained is movable. Hence place must be unmoved.

"Nature" includes both matter and form and so the physicist must know both (*Physics* II, 2, 194a13–27). Matter and form are both said to be contained, form as the limit and matter as the body of the contained.[88] If we look beyond the *Physics* to the *De Caelo*, there is *no* matter except that contained within the cosmos, i.e., in place; if we look to the *Metaphysics*, form, which is outside the cosmos, is absolutely immovable, is not a proper subject of physics, and is not in place.[89] Again, the same conclusion follows: because place is a limit of the containing body, place cannot be another thing contained, i.e., individuals constituted by matter and form and so capable of locomotion.

Matter, according to Aristotle, is the principle of motion and change in things, and things in motion are moved by virtue of their matter. So contained body changes place because of its matter. But as we saw earlier, place cannot be identified with matter. Therefore, place must be unmoved.

Place is not only unmoved, it is "first": "the first unmoved limit of what contains." As Ross and Hussey show, the meaning of "first" and of the river as "unmoved" seems problematic. The image of a movable vessel, which contains body, and the unmoved river contrasts two vessels, one movable and one unmoved. Both vessels "hold" what is in them and so again recall the problem of how one thing is "in" another: "Now we say we are in the heavens as in a place, because we are in the air and it is in the heavens" (*Physics* IV, 2, 209a33–35). So the contained body, A, is in the vessel, and the vessel is in the river. "The whole river" is place just as the whole heaven is.

Contra Ross, place is not the "first" in the sense of "nearest" to the contained body, i.e., the boat. Place is first as the whole heaven is what first

87 Cf. Homer *Il.* xviii, 607; xxi, 194; Herodotus, *Historiae* iv, 8; Mimnermus, fr. 10 (Diehl); Aristotle, *Metaphysics* I, 3, 983b27; for an excellent analysis of these sources, cf. Kirk, Raven, and Schofield, *The Presocratic Philosophers*, 10–17.
88 Cf. Aristotle, *De Caelo* IV, 3, 310b10, and 4, 312a12.
89 Aristotle, *De Caelo* I, 9, 278b8–279b2; *Metaphysics* XII, 1, 1069a37–1069b2; XII, 7, 1072a25; 1072b5–13; cf. also *Physics* II, 7, 198a35–198b4; VIII, 10, 267b17–26.

surrounds everything that is contained within the heaven. A full account of how the heaven is and is not in place remains to be seen. But clearly we are in the air as our "local" place, and we are "in the heaven" *because* the air is in the heaven. Likewise, the contents of the boat are in the boat as a "local" place and in the river *because* the boat is in the river. Hussey complains that place is "remote"; but place in the sense of the heaven (or the river) is the "common" place for all things and, so, the more proper object of the investigation.

Here is the point of the distinction between the vessel and the river: whenever what is contained is moved and changed, it uses "local" place as a movable vessel rather than the surrounding "common" place; but a vessel cannot be place in the fullest and most important sense precisely because it is moved, and place must be unmoved. Therefore, place is what surrounds and is unmoved, e.g., the whole river. Finally, *contra* Hussey, if a boat moves *up*stream or *down*stream, it has changed place relative to the river as a whole – we say that it has moved x miles precisely because we measure such motion relative to the "whole river" as unmoved. And the elements change place when they are moved upward or downward in exactly the same sense as a vessel moves relative to the whole river. When Aristotle concludes that place is the first unmoved limit, he turns to the heaven and the elements.

Without explaining his definition of place, Aristotle proceeds to the most important consequence of it: that the middle of the heaven and the last part relative to us seem most of all to be "the up" and "the down" respectively.[90] This constitution of the heaven in turn explains the motion of the elements:

> Therefore, since the light is what is carried upward by nature, and the heavy downward, the limit in respect to the middle, which is contained, is down as too [is] the middle itself and the [limit] in respect to what is last [is] up as too [is] the last itself. (212a24–28)

In short, the constitution of the cosmos as determinate, i.e., "up" and "down," is a direct consequence of defining place as a limit, and moved things are explained within the determinate cosmos. This consequence confirms the intimate relation between place and the cosmos and so provides a strong clue as to why place is a term without which motion in things seems to be impossible.[91] Here is the constructive moment of Aristotle's account of place, and it accounts for place as a common and universal term: place is a single principle that determines the cosmos as a

90 Aristotle, *Physics* IV, 4, 212a20–24; cf. *Physics* VIII, 10, 267b6–7, for an identification of the middle and the circumference of the heaven as its ἀρχαί.
91 It also shows again the strong sense in which place resembles form as a limit: form limits the contained and in so doing renders it determinate and the object of the definition. Cf. Aristotle, *Physics* II, 1, 193b-3.

whole and the elements from which all natural things and artifacts are composed.

In the *De Caelo* too Aristotle argues that place is the "first" [πρῶτον] limit that differentiates the cosmos. Here "first" is again unequivocally identified with what contains everything moved, i.e., the boundary of the heaven. Arguing against the Pythagoreans about the position of the earth and the importance of the center of the cosmos, Aristotle gives his own view:

> For the middle is a source and precious, but the middle of place seems last rather than a source; for the middle is what is bounded and the limit is the boundary. And the container, namely the limit, is more precious than the limited; for [the limited] is the matter while [the container or limit] is the substance of the system. (*De Caelo* II, 13, 293b11–15)

The first limit of the cosmos, which in *Physics* IV is place, is identified with substance, the first category of being, whereas the limited, or contained, is like matter. Matter is a principle of change oriented toward and requiring limit; thus, matter is the limited. But place "is more precious than the limited": it is the boundary and "substance" of the system, i.e., place is the first constitutive principle of the cosmos.[92] The first candidate for substance is form.[93] As we have already seen, place is like form insofar as it surrounds, being a limit and a determinate constitutive principle of what is contained.

What does it mean to call the surrounding limit the "substance," οὐσία, of the system? The first motionless limit is the substance of the cosmos because it renders the cosmos determinate rather than indeterminate. By rendering the cosmos determinate, the limit makes it a whole and defines every part within the whole as "up," "down," or "middle."[94] The surrounding limit, the substance of the cosmos, is place, which as the first unmoved limit of what contains determines "up," "down," and "middle" within the cosmos (*Physics* IV, 4, 212a20). These conclusions address the opening problem of *Physics* IV, 1: what is "the where" for things that are and are moved? Place is "the where": it renders the cosmos one and determinate in respect to the category "where" for all things that are within it, i.e., within the vault of the heavens. The limit is substance because it is a principle of determinacy for all things within the cosmos, and these things are themselves made determinate in respect to "where" by place as a limit.

---

92 Cf. also *De Caelo* II, 1, 284a4–6.
93 Cf. Aristotle, *Metaphysics* VII, 3, 1029a6–7; 27–32; 17, 1041b11–31; VIII, 1, 1042a24–31.
94 Cf. *De Gen. et Corr.* II, 8, 335a20: "Now each of them [the elements] tends by nature to be carried towards its own place; but the figure, that is the form, of them all is at the limits." In *Metaphysics* XII, Aristotle asks if the cosmos is of the nature of a whole – he answers yes – and if god, the unmoved mover, is a part of or separate from this whole. God is part of this whole only in a weak sense, as a general is part of the army, but in the stronger sense god is separate from this whole. The whole constitutes nature and things that are by nature, whereas god is separate from nature. Cf. *Metaphysics* XII, 10, 1075a13–24.

Again, the analogy with form is telling. Form is substance in the primary sense because it is the first determinative principle of anything in respect to its definition. Like form, place is a determinative principle making the cosmos as a whole determinate in respect to "where" with the result that every place within the cosmos can always be defined as "up," or "down," etc. Corelatively, as we shall see in the *De Caelo*, any and all things within the cosmos are determined in respect to place, i.e., where they are. Aristotle defines the natural place of each element, fire is up, earth is down, etc. Because all natural things are composed of these elements, all natural things have a natural place within the cosmos. (Artifacts, insofar as they are composed of the elements, also experience natural motion and so also have a natural place [*Physics* II, 1, 192b19–20].) In short, nothing within the cosmos escapes the "mark" of the surrounding limit: "where" within the cosmos can always be defined while all things always have a natural place.

This relation forms the background for the account of elemental motion developed in the *De Caelo*. Place is a cause of motion not as one of the four causes, but as the first unmoved limit, a determinative principle that by differentiating the cosmos into actually "up," "down," etc., renders the cosmos determinate. As determined in respect to up and down, the parts of the heavens constitute the actuality for the elements that are moved naturally insofar as they are potentially "up" or "down."[95] Here is the crucial issue: given Aristotle's definition, motion is impossible without place because as a limit place renders the cosmos determinate in respect to the six directions, up, down, front, back, left, and right, and so constitutes "the where" of all things that are and are moved. Every natural body is carried to and/or rests in its proper place (*Physics* IV, 5, 212b29–35). Because *all* body is composed of these elements, natural place is a common and universal cause of the motion of all body. In *Physics* IV, 1, Aristotle comments that the power of place must be something marvelous and prior to all things (208b34). And on his account, it most certainly is.

Machamer, who concludes unequivocally that because place cannot be one of the four causes it cannot be a cause at all, makes an important point:

Throughout this discussion I have been shifting back and forth, as Aristotle does, between two senses of place. The first sense is that defined in *Physics* IV, 4 (212a20), where place is defined as the innermost motionless boundary of what contains. This I shall call the relative definition of place. It is relative in the sense that it presupposes the existence of a body which is contained by the boundary. In this sense of place the natural place of earth is dependent upon the actual location of extended earth and is defined by what surrounds the earth.[96]

95 Cf. *De Caelo* IV, 1, which will be discussed. For the problem of elemental motion in *Physics* VIII, 4, cf. H. Lang, *Aristotle's* Physics *and Its Medieval Varieties*, 63–84.
96 Machamer, "Aristotle on Natural Place and Natural Motion," 381.

(1) Machamer is correct that there cannot be a limit without something that is limited and in this sense place depends upon the first containing body, just as form as a limit depends upon matter.[97] I shall consider the first containing body and its relation to place in Part II. But body, whether limited by form or place, cannot be at all without a limit, and the limit causes body in the more important sense (*Physics* II, 8, 198b10).

(2) Machamer is also correct that the natural place of earth (or any element) depends upon actual location within the cosmos, i.e., up, down, etc.; and he speaks of "extended earth." But Aristotle does not call earth or the cosmos "extended"; rather, he thinks of the cosmos as intrinsically directional and the elements as intrinsically oriented, i.e., fire upward, earth downward, and air and water toward the middle. As I shall discuss at greater length in the conclusion, the intrinsic directionality of the cosmos is a formal characteristic precisely because it is granted by place, which, like form, is a limit.

When Aristotle investigates void, he makes an important observation: those who posit the void as place when full and otherwise empty speak "as if the same being is empty and full and place, but 'the to be' for these things is not the same being." [*Physics* IV, 6, 213a18–19: ὡς τὸ αὐτὸ μὲν ὂν κενὸν καὶ πλῆρες καὶ τόπον, τὸ δ᾽ εἶναι αὐτοῖς οὐ ταὐτὸ ὄν.] By limiting the cosmos, place makes the six directions actual within the cosmos; each element in some sense – this point too lies ahead in Part II – depends upon its respective actual place. Rendering the cosmos directional is a formal determination and so entirely different from full and empty, which are material determinations.

(3) Finally, and this point will reappear in the examination of the void, the word αἰτία derives from the word "responsible." If in saying that *a* is dependent upon *b*, one means that *b* is somehow responsible for *a*, then one grants a causal relation. Clearly though, the exact nature of that relation remains to be determined.

But *Physics* IV, 4, is not quite complete. Two sentences remain, and they have been enormously important for the history of this argument.

And it is for this reason [that the middle is down and the extremity up] that place, i.e., what is surrounding, is thought [or seems] to be some surface and like a vessel. Again, place is together with the object [contained] for the limits are together with the limited. [καὶ διὰ τοῦτο δοκεῖ ἐπίπεδόν τι εἶναι καὶ οἶον ἀγγεῖον ὁ τόπος καὶ περιέχον. ἔτι ἅμα τῷ πράγματι ὁ τόπος ἅμα γὰρ τῷ πεπερασμένῳ τὰ πέρατα.][98]

---

97 For one example, cf. Aristotle, *Physics* II, 9, 200a-10.
98 Aristotle, *Physics* IV, 4, 212a28–30. Cornford thinks this line does not belong here, but his view is criticized and rejected by Ross. Cf. Wicksteed and Cornford, 315, n. b, and Ross, 576. Carteron labels his translation of these lines "Vérifications de la définition," which is correct.

Historically, place in Aristotle is often identified as a surface and this identification has affected not only interpretations of Aristotle's view but the very translation of *Physics* IV, 4. Because the claim that Aristotle identifies place with a surface has been so common and produces such serious difficulties for his account, it requires consideration.

Does Aristotle close his account of place by identifying it as a surface? Ross suggests that "[h]ere Aristotle notes the agreement of his account with two others of the ἔνδοξα."[99] But this suggestion is not quite right: Aristotle believes his account explains what others who consider the question think (usually with less precision). In effect, his own account closes with the assertion that place is the first unmoved boundary of what contains along with a reference to the contained. Having closed his account, he turns to the views of others to provide a pathology of why they think as they do. Place is often thought to be a surface and like a vessel because, as he has just concluded, it is a limit and the limit is together with the limited. *Because* place is a limit and together with the limited, it seems, or is thought [by others], to be a surface.

Aristotle regularly uses δοκεῖ in this sense – for example, in *Physics* III, 1, things without which motion is thought [or seems] to be impossible, place, void, and time. That these things "are thought" or seem such need not represent Aristotle's view; indeed, he ultimately rejects the void from this list. When he turns to the void, δοκεῖ introduces the views of his opponents, who [wrongly] assert it (*Physics* IV, 6, 213a16–17).

Ross also notes that Aristotle's usual word for surface is ἐπιφάνεια, although there is no problem with ἐπίπεδον here.[100] And Ross is surely correct, as far as usage goes. However, the use of ἐπίπεδον may yield ground for speculation as to its origin. In the *Timaeus*, we hear how god made the body of the all [τοῦ παντός] (*Timaeus* 31B7). The view that the body of the all is a surface [τὸ ἐπίπεδον] having no depth is explicitly rejected in favor of the view that the world must be solid (*Timaeus* 32A7–8). ἐπίπεδον may have been used by those claiming this view – and so is repeated by both Plato and Aristotle – or it may be a later technical term used anachronistically to express very early views.[101]

99 Ross, 576. He also thinks that this "agreement" produces a contradiction: "The place of a thing contains it; and the outer surface of a body coincides with its place, i.e. with the inner surface of its container, so that body and place have the same size. This is unfortunately incompatible with the definition of place in 212a20–21 [i.e., place is the first motionless limit of what contains]." I shall argue that there is no contradiction here. For an example of how casually and readily this view is taken to be Aristotle's, cf. Feyerabend, "Some Observations on Aristotle's Theory of Mathematics and of the Continuum," 222.

100 Ross, 576; ἐπιφάνεια is the term at *Physics* IV, 1, 209a8. However, in his commentary on Euclid's Elements (*In primum Euclidis Elementorum librum commentarii*, p. 116) Proclus comments that both Plato and Aristotle used these words interchangeably. In his rejection of Aristotle's account of place as a limit Philoponus regularly uses ἐπιφάνεια and not ἐπίπεδον; cf., e.g., Philoponus, *In Phys.* 564–565 *passim.*

101 Cf. Homer, *Il.* xviii, 607, where the river Okeanos is put on the "outer rim"; so, "outer rim" may be interpreted as surface.

In either case, the reference to Plato's view forms a fitting conclusion to Aristotle's account of place – an account that began by praising Hesiod and that even praises Plato because he alone tried to say *what* place is.[102] After citing Hesiod, Aristotle connects place to the motion of the elements and the intrinsic determination of the cosmos, commenting that the power of place must be marvelous indeed (*Physics* IV, 1, 208b30–209a2). He alone accounts fully for what others have claimed, provides a pathology of those claims, and shows why place is marvelous. Hence these lines do not identify place as a surface, and, as I shall argue, place for Aristotle cannot in fact be a surface.

To make my case I must first consider both the claim that Aristotle identifies place as a surface and the possible textual support for it. The claim apparently begins immediately with Theophrastus. An *aporia* concerning place asserts: "A body will be in a surface."[103] That is, if place is a surface and body is in a place, then body will be in a surface. The attribution of this view to Aristotle is now entrenched in the literature. Theophrastus is followed by many others, including Simplicius, Philoponus, Avicenna, and Averroes (but not Thomas); and among modern commentators, Ross, Sorabji, Hussey, Furley, and Apostle to name but a few.[104]

This interpretation of the *Physics* is often reflected in the translation, which in addition to being important in its own right, bears witness to the important substantive issue at stake here. First, the concluding lines of *Physics* IV, 4, present the only occurrence of the word (or notion) "surface" [ἐπίπεδον, or ἐπιφάνεια] in Aristotle's constructive account of place.[105] Other translations or references to a surface in this account are interpretive and involve a serious misunderstanding of Aristotle's position: for Aristotle, all surfaces [of body] are limits, but not all limits [of body] are surfaces; hence that place is a limit need not entail its being a surface. In fact, *contra* Hussey, who writes "[b]ecause place is a limit, it is a surface," on Aristotle's view, place cannot be a surface.[106]

One issue of translation involves "τὸ ἔσχατον." Hardie and Gaye often, although not always, translate it as "surface" and in one case

---

102 On references to the ancients as "markers" or "punctuation marks," cf. H. Lang, "The Structure and Subject of *Metaphysics* Λ," 260, 267, 278.
103 Simplicius, *In Phys.* 604.5–11, quotes Theophrastus here. On the relation of Theophrastus to Eudemus and the problem of place for Aristotle's immediate successors, cf. Sharples, "Eudemus' *Physics*," (in press).
104 Simplicius, *In Phys.* IV, 518.17–18; Philoponus, *In Phys.* 563.27–564.3; Thomas, *In Phys.* IV, lect. 6, par. 471; Hussey, 118. Sorabji asserts this view in a number of places. For one example, cf. his introduction to Simplicius, *On Aristotle's Physics 4.1–5, 10–14*, 1. Furley attributes this view directly to Aristotle in "Aristotle and the Atomists on Motion in a Void," 88. In Apostle, cf. 247, n. 47.
105 Urmson cites this text in defining ἐπίπεδον. Urmson, *The Greek Philosophical Vocabulary*, 58. The only other occurrence of the notion of "surface" in *Physics* IV is at *Physics* IV, 1, 209a8.
106 Hussey, 118.

"innerside."[107] In addition, when Aristotle defines place as "the first un-moved limit of what contains," they translate "first" [πρῶτον] as "inner-most" – undoubtedly a reflection of the translation of τὸ ἔσχατον as "sur-face" (212a20). Wicksteed and Cornford also translate τὸ ἔσχατον as surface here.[108]

The point is obvious: τὸ ἔσχατον does not and cannot mean "surface." It means "the extreme" or simply "the last." Hussey sometimes translates τὸ ἔσχατον as "limit," which is the sense here; however, "limit" should be reserved to translate πέρας and is identified with the very definition of place.[109] In other instances, he translates it more literally as "the ex-treme."[110] And his correction of older translations is important.

What other textual evidence is cited by those who claim that for Aris-totle place is a surface? We have now seen both texts that are generally cited: (1) At the end of his account of place, Aristotle explains why place is thought, or seems, to be a kind of surface (212a28). But as we have seen, the claim that place is a surface need not – and probably does not – rep-resent his view. (2) More importantly, Aristotle speaks of place as "the limit of the containing body [by virtue of which it is conjoined to the con-tained]" (212a6). If we accept the bracketed words, "limit" here might mean surface, which would seem to be required as the point of contact between container and contained. As I have argued, however, these words are almost certainly spurious; but even if one *were* to accept them, "con-joined" need not, indeed could not, mean contact.

Furthermore, the substantive problems entailed by claiming that place is a surface appear immediately. In *Physics* IV, 1, when formulating prob-lems related to place, Aristotle says that place has three dimensions, length, breadth, and depth, by which all body is bounded (209a5). As Urmson points out, Aristotle never speaks of place as two-dimensional.[111] But by definition surface can have only two dimensions. How can place have three dimensions yet be a surface having only two dimensions? Fi-nally, surface itself must be moved when the body of which it is a surface is moved; but Aristotle has specified that place is unmoved.

The problem lies in the nexus of place, limit, surface, and body. Sur-face is the limit of contained body, and place is the limit of the first con-

107 Cf. Aristotle, *Physics* IV, 4, 211a26, 32 (at 34 τὰ ἔσχατα is translated as "extremities"), 212a21–22 (at 24 τὸ ἔσχατον is translated "inner side" and in two occurrences, at 27 and 28, "the extremity"). These are unchanged in Barnes.

108 Indeed, at 211a28 they translate the phrase τοιοῦτος δ' ὁ πρῶτος ἐν ᾧ ἐστιν as "such then – the inner surface of the envelop, namely – is the immediate place of a thing."

109 Cf. Aristotle, *Physics* IV, 4, 211a26, 212a24, 26, 27.

110 Cf. Aristotle, *Physics* IV, 4, 211a32, 34; at 212a21–22 Hussey translates "the extreme limit"; 212a28.

111 Urmson makes this point in a note to his translation of Simplicius, *On Aristotle's Physics 4.1–5, 10–14.*, 22. Mendell speaks of place as "a two-dimensional locus" in *Physics* IV, "Topoi on Topos," 210.

taining body. Thus, although both place and surface are limits, there are two essential and irreducible differences between them: (1) Surface is the limit of bodies that occupy place and are moved, whereas place is the unmoved limit of the first containing body, which is not itself essentially in place and does not change place. (2) Surface is contained and must itself have a limit; but place is the limit of the *first* containing body and so cannot itself be contained or have a limit. Place cannot be a surface.

Surface is itself a complex topic, historically related to the notion of a "plane," and even in ancient times there was disagreement concerning its definition.[112] This disagreement, which has not been sufficiently explored in the current literature, may underlie some objections raised by Aristotle's successors. Theophrastus, Philoponus, and other commentators on Aristotle may assume a different definition of "surface" than does Aristotle and, obviously, such an assumption would make considerable difference to their arguments.[113] Furthermore, these commentators are not necessarily representing or criticizing Aristotle's views. Consequently, we cannot presuppose a common definition necessarily at work for both Aristotle and his various commentators.[114]

Because, according to Aristotle, a surface is a limit, the characteristics of a surface are also those of a limit. And insofar as the characteristics of surface apply to any limit, they must apply to place because it too is a limit. But, I shall argue in a moment, place is unique and so possesses characteristics not shared by any other limit, including a surface.

Natural bodies have surfaces and volumes as their limits and the mathematician considers these.[115] However, although they are "attributes" of body, the mathematician considers them apart from body (and hence apart from natural motion); but this separation does not falsify the mathematician's considerations (*Physics* II, 2, 193b32–35). Herein lies the first conclusion about surface – a conclusion that applies to all limits and so to place as well: although surface (like any limit) is an attribute of body, it may be considered apart from body and in this sense is prior to body. Thus, surface is "formal" and so resembles number or curved in contrast to flesh or a snub nose.[116] Such things do not exist apart from (or prior

112 For a helpful introduction to this topic, cf. Euclid, *The Thirteen Books of the Elements*, trans. with introduction and commentary by Sir Thomas L. Heath, I, Def. 5, "surface [ἐπιφάνεια] is that which has length and breadth only" (notes by Heath, 169–170).
113 Sandbach argues that Theophrastus (and the later Greek tradition, especially the Stoics) was primarily interested "in the prosecution of his own studies" and was not "the guardian of an orthodox tradition, but an independent thinker," *Aristotle and the Stoics*, 2; cf. also 33.
114 Sorabji seems to assume that everyone shares a common set of terms and conceptions, "Theophrastus on Place," 143 ff.
115 Aristotle, *Physics* II, 2, 193b23–25; cf. *Topics* VI, 4, 141b22. Also cf. Mueller, "Aristotle on Geometrical Objects," 159.
116 Aristotle, *Physics* II, 2, 194a4–6; the reference here is to *Metaphysics* VII, 10, 1035a5–10. The point of the argument of the *Metaphysics* is that when parts are prior to the

to) the bodies of which they are attributes; rather, because of their formal natures, they may without distortion or falsehood be considered apart from and independently of body.[117] Hence, even though surface is always found conjoined to body, it is not in and of itself material, but formal. As formal, surface, like any limit, may be considered apart from body and motion, and such is the task of the mathematician.

A limit is that which contains and surface may be said in a sense to contain the body of which it is the surface (*De Caelo* II, 1, 284a6). The notion of containing is thus fundamental to surface as well as to place; indeed, *Physics* IV, 5, begins with specialized problems concerning place as a container. A limit, including place and surface, may be thought of as the last [τὸ ἔσχατον] and first [τὸ πρῶτον] container of each thing: "the last of each thing, i.e., the first beyond which it is not possible to find any part, and the first within which every part is" (*Metaphysics* V, 17, 1022a4–6). For this very reason, the notion of limit applies to the form of a thing that has magnitude (1022a6–7). Thus the limit does not contain as the last physical part; it contains as a figure or form within which all the material parts are contained. In this sense, no limit is itself body, or in any sense material, but is like the form or figure of a body and in this sense is its container.

These observations may seem to support the view that place is a surface because both place and surface are limits. Any surface contains the body of which it is a surface; in *Physics* IV, 5, place is shown to contain the parts of the heaven, which as parts are in place, although as a whole the heaven is not in place. Furthermore, Aristotle calls place the first limit of what contains and so it would seem to be the surface of the first container. In fact, the characteristics shared by place and surface as limits may explain why the commentators so often identify place for Aristotle as a surface.

But surface and place also exhibit important differences. Surface limits, i.e., contains, body as its first "part," because it divides the contained body from what is outside, e.g., air or water, and is not part of the body.[118] According to Aristotle, all mathematical limits, e.g., points, lines, and surfaces, serve as limits in this way: they divide what is contained from what

---

whole – not in the sense of actually being prior to the whole, but in the sense of being formally separable from the whole – they may be defined independently of any reference to the whole, e.g., a circle may be defined independently of semicircles; but when parts are not prior to the whole, e.g., snub is only in a nose, the parts cannot be defined separately or considered apart from the whole.

117 Cf. Aristotle, *Metaphysics* XI, 2, 1060b16–17.

118 Cf. Aristotle, *Cat.* 6, 5a1–6. Sorabji's comments on this issue seem confused. "Moreover, it [place] cannot have a place, if a place is defined as the two-dimensional surface of a thing's *immediate* surroundings. For the two-dimensional surface of the banks and bed immediately surrounding a river cannot itself be *immediately* surrounded by a further two-dimensional surface" (his italics), "Theophrastus on Place," 143.

is outside the contained.[119] However, place cannot function as a limit by dividing what is contained from what is outside the contained because outside of place is nothing, i.e., neither body nor place nor anything that is in place. (The first unmoved mover is separate from the world but is neither among things that are by nature nor composed of things that are by nature and so is neither in nor out of place.) Place is unique as a limit because whereas all other limits bound something contained in place, place itself bounds that which contains everything else but as a whole is not itself contained.

This difference entails another distinction between place and surface. Surface must have a limit, whereas place is the first limit and so cannot have a limit: place constitutes the limit so that all else is in a container, i.e., the containing body that place limits.[120] Surface bounds (or limits) three-dimensional body because it is itself two-dimensional (just as a line having one dimension is bounded by a point, which has no dimensions, and a plane, having two dimensions, is bounded by a line, which has one dimension). To be two-dimensional is to be limited, and the word "surface" [ἐπίπεδον] means "plane" in the sense of restricted to two dimensions. Indeed, surface is able to limit three-dimensional body precisely *because* it is limited in this sense. Place, however, cannot have a limit because it itself is the first limit.

Aristotle never calls place either a surface or two-dimensional, and here is the reason why: place does not limit because it is limited, place is not a surface, and place is not two-dimensional. Unlike surface, place is not among the objects of mathematics; place is among the common and universal terms of physics.[121] Place, which is three-dimensional, must be the *first* limit of the containing body. As first it constitutes the limit of all other things and cannot itself have a limit. It limits the container within which all other things act as both limits and limited things. Thus, although place and surface are both limits and are both "attributes" of body, they are so in very different ways.

Here we face the most important question for Aristotle's account of place. What does "limit" mean in the unique case of place, i.e., the first limit of the containing body? Place cannot limit the containing body by

---

119 Cf. Aristotle, *Metaphysics* XI, 2, 1060b15–16; *Topics* VI, 6, 143b10–20; *De Anima* III, 6, 430b20–21. Heath, *Mathematics in Aristotle*, 193–194, refers this issue to Euclid (I, Def. 1). On mathematical abstraction as a positive process and not just division, cf. Mueller, "Aristotle on Geometrical Objects," 160–171.

120 On surface as itself limited, cf. *Metaphysics* V, 13, 1020a13–14; *De Caelo* I, 5, 272b17–24.

121 Modrak, "Aristotle on the Difference between Mathematics and First Philosophy," 121–125. Also, Lear, "Aristotle's Philosophy of Mathematics," 161–192. Supporting and further developing Lear's view, cf. Jones, "Intelligible Matter and Geometry in Aristotle," 94–102. Also (largely) in agreement and giving a fuller history of the problem of the status of mathematical objects in Aristotle, cf. Wildberg, *John Philoponus' Criticism of Aristotle's Theory of Aether*, 31–36.

dividing it from what is outside place; rather, place must limit as a constitutive principle: place renders the cosmos determinate in respect to "where" with the result that all things that are are "somewhere." Arguing that there cannot be an infinite body, Aristotle, assumes his own view of natural place and makes the relation between place, limit, and the directionally determinate cosmos clear:

In general, it is clear that it is impossible to say both that there is an infinite body and that there is some [proper] place for each of the bodies, if every sensible body has either weight or lightness, and if heavy [body] has locomotion by nature towards the center and if light [body has locomotion by nature] upwards. . . . Further, every sensible body is in place, and the kinds, i.e., differences, of place are up-and-down, and before-and-behind, and right-and-left; and these distinctions hold not only in relation to us and by convention, but also in the whole itself. . . . Surely what is in a place is somewhere, and what is somewhere is in a place. If then the infinite cannot be quantity – that would imply that it has a particular quantity, such as two or three cubits, for quantity just means these – thus too a thing is in place because it is somewhere and this is up or down or in some other of the six different positions; and each of these is a limit.[122]

Place is the limit of the first containing body because it renders the entire cosmos determinate in respect to the six directions. This determination differs from that granted by form because form is the limit of the contained and so gives the contained its very definition. Place resembles form because both are principles of determination; but unlike form, place is the principle of determination in respect to the container, not the contained. Determination in respect to "where" is granted by place and is nothing other than the six directions – up, down, left, right, front, and back – that define the "where" of any thing that is in place.[123]

Simplicius sums up the puzzle efficiently: but if that which has three dimensions is not body, an explanation is needed.[124] Only bodies seem to be three-dimensional; place is three-dimensional, but cannot be body. How can this be? Place is three-dimensional not as another body *determined* to be three-dimensional, but as a principle, a limit, that *determines* the cosmos in respect to "where."

Aristotle says that place is difficult in part because of its relation to form and matter. The issue with form now seems clear, but the relation

122 Aristotle, *Physics* III, 5, 205b24–206a7; on the six directions as given in nature and not just relative to us, cf. also, *Physics* IV, 1, 208b10–15, and *De Caelo* II, 2, 284b15–286a.

123 Preus argues that "the idea of cosmic direction (up and down, right and left, front and back) is more prominent in the *Physics* (especially IV.1) and *De Caelo* than in *Metaphysics* Λ, but the results of those inquires seem presupposed by the argument in Λ. It may be too that the idea of cosmic directions is a relatively 'primitive' element in Aristotle's thought; at any rate passages referring to the notion of cosmic orientation are the most pervasive sort of 'cosmic' passages in Aristotle's biology. They occur in all of the biological works, as well as elsewhere," in "Man and Cosmos in Aristotle," 473.

124 Simplicius, *In Phys.* IV, 531.9–11.

to matter remains obscure. When Aristotle defines place as the first limit of the containing body, we might expect him to say more about the containing body. But he does not, and the extensive speculation about surface and the relation between the containing body and its two-dimensional container may in part stem from this silence.

In fact, Aristotle *does* define the containing body. But he does not do so in *Physics* IV. Rather, his account of the containing body appears in the *De Caelo:* the heavens must be spherical because the complete is prior to the incomplete and the sphere is first among solid figures (*De Caelo* II, 4, 286b13–31). So the question becomes methodological: why does Aristotle omit any discussion of the containing body from *Physics* IV and include it only in the *De Caelo?*

The answer to this question, I would submit, returns us to the broader methodological issue of subordination. In *Physics* III, 1, when posing the list of topics, Aristotle asserts that we must first consider what is universal and common. And the examination of place reveals what "universal and common" means, a single principle of determination for all things that are and are moved. An examination of "proper terms" will follow later. And although Aristotle never identifies any accounts as considering such "proper terms" – or explains what "proper terms" means for the physicist – I shall argue that the *De Caelo* constitutes this examination.

Place renders the entire cosmos determinate in respect to "where" and so is clearly a universal and common term required by all things in motion. Hence, place is examined in the *Physics*. The *De Caelo* extends the investigations of the *Physics* by taking up more specialized problems and more proper terms, including "inclination."[125] Here we find an account of the fifth element (or first body) constituting the heaven. Several points follow.

(1) The heaven is a body, properly speaking, and as such is moved. It cannot be universal and common as is place, but it is a unique individual. Hence the investigation of its characteristics is a study of "proper terms" in the sense of an individual that is moved and in this sense caused, or constituted.

(2) The heaven is not a limit, but is the containing body limited by place. Hence, although we may be tempted to think of it as the first container, it too is contained in the sense that its parts are contained in place. The heaven is not, as is place, a constitutive determining principle, but that which is first determined by place. In short, the heaven contains everything that is in place not because it is a determining principle but

---

125 Solmsen contrasts the *Physics* and the *De Caelo:* "We do not maintain that the doctrine of place and that of natural places stand in flagrant contradiction to one another. . . . But it remains true that *De Caelo* establishes a more organic connection between the elements and their places," in *Aristotle's System of the Physical World,* 128–129; I shall consider this relation at length in Part II.

because it is itself the first part of the cosmos rendered determinate by place. Hence an examination of the heaven is not found among the common and universal terms but is properly postponed until after the examination of these primary terms is complete. Indeed, I shall argue in the examination of the *De Caelo* that "proper terms" for the physicist are just this: parts of the cosmos, each unique and specific of its kind, rendered determinate by a first principle.

(3) The heaven is both limited and spherical and so possesses a surface. Since surface must be the limit of a sphere and itself limited, the heaven, as a sphere, should exhibit a surface.[126] We should think of the surface of the heavens as the first visible effect of place as a limit, i.e., the constitutive principle rendering the heaven determinate.[127] For, as we have seen, in *Physics* IV, 4, when Aristotle concludes that place is a limit, he immediately indicates that this explains why it seems to be a kind of surface.

This point returns us to the primary issue: is place a surface according to Aristotle? No, it is neither a surface nor two-dimensional. It may be confused with surface because both are limits and both are attributes of body without being body *per se*. But surface is itself limited and limits by dividing, whereas place is the first limit, not itself limited, and limits as a constitutive principle rendering the cosmos determinate in respect to the six directions.

This view of place as a principle of determinacy returns us to the methodological relation between the definitions of nature, motion, and place. Nature is an intrinsic principle or cause of being moved and being at rest; motion is the actualization of the potential *qua* potential by its proper actuality. Thus, potency is the intrinsic principle of motion in the moved, and actuality is that same principle in the mover. On contact (and in the absence of hindrance) motion must occur because potency is oriented toward actuality, and the actuality of the mover and the moved are identical.[128] Consequently, the identification of proper actuality always explains why potency is moved.

Place is the first unmoved limit of what contains, differentiating the cosmos as contained into actually "up" at the outermost limit and actually "down" at the center; each element is by definition potential for its natural place and, unless hindered, will be carried to this place, e.g., fire toward the outermost limit, earth toward the center. In this sense, respective natural place provides the actuality of each element as movable

---

126 For ἐπιφάνεια as meaning visible surface, cf. *Hist. An.* I, 16, 494b19–20.
127 The word ἐπιφάνεια is clearly related to the notion of visible appearance and so may in part express this relation.
128 Aristotle, *Physics* III, 3, 202a19. Aristotle argues in a number of places that mover and moved must be together; cf. Weisheipl, "The Specter of *Motor Coniunctus* in Medieval Physics," 101–108 in the reprinted edition, for an argument that a mover in contact with the moved is not always needed.

by nature. In short, Aristotle's definition of place operates within a domain defined by his definition of motion as the actualization of the potential *qua* potential by what is actual. Being movable implies being potential; potency by definition is always oriented toward actuality. Place, by being a limit, determines the cosmos as "up" or "down," which serves as the actuality of the elements, themselves potentially up or down.

By rendering the cosmos determinate in respect to "where," place causes the motion of the elements. All movable body is composed of the elements (*De Caelo* III, 2, 301b30). Thus, to be movable on Aristotle's definition implies to be in place as he defines it, and "place" must be "where" natural things are. "Place" as the "first motionless limit of that which contains" causes motion as the actuality of the elements. Without place as the limit rendering the cosmos determinate, elemental motion would be impossible.

Aristotle Concludes His Account

In *Physics* IV, 5, Aristotle completes his account of place by returning to the special problems raised at the outset. Assuming that place is the first unmoved limit of what contains, he turns to the most serious problem: how the heaven is (and is not) in place; further problems are then briefly resolved, and the chapter ends with an account of the elements moving toward and resting in their respective natural places. The account of elemental motion completes the examination of place by showing that it is, what it is, and that it possesses the characteristics rightly thought to belong to it (213a10–11).

The most important problem, the relation between the heaven (the first contained body) and place (the unmoved limit of the first containing body), is implicit in the question of how one thing is "in" another and in the boat/river metaphor. Aristotle argues that as a whole the heaven is not in place and does not move, but that its various parts are in place and do move.[129] The immediacy with which he introduces bodies in motion and locomotion evidences the close relation between natural things, motion, and place; the resolution of this problem shows both how place operates for the cosmos and the constitution of the cosmos itself.

Aristotle returns to the problem of how one body is "in" another. If one body were outside and surrounding, i.e., distinct but touching, another body, then the latter would be in place – and if not, not (212a31–32). But what about the first containing body: is it too in place or not? For example, if there were water with no surrounding body, its parts would be moved because they are surrounded by one another, but the whole would remain unmoved (212a32–34).

---

129 These claims answer the question raised at *Physics* IV, 4, 211a13.

Hussey comments that "Aristotle begins by generalizing the problem (to test his theory) to the case where the body not in place is not even rigid, but composed of water, and so homogeneous and capable of internal movement of parts."[130] Ross refers to the body of water as an "extravagant hypothesis."[131] Both accounts assume that Aristotle is positing the surrounding water as a serious working hypothesis. But Aristotle gives no account of this water and certainly no mention of its "rigidity" or the homogeneity of its parts. It appears and disappears in a single sentence. What is its function? Fifteen lines above, Aristotle identifies "the whole river" as place because it is motionless; now he returns to the image of water to suggest that his position agrees with the wisdom of the ancients in the myth of Okeanos: in the outermost body, the whole will not be moved as a whole, but the parts are moved.

First Aristotle considers the whole, which represents the heaven in the sense of a circular vault containing all natural things, and the relation between the heaven and place. The account of how the heaven is and is not in place is brief. A whole that is not surrounded by another body would not, as a whole, be moved because it does not simultaneously change its place, even though it moves in a circle: the place in which the rotary locomotion occurs is the place of the parts; thus the parts, not the whole, change place (*Physics* IV, 5, 212a35–212b). By definition, the whole is not in place because it is not surrounded by another body – the primary requisite for being in place; all movable things are in place; because the heaven is not in place, it cannot be moved as a whole. However, as Aristotle explains in a moment, the parts of the heaven are moved because each is contiguous to the next and one part contains another (212b11, 13). But before turning to the parts, we must consider the immobility of the heaven.

The problems of the motion of a whole as such and of the heaven as a whole of parts are articulated more fully in *Physics* VII and *De Caelo* II, respectively. Their arguments fully support the conclusion of *Physics* IV, 5. In *Physics* VII, 1, Aristotle argues that everything movable is divisible, and if a thing is in motion as a whole, then all of its parts must be in motion: if even one is not in motion, then the whole as a whole is not in motion.[132] Furthermore, the respective motions of the parts must all be simultaneous.[133] The model here is a whole in which all the parts are connected by being either contiguous or continuous. So, for example, it is impossible to move the upper arm without also moving the forearm, hand, and fingers.

130 Hussey, 119.
131 Ross, 577.
132 Aristotle, *Physics* VII, 1, 242a40–45. Cf. H. Lang, *Aristotle's* Physics *and Its Medieval Varieties*, 35–62.
133 Aristotle, *Physics* VII, 1, 242a60–62.

It is easy to see why the heaven as a whole cannot be in motion as a whole. As Aristotle argues in *De Caelo* II, 14, the center of the earth, which coincides with the center of the whole, is at rest (296b10, 21–27). Because there is a part at rest while another part moves, by definition the whole cannot as such be moved. Therefore, as a whole, the heaven cannot be in motion.[134]

Before proceeding to the motion of the parts, Aristotle recalls several relevant points. First, some things are moved in a circle whereas others, those capable of rarefaction and condensation, are moved up and down.[135] The heavens, which are moved in a circle, are composed of the fifth element, i.e., aether. The other four elements are transformed into one another and so are capable of rarefaction and condensation. These two categories taken together exhaust *all* things exhibiting motion; hence an account of them must be the most general and inclusive possible.

Second, some things are in place potentially, others actually (*Physics* IV, 5, 212b3–4). This distinction opens another approach to the problem of how one thing is "in" another. If a thing is continuous, its parts are in place potentially, because in a continuous substance the conditions of being in succession and touching are not met. But something that is continuous may be divided with the parts touching; when they are divided and touching, they are in place actually (212b4–6). If we suppose a vaulted heaven, it can be in place only if surrounded completely by another body outside it and touching it; but there is no such body; thus the whole as a whole fails to meet the necessary condition for being in place. However, the parts of the heaven are contiguous, each to the next (212b11). As Aristotle says in a moment, they are contained in, or surrounded by, the whole. Hence the parts of the heaven must be in place actually.

The problem of the parts of the heaven recalls a distinction from *Physics* IV, 4: a body can be in place either (a) *per se* or (b) accidentally

---

134 Hussey, 119, gives quite a different account: the whole body "cannot change its place as a whole, because it has no place to change. . . . For it is surrounded neither by another body nor by a void (since it is assumed there is no such thing); so there is nothing for it to move away through. The only possible movement of the body as a whole, then, is one involving no change of place of the body, but only exchange of places between parts in a concerted way, e.g. rotation." This view involves a serious problem. The entire argument is cast as a contrast between the whole body and its parts. Place as a whole is without place (Ross too suggests this point, 577) because it has no body outside and surrounding it, not because there is nothing for it to move through (a point that is never mentioned and has nothing to do with Aristotle's formal definition of place). Void too is not mentioned here; its discussion begins as analogous to that of place in a little more than a page. Here Aristotle denies – this is the point of the argument – that the body as a whole can be in motion. The parts, on the other hand, are moved as parts, and their motion is not a motion of the body as a whole. (Aristotle never speaks of these parts as "exchanging places," which is an expression that may be incompatible with his frequently expressed view of natural place.)

135 Aristotle, *Physics* IV, 5, 212b-3. I follow Hussey's punctuation here, which must be right (cf. Hussey, 120). Some commentators speculate that this line may be a later addition.

(211a19–22). (a) Being in place *per se*. Every body that is movable either in respect to locomotion or in respect to increase (and decrease) is in place *per se;* but the heaven, i.e., the circular vault encompassing the cosmos, is not in place as a whole because no body contains it (*Physics* IV, 5, 212b8–10). Correlatively, it is not movable *per se* and does not *per se* change place; but on the line on which the circular locomotion of the heaven occurs, the parts are in place *per se* because each is divided from and contiguous with the next (212b10–11).

The relation between locomotion and place is striking: the heaven as a whole is not in place and as a whole does not move; but its parts, which are actually divided and touching, are both in place and moved, i.e., they exhibit circular locomotion. The conditions that make the parts of the heaven occupy place are the very conditions that allow them to be moved *per se* (although we do not get a full account here of why they are moved).

(b) Being in place accidentally. A thing, e.g., the soul or the heaven, is in place accidentally if it is in place not because of what it is but because it happens to be contained in something that is in place essentially (212b11–12). The heaven is in place accidentally because all its parts are in place (212b13). In effect, the whole heaven is present in all its parts, and because all the parts are in place essentially, the heaven present in them is in place accidentally. That is, the heaven is in place not because of what it is but because it happens to be present in its parts, all of which are in place. For this reason, the upper part of the heaven is moved in a circle (and is in place) whereas the whole as a whole is neither in place nor moved.

The argument about the heaven now concludes:

> For what is somewhere is itself something, and again there must be something else outside this in which it is [and] which surrounds it. But in addition to the all, there is nothing outside the all, i.e., whole, and on account of this all things are in the heaven. For the heaven is, perhaps, the all. And place is not the heaven, but of the heaven what is last and touching the moved body. And on account of this, the earth is in the water, this in the air, this in the aether, and the aether in the heaven, but the heaven is not in another. [212b14–22: τὸ γάρ που αὐτό τέ ἐστί τι, καὶ ἔτι ἄλλο τι δεῖ ἔιναι παρὰ τοῦτο ἐν ᾧ, ὃ περιέχει· παρὰ δὲ τὸ πᾶν καὶ ὅλον οὐδέν ἐστιν ἔξω τοῦ παντός, καὶ διὰ τοῦτο ἐν τῷ οὐρανῷ πάντα· ὁ γὰρ οὐρανὸς τὸ πᾶν ἴσως. ἔστι δ' ὁ τόπος οὐχ ὁ οὐρανός, ἀλλὰ τοῦ οὐρανοῦ τι τὸ ἔσχατον καὶ ἀπτόμενον τοῦ κινητοῦ σώματος [πέρας ἠρεμοῦν]. καὶ διὰ τοῦτο ἡ μὲν γῆ ἐν τῷ ὕδατι, τοῦτο δ' ἐν τῷ ἀέρι, οὗτος δ' ἐν τῷ αἰθέρι, ὁ δ' αἰθὴρ ἐν τῷ οὐρανῷ, ὁ δ' οὐρανὸς οὐκέτι ἐν ἄλλῳ.]

The details of this view emerge in the *De Caelo*. There Aristotle argues that *all* sensible body is included within the heaven; there is neither body, nor place, nor motion outside the heaven (*De Caelo* I, 9, 278b23–279a13). So, to occupy place and be a "something," a thing must be "in" another and, ultimately, must be within the circular vault of the heaven.

What does "heaven" mean here? In *De Caelo* I, 9, Aristotle gives three meanings for "οὐρανός": (1) the outermost circumference, or the body that is there, (2) the body continuous with the outermost, which contains the moon, the sun, and some of the stars that are "in" the heaven, and (3) all body contained within the outermost circumference, "since we habitually call the whole or totality" the heaven (278b11–21). "Heaven" is used in *Physics* IV, 5, in the last sense: it is not itself another body – indeed, it is compared to soul – but it is the "all" within which all body is contained. And place is its first motionless limit. Thus the heaven itself is not in place, but all body must be in place within it.

As we have already seen, the parts of the heaven are actually divided and touching and so must be in place. The notion of them as "surrounded by one another" may refer to the fact that a circle has no beginning or end. But because there is nothing that surrounds and touches it, the heaven as the all or the whole is neither in place nor movable. Its first limit, which is not itself a body, is place – place that renders the cosmos determinate and thereby constitutes place for everything within the circular vault of the heaven. Consequently, the all or the whole is rendered determinate by place as its first limit. The center is down and the limit is up: the heaven as a whole is neither form nor matter but a whole determinate in respect to "up," "down," "left," "right," "front," and "back" because place constitutes its first motionless limit. Hence working out from the center, i.e., the earth, each part of the whole can be said to be in that which surrounds and touches it, until we see that the aether is in the heaven.[136] But the heaven as a whole is neither surrounded nor touched by another containing body: the heaven as a whole is not in place.

It is clear, Aristotle tells us, that the various problems raised concerning place are solved on this explanation (*Physics* IV, 5, 212b22–23). He lists seven, and an examination of them both returns to problems raised earlier and provides, as it were, a summary that completes his constructive account of place. (1) There is no necessity for place to grow with the body in it (212b22–24). The ultimate defining place of all body is the first motionless limit, and this limit obviously does not change as body within it changes. (2) There is no necessity that a point have a place (212b24). A point, on Aristotle's definition, is the terminus of a line; as such, it is neither divided from the line nor contiguous with it.[137] Therefore, a

---

136 For the [persuasive] claim that the establishment of a geocentric (and homocentric) model for the cosmos – a model designed to explain planetary motion – was a major scientific achievement that inaugurated a new phase in Greek astronomy, cf. Goldstein and Bowen, "A New View of Early Greek Astronomy," 330–340.

137 Aristotle, *Physics* VI, 1. That there is no such thing as a "free-standing" point (which would have place) is itself the source of an interesting controversy. Cf. Konstan, "Points, Lines, and Infinity," 2–5. Hussey, 121, claims that points do have "locations," by which he seems to mean places, and cites *Metaphysics* V, 6, 1016b25–26. But this argument establishes that points must have position (θέσιν), not place (τόπος).

point fails to meet the necessary conditions for being in place. (3) Again, there is no necessity that two bodies be in the same place (212b25). Place is not a body, but a limit; therefore, place and what is contained in place can be in the "same place" just as form and matter may be said to be in the same place.

(4) Aristotle repeats his claim that place cannot be a material interval because between the boundaries of the place is whatever body happens to be there, not an interval in body (212b26–27). Because place is a limit, it cannot be a corporeal extension of a body.[138] Rather, the cosmos is determinate with a center and circumference, down and up. Hence, whatever is in place is whatever body happens to be there either naturally or unnaturally.

(5) Place is also "somewhere" [πού], not in the sense of being "in" place, but as the limit is in the limited – for not all being is in place, but only movable body (212b27–29). Place constitutes the limit of the cosmos so that everything contained within it is in place; but place is not in place essentially, i.e., by virtue of what it is – only movable body is in place essentially, and place is not a movable body. However, place is accidentally in place just as the heaven is: place is "somewhere" accidentally because as a limit, it happens to be contained in that which is in place essentially.[139] (Only the unmoved mover is not in place either accidentally or essentially.[140])

The last two problems are closely related and concern the elements. Neither is clearly explained. (6) Each element is carried by nature into its own place (212b29–31). A body next in the series and in contact is "akin" to its predecessor so that they may act and be acted upon by one another (212b31–33). A full account of this motion appears only in the *De Caelo*. But we can see here that, on Aristotle's definition of motion and mover/moved relations, the motion of a body to its own place must be natural, rather than violent, and produced by a mover that is "together" with it. And, as I shall argue, these conditions are met by Aristotle's account of elemental motion.

Finally, (7) each element rests in its proper place because the elements have the same relation to place as does a separable part to its whole

138 Hussey, 121, claims that places cannot be bodies because "places turned out to be two-dimensional" and refers the reader to 209a4–7 where Aristotle argues that place seems to be three-dimensional, and as this is a characteristic of body, whenever body is in place, two bodies will be in the same place. Aristotle never suggests that place is two-dimensional; rather, he denies that it is a body. It is a limit of body. Hence it is in a sense three-dimensional, although not a body. That all body is three-dimensional does not require that everything three-dimensional be body.

139 Hussey, 121, repeats here that place is two-dimensional and adds that "it defines a three-dimensional extension inside itself." There is no evidence for this view in Aristotle.

140 Aristotle, *Physics* VIII, 6, 258b13–16; 259b23–31; *Metaphysics* XII, 7, 1072b5–13; 10, 1075a12–15.

(212b33–213a). The elements are related to one another as matter and form; for example, water is the matter of air and air some sort of actuality for water. Again, this point is not explained here – Aristotle promises a fuller account later.[141] But he defines the elements here as potency and actuality of one another, and he insists that as such there is contact between them.[142] They have "organic union" [σύμφυσις] only when both come to be one in actuality, i.e., no potency remains; when they are related as potency and actuality, there is not unity, but contact between them. "And" Aristotle says, turning to the problem of the void, "concerning place, both that it is and what it is has been said" (213a10–11).

The problem of elemental motion has not been solved. A full account of it, as I shall argue, does not belong in an argument dealing with common and universal terms, but must be reserved for special terms. A consideration of these terms will concern us in the second part of this study.

Final Considerations of Place

Because Aristotle's account of place is complete, we may consider its results and implications before turning to "void," the second candidate for explaining "where" things are. I turn first to the unity between the logic and content of Aristotle's account of place as "where" things are and are moved.

Aristotle begins with the most important topic, i.e., nature. Nature is a principle of motion: an intrinsic ability to be moved or to be at rest in that to which it belongs primarily in virtue of itself and not accidentally. Thus as principles, nature and motion are radically linked from the outset. *Physics* III, 1, confirms this point: if we are to understand nature, we must understand motion. Because motion appears as the crucial principle for things that are by nature, the domains of nature and motion are identical.

An account of motion as a relation between movers and moved things follows immediately: motion is the actualization of the potential *qua* potential by that which is actual. The potential has its definition in the actual and is actively oriented toward its proper actuality. Hence what is potential cannot fail to be actualized by its proper actuality, and identification of proper actuality always explains why, in the absence of hindrance, potency is actualized by actuality. Having defined motion, Aristotle turns immediately to other universal terms, including place, without which motion seems to be impossible.

The examination of place appears at *Physics* IV, 1–5 (void follows in 6–9), and this examination exhibits the same structure. The problem is

141 Aristotle, *Physics* IV, 5, 213a4–5; cf. *De Gen. et Corr.* I, 3.
142 Aristotle, *Physics* IV, 5, 213a6–9. Cf. *Metaphysics* IX, 7, 1049a19–27. Cf. H. Lang, "Why the Elements Imitate the Heavens," 335–354.

posed in its most universal form: where are all things that are and are moved? Aristotle offers two candidates for "where" [ποῦ], place [τόπος] and void [κενόν], and he takes up place immediately. As a problem, place involves all things that are and the most important kind of motion, locomotion, exhibited by them. After enumerating the problems associated with place, Aristotle identifies four characteristics rightly thought to belong to it. (1) Place first surrounds the thing contained without being part of it. (2) Place is neither less than nor greater than the thing contained. (3) Place must be separable from what is in place, and (4) place has "up" and "down": each of the four elements is carried by nature to its proper place "and [they] remain in their proper places and this makes either up or down." These characteristics constitute the broadest possible criteria for any account, and Aristotle clearly believes that his account alone meets and explains them.

A series of distinctions concerning motion and things in motion follows. Even though he specifies place as involving two kinds of motion, locomotion and increase and decrease, Aristotle considers locomotion, the first and most important kind of motion in things. Motion must be either *per se* or accidental; accidental motion can occur in things that are also capable of *per se* motion or in things that can only be moved accidentally. Thus motion raises the problem of how to define one thing as "in" another. Whenever one thing is "in" another, it can be "in" either as a part is in a whole, i.e., the part and whole are undivided and continuous; or as one thing is simply "in" another, as wine is in a cask or water a jug, i.e., container and contained are divided but touching. According to Aristotle, things are in place in the latter sense.

With these points established, Aristotle takes up the specifics of place as a problem. He lists four possible candidates for place: form, interval, matter, and limit. Rejecting the first three, he asserts that place must be a limit: it is the first unmoved limit of what contains, an immaterial constitutive principle that renders the cosmos determinate. As such it is the οὐσία of the all, or whole, and accounts for the motion of all body – both that which is moved in a circle and that which is moved up and down.

The account concludes with the most specialized problems that any account of place must address. Most importantly, Aristotle explains how the heaven and its parts both are and are not in place. Because there is no body outside, surrounding and touching the heaven, it cannot be in place essentially. But its parts are actually divided, touching each other in succession, and contained in the whole circle; therefore the parts of the heaven are in place essentially. And because the heaven is contained in these parts, it is in place accidentally.[143] Five common problems con-

---

143 On Eudemus's solution to this problem, cf. Sharples, "Eudemus' *Physics*" (in press).

cerning place are raised and quickly dismissed, e.g., place need not grow with what is in place, a "point" need not be in place, etc.

The account of place concludes, so to speak, with two problems about elemental motion; namely, why the elements are moved to their respective natural places and why each element rests in its natural place. Aristotle does not solve these problems here and promises a better account later – that of the *De Caelo,* as I shall argue in Part II. This, he says, concludes his account of place – that it is and what it is.

But the account of "where" things are is not yet complete. The void is often proposed as place with nothing in it and so stands as a serious answer to the question "'where' are things that are and are moved?" Thus void is an important competitor to Aristotle's concept of place, and rejection of it supports his account of place. Indeed, this rejection completes Aristotle's account of "where" things are. Turning to his analysis of void, I shall again argue that it presupposes his accounts of nature, motion, and place.

4

Void

Aristotle's arguments about the void present special interests and problems because of their long history in Aristotle's commentators. A full accounting of the responses to these arguments lies beyond the scope of my analysis, although some special cases will be taken up. One point however has been crucial for interpretations of these arguments: Euclidean geometry requires a three-dimensional infinite space.[1] Euclid flourished (probably) one generation after Aristotle, and his geometry was enormously influential. Because Aristotle defines place as the limit of the first containing body, place was often thought to be both finite and, as we have seen, a two-dimensional surface. Hence on this view, Aristotle's notion of place fails on both counts to meet the requirements of Euclidean geometry. Taken together, the requirements of Euclidean geometry and the apparent failure of Aristotle's account of place to meet them often motivated first a commitment to the void and then criticism of Aristotle's arguments rejecting the void. Therefore, these criticisms derive from the conjunction of Euclidean geometry and a common misunderstanding of Aristotle's account of place. My interest lies in a direct analysis of the arguments concerning the void in *Physics* IV.

Place has not been established as the *exclusive* answer to the question "where are things?" because void [κενόν, i.e., "empty"] presents an alternate answer. Consequently, void must be examined, and, Aristotle says, the same questions must be posed for it as for place, namely, "if it is or not, and how it is and what it is."[2] Indeed, arguments about void resem-

---

1 For one example of an analysis of the impact of Euclid on Byzantine, Arabic, and Christian physics, cf. Pines, "Philosophy, Mathematics, and the Concepts of Space in the Middle Ages," 84–90.
2 Aristotle, *Physics* IV, 6, 213a12–14; Ross comments, 581, that given that Aristotle denies the existence of the void, the phrase καὶ πῶς ἔστι καὶ τί ἐστιν may "excite suspicion. We may suppose (1) that Aristotle means 'and *if* it exists, how and what it is.' But (2) . . . it is better to suppose that he means 'in what sense there exists something which is what the supporters of a void mistakenly describe as a void, and what is its nature.'" I shall argue in a moment that Aristotle does not deny the existence of a void directly; rather, his re-

ble those concerning place because people treat void as a kind of place –
place that is full when it receives body and empty when no body occupies
it.[3] So we must consider first what people say who assert the void, then
what those say who deny the void, and, finally, common opinion on these
problems (*Physics* IV, 6, 213a19–21). Solmsen, labeling the arguments
about void "a digression," comments that "[t]here is no evidence that it
had ever before him been brought into relation with place."[4] But Aristotle expressly claims that void is used to account for the cosmos (τὸ πᾶν)
and to explain "differences in position, shape, and order . . . ," and he
names several proponents of the void.[5] Hence he clearly takes void as a
serious competitor of place in accounting for "the where" and, ultimately,
for nature.

The analogy between place and void is problematic. Those proposing
the void as place when it is full and empty when it is not think of the
empty, the full, and place as the same thing, when in fact they are different (*Physics* IV, 6, 213a18–19). But "the full" is not examined in *Physics* IV
(or in physics more generally), perhaps because no one proposes it as
"where," or place – it clearly cannot receive body. The confusion, however, between void and place is serious.

In one sense, the investigation of void, like place, belongs to the
physicist and should be conducted as was that concerning place
(213a12–14). But void is a special case of place, i.e., empty place, and so
presupposes place. Consequently, although void is proposed as an alternative to place as "the where" – and so presumably is an independent

jection of it follows from the main point of the argument, namely, to show that a void fails
to serve as that without which motion seems to be impossible.
3 Aristotle, *Physics* IV, 6, 213a14–19. Commenting on 213a16, Hussey says, 122, "Aristotle
claims that in theories of the void, the void is a kind of place. The content of the claim
seems to be that the void, as generally hypothesized, is a (non-bodily, three dimensional)
spatial extension." Although Hussey is clearly right that void is proposed as a candidate for
place, there is no indication here either that the void is "spatial extension" or that it is three-
dimensional. Furley, *The Greek Cosmologists*, 190, seems to give a similar account, claiming
that a void place is possible only if place is thought of as an interval. He also claims, 189,
that "[i]n the history of Greek philosophy, Aristotle was an extremist in his refusal to accept the existence of void in the universe. In defense of his own view that the universe is
finite and filled to its limits with a continuous quantity of matter, he mounted several at-
tacks on the void." Whereas Aristotle often seems to speak of the world as a plenum (for
an example in *Physics* IV, cf. 1, 209a26–27), he proves that it is rendered determinate in respect to "where" by place, not that it is full of a continuous quantity of matter.
4 Solmsen, *Aristotle's System of the Physical World*, 140.
5 Aristotle, *Physics* I, 5, 188a23; cf. *Metaphysics* I, 4, 985b5. Ross seems to think that atoms
are being accounted for here, 488. Sedley argues that given his definition of void "Aristotle's evidence on this question is of little historical value," in "Two Conceptions of Vacuum," 179. For proponents of the void, cf. *Physics* IV, 6, 213a25: Anaxagoras; 213a33: Democritus and Leucippus; 213b: "many other physicists"; 213b3: those who say that the void
exists; 213b12 and 214a29: Melissus; 213b24: the Pythagoreans; 9, 215b25: Xuthus. For a
survey of possible targets for these arguments, including Parmenides, cf. Malcolm, "On
Avoiding the Void," 75–94.

and equal competitor – an examination of the void can be neither independent of nor parallel to that of place.[6] And as Aristotle concludes in *Physics* IV, 7, however the term "void" is construed, it entails a contradiction.

Analysis of the term, however, does not complete the investigation. Two further chapters treat the void as if it were a coherent alternative to place as "the where." Historically, these difficult arguments have been the primary focus for analysis of "Aristotle's rejection of the void." If "void" is an incoherent term, then why does Aristotle list it as a term without which motion seems to be impossible, treat it as parallel to place and pursue it so rigorously?

As terms presumably involved in motion (and hence physics), place and void must each be examined because each is proposed as "the where." Considered according to its definition, however, void is a subset of place, and so its investigation presupposes the concept of place. Hence – and I shall argue this point – Aristotle's analysis (and rejection) of void assumes important features of his account of place.

Here we return to the methodological issue of subordination: Aristotle's account of place presupposes his definitions of motion and nature. Hence insofar as the analysis of the void presupposes that of place, it must also presuppose the definitions of motion and nature: there is no independent analysis (or rejection) of the void taken in and of itself.[7] Like his opponents, Aristotle treats the void as some sort of "cause" of motion in things, and he does so by presupposing his own definition of motion (and nature).[8]

Aristotle works within the science of physics as he defines it. And within this physics, rejection of the void further supports Aristotle's account of place by leaving it as the true, exclusive, and unique answer to the question "where are things that are?"[9] But the rejection of the void does more

---

6 Aristotle, *Physics* IV, 1, 208b25–26; 6, 213a15–19.

7 The common view, of course, is that Aristotle rejects the existence of a void. For a clear statement of this view along with a sense of its history, cf. Weisheipl, "Motion in a Void," 121–142. For a recent example, cf. Lettinck's summary of *Physics* IV, 6, which begins, "The arguments for and against the existence of the void are about the same as those in connection with place," *Aristotle's* Physics *and Its Reception in the Arabic World*, 317.

8 The causal force of these arguments emerges very clearly as Aristotle begins his fourth argument against the void: "For if there is some motion for each of the simple bodies by nature, for example fire upward and earth downward and toward the middle, it is clear that the void would not be a cause of the motion. Of what therefore will the void be a cause? For it seems to be a cause of motion according to place but it is not a cause of this," *Physics* IV, 8, 214b13–17: εἰ γὰρ ἔστιν ἑκάστου φορά τις τῶν ἁπλῶν σωμάτων φύσει, οἷον τῷ πυρὶ μὲν ἄνω τῇ δὲ γῇ κάτω καὶ πρὸς τὸ μέσον, δῆλον ὅτι οὐκ ἂν τὸ κενὸν αἴτιον εἴη τῆς φορᾶς. τίνος οὖν αἴτιον ἔσται τὸ κενόν; δοκεῖ γὰρ αἴτιον εἶναι κινήσεως τῆς κατὰ τόπον, ταύτης δ' οὐκ ἔστιν.

9 When at *Physics* IV, 4, 211b5 Aristotle gives his list of four candidates for place, i.e., form, matter, interval, or limit, Hussey comments that this list is odd, in part because there is no effort to show that it is exhaustive (115). Although it is certainly true that Aristotle ar-

than indirectly support place. The void, as I shall argue, makes disorder prior to order, the indeterminate prior to the determinate. Hence as a competitor to place, void challenges the view that nature is always and everywhere a source of order, and its rejection is crucial not only to Aristotle's account of "the where" but also to his view of nature itself. He certainly believes that his arguments inflict a defeat, in fact a complete rout, on proponents of the void. And thus their view of nature – opposed as it is to his own – is also refuted.

The overall structure of Aristotle's analysis is important. He first takes up accounts given by those who assert a void. Their other opponents, e.g., Anaxagoras, give silly arguments against the void because they do not consider what those asserting the term "void" mean (*Physics* IV, 6, 213a24–29). Four brief but serious arguments in favor of the void occupy the remainder of *Physics* IV, 6. (1) Both locomotion and increase require a void because if something is full, there can neither be motion nor addition of anything further (213b4–6). The assumption of motion in a plenum produces various absurdities, e.g., many bodies in the same place (213b6–12). Therefore, we must assume a void. (2) When things appear to contract or be compressed, some assume the presence of a void into which the contraction occurs (213b6–18). (3) Increase is thought to occur on account of a void because nutriment is body, and two bodies cannot be together; likewise a vase full of ashes will also hold its own volume again of water; both examples imply void space being filled (213b19–21). (4) Finally, the Pythagoreans argue for a separate void, entering the world from without and distinguishing, or separating out, the terms of a series, primarily numbers (213b24–27). By his own lights, Aristotle will refute these arguments later; he takes them seriously because they present the most important challenges to his own view.

Physics IV, 7, begins the analysis of void. The overall force and structure of this analysis is crucial to the construal of its particulars. The analysis begins not with nature or with a void occurring in nature, but with the meaning of the term "void." According to Aristotle, the meaning of "void" entails several contradictions, e.g., it is both empty and contains what is heavy and light; it is both separate and not separate (213b30–34). Next he refutes (from his own point of view) three of the four arguments from *Physics* IV, 6. The fourth, quite different, and most serious argument for the void is addressed in *Physics* IV, 8–9. Each of Aristotle's complex arguments concerning motion in a void results in a contradiction and consequent rejection of void as a cause of motion.

rives at limit only after eliminating the other three, his assertion of the limit does not rest solely on its being the only remaining possibility. He provides constructive arguments for it. Here there are two points: (1) the list may not be complete, and (2) it is not immediately clear here that void means "interval" (often translated "extension"), which has been rejected earlier. For such an identification, cf. Hussey, 123.

Historically, these arguments are among the most important in the Aristotelian corpus.

Aristotle concludes with an unequivocal rejection of the void: there is neither a distinct void nor a potential void.[10] Indeed, the void fails in every way. First and foremost, it is an incoherent concept. Second, were we nevertheless to assume it as a possible cause of motion, it would not only fail to cause motion but would render motion impossible.

Aristotle shows that as a concept void is incoherent: "empty place" is self-contradictory. In rejecting void, Aristotle refutes serious challenge to his own account of nature. He both attributes arguments for the void to its "proponents" and criticizes opponents of void other than himself for failing to take the term as intended by its proponents. Hence even though he concludes that the void is an incoherent concept, Aristotle treats it as if it were a meaningful term in order to refute the arguments of its proponents and he concludes that assumption of a void would render motion impossible.[11]

Several methodological conclusions follow. (1) Within physics as a science, analysis of a constructive term such as place exhibits both the order of the cosmos and nature as a principle of order.[12] But the physicist cannot consider those things that would render physics as a science impossible, including accounts that render motion impossible.[13] Consequently, when analysis leads to the conclusion that void renders motion impossible, constructive treatment of void is impossible within physics: an analysis of void can produce no constructive conclusions for physics as a science.

(2) Aristotle's arguments concerning the word "void" allow it to stand as if it had a coherent meaning. Thus their starting point in this concept originates exclusively with his opponents. As a result, *Physics* IV, 6–9, contributes directly neither to Aristotle's account of the world nor to his accounts of motion and nature.

(3) Aristotle rejects the void (as I shall argue) because arguments for it invariably produce a contradiction. His rejection is warranted if and only if the contradiction originates solely with the assumption of the void. Hence, premises other than the void, even if assumed only for the sake

---

10 Aristotle, *Physics* IV, 9, 217b20–21; this conclusion echoes the opening of Aristotle's constructive account. Cf. *Physics* IV, 7, 214a18–19.

11 Cf. Aristotle, *Metaphysics* IV, 4, 1006a15–18. Thorp, "Aristotle's *Horror Vacui*," 149–166, argues that because Aristotle only criticizes the arguments of other thinkers "it seems almost as though he suffers from an irrational aversion to the void, a neurotic *horror vacui*," 150, and he argues that Aristotle indeed denies the void. I suggest that this view is wrong on both counts.

12 On the cosmos as orderly, cf. Aristotle, *De Caelo* II, 14, 296a34 (also *Metaphysics* XII, 10, 1075a13, 14); on the order of the whole, cf. *De Caelo* III, 2, 300b23 and 301a10; on nature as a cause of order, cf. *Physics* VIII, 1, 252a12; *De Caelo* III, 2, 301a5; *Gen. of An.* III, 10, 760a31.

13 Aristotle, *Physics* I, 2, 184b25–185a4; 185a13–19.

of these refutations, must be taken as true. Can we conclude that these other premises present constructive commitments for Aristotle's account of nature, even though they appear within a refutation? No. One may allow any number of premises as true for the sake of an argument; so, for example, a materialist account – and some supporters of the void are materialists – holds various assumptions that a critic of materialism might also assume for the sake of an argument. The same assumptions might turn out to be totally irrelevant (irrespective of the question of their truth) for alternate accounts. Hence even premises that must be true within the refutations of arguments for a void cannot be understood as operating constructively within Aristotle's physics without independent evidence from arguments like those concerning place that are constructive.

I turn now to Aristotle's rejection of the void as a cause of motion, considering first Aristotle's analysis of the meaning of the name "void" (*Physics* IV, 7) and then the arguments supporting the void as a cause of motion (*Physics* IV, 6) and Aristotle's replies to them. The direct and *ad hominem* relation between the arguments for the void in *Physics* IV, 6, and Aristotle's replies to them in *Physics* IV, 7–9, will thus be clearly established.

The Name "Void"

Aristotle opens *Physics* IV, 7, saying that it is necessary to determine what the word "void" signifies: "the void seems to be place in which there is nothing" (213b30–31). There is a reason why proponents of the void think of it in this way. They identify "what is" [τὸ ὄν] with body and think that every body is in place; void is place in which there is no body "so that if somewhere [ποῦ] there is no body, then there is nothing there" (213b31–34).

Aristotle's first set of opponents are materialists who identify "what is" with body and define place as empty or full *of body*, i.e., void or plenum.[14] This conception of "what is" (and consequently place) is by definition at odds with Aristotle's definitions of place and motion. Place as a limit renders the cosmos determinate as actually up and down, while motion is the actualization of the potential *qua* potential by what is actual. Aristotle requires these principles in part because he conceives of the world as determinate, and he identifies things that are by nature primarily with form, explicitly rejecting the view that identifies them primarily with matter (*Physics* II, 1, 193b7). But materialism excludes both limit and actuality as concepts because they are immaterial.

This systematic incompatibility lies at the heart of Aristotle's "refutations" of materialist arguments for void. Because materialists identify "what is" with body, "void" means place in which there is no body; but body is thought to be tangible, which means to have weight or lightness;

---

14 Cf. again the beginning of the account, *Physics* IV, 6, 213a17–19.

therefore, a void is that which has nothing heavy or light in it (*Physics* IV, 7, 213b33–214a3). Emphasizing that this conclusion follows from a syllogism, Aristotle derives several contradictions from it.

The key here lies in the identification of body with what is heavy or light.[15] For Aristotle, heavy (earth) and light (fire) are immediately associated with the cosmos as differentiated by place into actually up and actually down because up and down are "where" the light and the heavy are carried naturally and immediately if nothing intervenes. That is, they are the respective proper places that naturally contain light and heavy. Consequently, the introduction of "heavy" and "light" into the definition of void shows that for Aristotle the void presupposes a determinate world in the sense that bodies are determinate as heavy and light; in their turn, heavy and light are immediately linked to an inherently directional world. Because the void is by definition indeterminate but (according to Aristotle) body as heavy or light presupposes a determinate world, there will be little difficulty generating contradictions.

By definition, void is place (which means "contains light and heavy") that is empty, i.e., does not contain light and heavy. From this flawed definition, several contradictions follow. (1) This definition is absurd if a point is a void. A void must be place, i.e., empty place, in which there is an interval for sensible body.[16] But a point (as Aristotle defines it) is the terminus of a line; a line is that which has one dimension only, length, and as its terminus, a point must be without dimension.[17] Sensible body is obviously three-dimensional, and if void were place in which there were an interval for sensible body, it too would presumably be three-dimensional. Therefore, if a point is a void, it both is and is not three-dimensional.[18]

(2) Some thinkers (again presumably materialists) define void as "what is not full of body perceptible according to touch" (*Physics* IV, 7, 214a6–7). But if an interval contains color or sound, is it void, i.e., empty, or not? Either answer produces a contradiction. If it is not empty, then the interval meets the definition of a void because it does not contain body sensible to touch, yet it is "occupied" by color or sound; if such an interval were empty, then an empty interval can have "something" (color or sound) in it.

Aristotle suggests that proponents of this view might reply that if the interval with color or sound could receive tangible body, then it would be a

---

15 Hussey, 125–126. Hussey rightly argues that Aristotle intends to make the definition of the void more precise.

16 Aristotle, *Physics* IV, 7, 214a5–6: δεῖ γὰρ τόπον εἶναι ἐν ᾧ σώματος ἔστι διάστημα ἁπτοῦ. The word "διάστημα" appears here for the first time. It is translated as "extension" by Hussey, but as "interval" by Hardie and Gaye.

17 Aristotle, *Metaphysics* XI, 2, 1060b15, 18; *Physics* VI, 1, 231b9; cf. Konstan, "Points, Lines, and Infinity," 1–32, and H. Lang's, "Points, Lines, and Infinity: Response to Konstan," 33–43.

18 Ross gives several other interpretations, none of which is very helpful, 584–585. Once Aristotle introduces "point," it is best to assume his definition. And with this definition the argument is less obscure.

void (214a10). This answer avoids the dilemma of answering "yes" or "no" to the question of whether or not such an interval is empty; nevertheless, it fails to answer the question of whether intervals having sensible quality but not sensible body are voids. In any case, as the next objection shows, calling the void "ability to receive tangible body" is also problematic.

(3) Some say the void is that in which there is no "sensible substance" and so call the void "the matter of body" (214a12–13). They contradict themselves by speaking of the void as both separable and matter. Here Aristotle assumes his own doctrine of matter, i.e., potency that is not separable from form.[19] The void is "the matter of body" in the sense of "a capacity to receive body"; however, being defined in relation to body, it is not altogether separate from it. But it must be separable from that body because there is no body in it. Hence a void is both separate and not separate from body, which is clearly contradictory (214a13–16).

Aristotle concludes his analysis of the name "void," by rejecting the definition of void as place with nothing in it:

And since [things] concerning place have been distinguished and the void must be place, if it is, deprived of body, and place, both how it is and how it is not, has been said, it is clear that thusly a void is not, neither separated nor unseparated (214a16–19).

Defining void as "empty place" always produces a contradiction. In fact, people think there is a void only because place is and the void is "a special case of place"; thus, the reasons that support place as a term required by motion seem [mistakenly] to support void (214a21–22). Consequently, the fact of locomotion appears to support both place and void (214a22–23). But as a concept "void" always and necessarily produces a contradiction and the only reason for asserting it is a confusion between the concepts "void" and "place." Aristotle concludes unequivocally: "there is no necessity, if motion is, for a void to be" (214a26). "Void" is not a term without which motion seems to be impossible.

Only the responses to particular arguments proposed for the void remain. The claims of Aristotle's opponents provide the starting point of each of his arguments. Consequently, they are not proofs properly speaking, but are refutations of the [false] assumption of the void. Thus, they provide no information about Aristotle's own views.

Refutations of Opposing Arguments

A general claim precedes Aristotle's refutations of his opponents: proponents of the void "think that [it is] a cause of motion as that in which [a thing] is moved" (*Physics* IV, 7, 214a24–25). This claim anticipates his

---

19 Aristotle, *Metaphysics* VII, 3, 1029a11–33. Ross says that this statement is unhistorical because no thinker prior to Aristotle held this view of matter, 585.

refutations of the void: each concludes that the void fails to serve as a cause of motion. Consequently, Aristotle neither argues that a void does not exist nor provides a constructive account of the phenomena cited as evidence by proponents of the void. His arguments (and conclusions) are strictly limited: the void fails to serve as a cause of motion. His refutations of four of the five arguments raised in *Physics* IV, 6, now follow. They are brief and will be considered in order, first the argument and then its refutation. In conclusion, I shall consider the question of how each pair relates to Aristotle's own constructive position.

(1) Its proponents claim that a void is necessary for motion.[20] Without a void there can be no change of place (or increase) because (1) what is full cannot contain anything more (213b5–7), and (2) if something more were added to a plenum, then two bodies would be in the same place at the same time, and a variety of problems would follow (213b7–12). But, Aristotle replies, in at least one case (which seems to have escaped Melissus) motion clearly does not require a void: a plenum can undergo qualitative motion because it does not involve change of place and so need not involve a void.[21] Even locomotion does not require a void because moving bodies simultaneously make room for one another, as is clear in the rotation of continuous things such as liquids.[22] Hence the claim that a void is *necessary* for motion is obviously false.

(2) From the fact that some things seem to contract, proponents of the void conclude that the compressed body contracts into void present in it (213b15–19). Thus they explain contraction as a change of place requiring a void. But they are wrong: things can be compressed not because there are "void places" in the thing, but because what is contained, e.g., air, is being squeezed out.[23] Again, the argument fails to show that a void is *required* for motion.

(3) Increase too, proponents of the void claim, occurs because void is that into which a body expands. But increase can also be explained by qualitative change, as when water is transformed into air. Presumably, air and water occupy different places and different extents; thus alteration (or qualitative change) could make place for growth. Again, Aristotle does not give his own account, but provides a reasonable alternate account to prove that proponents of the void fail to show that it is *necessary* for motion.

Finally, (4) a vessel full of ashes will hold its own volume again of wa-

20 Hussey, 124, attributes this argument (and others for the void) to the ancient Atomists. Ross, 582, refers it to *De Gen. et Corr.* I, 8, 325a23–32, b4, which identifies Leucippus as holding this view.

21 Aristotle, *Physics* IV, 7, 214a26–27; on Melissus, cf. Diels and Kranz, *Die Fragmente der Vorsokratiker,* B7 (7); Sedley, "Two Conceptions of Vacuum," 178–179.

22 Aristotle, *Physics* IV, 7, 214a28–32. For Plato's argument that the revolution of the universe will not allow any place to be void, cf. *Timaeus* 58A5-B. Hussey, 127, introduces a problematic example not found in the argument here and best left where it does occur.

23 Aristotle, *Physics* IV, 7, 214a33–35; for a partially parallel argument, cf. Plato, *Timaeus* 60C-5.

ter without overflowing; there must be empty spaces in the ashes that absorb the water (213b19–22). Aristotle gives essentially the same reply as before: things are not compressed into a void; rather, what was formerly present, e.g., air, is squeezed out and so the vessel now holds "more."[24] This argument too fails to demonstrate the *necessity* of void.

These arguments for void are, Aristotle says, "easy to refute" (*Physics* IV, 7, 214b10–11). In each, void supposedly required by motion is "the where" into which things move. Aristotle rejects these arguments either with a counterexample, e.g., the rotation of a liquid is an example of movement in a continuous body resembling a plenum, or with an alternate explanation, e.g., air is replaced by water. Because the burden of proof lies with proponents of the void, the counterexample or alternate explanation is sufficient to show that these arguments do not demonstrate the *necessity* of a void for motion to occur.

The logical limit of these arguments is clear. In each case, Aristotle rejects the argument for the void because the conclusion does not follow necessarily. And the purpose of his responses is strictly limited: rejection of the claim that void is necessary for motion. Hence his responses do not imply that he is committed to a plenum or to a view of locomotion as mutual replacement within a plenum. We know only that the arguments do not *necessarily* establish void as a term required by motion.

A Separate Void

A more serious argument for the void is initially presented in *Physics* IV, 6, but is not addressed by Aristotle until *Physics* IV, 8. It differs from the first three arguments in its structure, in its substance, and, finally, in the length and complexity of Aristotle's reply to it.

The Pythagoreans, too, said void is and that it enters the heaven from the infinite air, as if it [the heaven] also inhaled the void which distinguishes the natures [of things], as if the void were something that separates and distinguishes things that are in succession. And this it does primarily for the numbers, for the void distinguishes their nature. (*Physics* IV, 6, 213b23–27)

Although their position remains unclear, the Pythagoreans presumably were not materialists in the sense of identifying "what is" with body.[25] Their concept of number, according to Aristotle, is in some sense formal.[26] Furthermore, unlike the materialists, for whom void is an interval

---

24 Aristotle, *Physics* IV, 7, 214b3–9; for a clearer account of this problem, cf. the spurious *Problemata* 8, 938b25–939a9.

25 On the views of Pythagoras and the Pythagoreans, cf. Heidel, "The Pythagoreans and Greek Mathematics," 2, who suggests some historical evidence for rejecting Aristotle's account of Pythagoras here.

26 Cf. Aristotle, *Metaphysics* I, 5, 985b23 ff; I, 6, 987a22 ff; Kirk, Raven, and Schofield suggest that Aristotle was interested in emancipating Pythagoras from efforts to Platonize his philosophy and refer to "his full and comparatively objective accounts of fifth-

in body, the Pythagoreans seem to suggest that there is a void that enters the heavens from without and distinguishes the nature of both things and numbers. Although here too void may be understood as empty place, it functions as a principle of the unlimited that separates out the limited.[27] In short, it is formal, differentiating numbers and perhaps the heaven too. Because it is a formal principle, Pythagorean void constitutes a more serious challenge to Aristotle's concept of place than does the void of the materialists.

The Pythagorean void (which is also rejected by Plato) requires a separate set of considerations, and these occupy *Physics* IV, 8.[28] Aristotle's intention could not be clearer. He begins: "that there is not a void thus being separate, as some say, let us say again" and concludes "that, then, there is not a separate void, from these things is clear" (214b13; 216b20).

Two sets of short arguments – I shall designate them 1a through 1d and 2a through 2e – precede a longer and more complex argument that occupies most of *Physics* IV, 8. In the first set, 1a–1d, each argument concludes that the void cannot be a cause of motion and so, presumably, cannot be a "special term" required by motion in things. In the second set, 2a-2e, the conclusion is even stronger: if there were a void, it would not be possible for anything to move (214b30–31). Hence, these arguments must presuppose, and so are subordinated to, Aristotle's definitions of motion and things in motion; furthermore, they unambiguously establish the main force of his arguments against the void: the void fails to cause motion. I turn to these two sets of arguments.

(1a) Aristotle asserts his own view of natural elemental motion, which he claims the void clearly cannot cause:

> If there is for each of the simple bodies some natural locomotion, e.g., for fire upward and for earth downward and towards the middle, it is clear that the void cannot be a cause of the locomotion. Of what, therefore, will the void be a cause? It is thought to be a cause of motion in respect of place, but it is not [a cause] of this. (214b13–17)

Posed as a hypothetical, natural elemental motion clearly functions as the natural motion in things, and it requires a cause. Place, as we have seen, causes elemental motion by rendering the cosmos determinate.[29]

---

century Pythagorean metaphysics and cosmology," *The Presocratic Philosophers*, 215–216.

27 On Pythagorean void as the "unlimited" that differentiates the "limited," cf. Guthrie, *History of Greek Philosophy*, 277–280, 340.

28 Plato denies the Pythagorean void throughout the *Timaeus;* cf. 30C-31B2; 32C3–34B5; 59A; 60C; 79B.

29 Ross, 587–588, complains that this argument is worthless because "cause" is used ambiguously. Hussey, 128, claims that Aristotle must be thinking of the four causes and asserts that place is a final cause. But as we have noted earlier, Aristotle's notion of "cause" can operate quite broadly and is not thereby either ambiguous or restricted to the four causes; the most likely meaning of cause here is actuality relative to potency, as required

Void, i.e., empty place, is by definition indeterminate and so clearly cannot cause motion, especially the motion of the elements (as Aristotle defines it).

(1b) If there were a void, where would a body placed into it move, because it cannot be moved into the whole of the void (214b17–19)? The elements (of which all bodies are composed) are by definition light and heavy; therefore, each is moved naturally toward or rests in its proper place, up, down, or middle respectively. But such natural motion would be impossible in a void. The void is undifferentiated and so without "up," "down," or "middle"; but each element is naturally moved to or rests in its natural place "up" or "down"; consequently, the void fails to explain natural motion or rest in things and cannot serve as place or cause motion.[30]

(1c) In what sense would a body be "in" the void (214b24–25)? Motion does not result when some body is in a separate and persisting place because the part will not be in place but in the whole (214b25–27). This argument presupposes the earlier account of how a thing is in place: place must both surround and be divided from and touching that which is in place. If a void were separate and persisting, it could not be touching the body in place. Therefore, a body would be in a void as a part in a whole, not as one thing is "in" another. Because motion requires that mover and moved touch, the void fails to meet this condition and cannot serve as a cause of motion.

(1d) Finally, if place is not separate, neither will the void be separate (214b28). And place, as we have seen, is not absolutely separate; as a limit, it is accidentally in place because it is "contained" in the first containing body. Because void is a subset of place, it too cannot be absolutely separate.[31]

These arguments are striking for their brevity and for the claim that each is obvious – a claim that results from assuming conclusions established earlier concerning place and motion. The same assumptions operate in the second set of arguments against the void. Here Aristotle argues that far from being a term without which motion seems to be impossible, the assumption of a void makes it impossible for anything to be moved (214b28–31). Again, the arguments are very brief.

(2a) In a void, a body would necessarily be at rest because "a void . . . admits no difference," and so a body would not be moved here rather

---

by Aristotle's definition of motion. Place, too, as a cause may best be thought of as actual relative to the potency of the elements; Aristotle nowhere calls it a final cause. This problem will be discussed at length in the second half of this study.

30 Hussey, 128, refers to Aristotle as running a "thought-experiment: what will be the motion of a 'test particle' placed in a void?" Aristotle is opposed to the Atomists and (as we shall see at length later) rejects the notion of particles in favor of his own account of the elements. Therefore, he is not likely to be thinking of a particle here.

31 Hussey, 129, connects this argument to that concerning extension, but there is neither indication nor need to do so.

than there.[32] On Aristotle's view, the cosmos is determinate as actually up, down, and middle, while each element is determined toward its proper place.[33] Hence, in the absence of hindrance, fire by nature is always moved up rather than down, while earth is moved down rather than up. But void is by definition indeterminate; thus, body that is determined in one direction rather than another will not be moved naturally in it.

(2b) All motion is either compulsory or natural, and natural motion is prior because compulsory motion presupposes natural motion; therefore, each element must have a natural motion without which it cannot have compulsory motion either (215a-6). Natural motion requires that the cosmos be determinate as actually up, down, and middle – a determination that an indeterminate void fails to provide (215a6-b). Without determination, nothing is ever moved with a natural motion; thus either there is no natural motion or there is no void.[34] The conclusion goes without saying: there is motion, and so there cannot be void.

(2c) Things that are thrown move even after they no longer touch the thrower. Either "mutual replacement, as some maintain" or the air through which they move must be moving them.[35] Whichever explanation one chooses, a void would fail to provide a mover for things in motion and so cannot cause it. Here Aristotle presupposes a conclusion established only in *Physics* VII and VIII: "everything moved must be moved by something."[36] So the problem of explaining motion requires that one identify a mover in contact with the moved.[37] He identifies the two possible movers during projectile motion and concludes that in a void neither would be available. Hence, a void fails to account for projectile motion. I shall discuss Aristotle's definition and account of projectile motion in Part II. But these brief refutations tell us nothing of Aristotle's own view.

(2d) Because the void is undifferentiated, it cannot explain why things rest "here" rather than "there"; hence a moved thing would be moved forever unless it were blocked by force from the outside (215a19–22). Na-

---

32 Aristotle, *Physics* IV, 8, 214b33–215a. On different grounds Plato too concludes that what is uniform cannot cause motion. Cf. *Timaeus* 57C-58. Ross, 588, refers this passage to *Phaedo* 109A and *Timaeus* 62D.

33 Hussey, 129, again treats this argument as concerning a "test particle."

34 Aristotle, *Physics* IV, 8, 215a-14; cf. Plato, *Timaeus* 63A5–64.

35 Aristotle, *Physics* IV, 8, 215a13–19. Plato seems to argue for mutual replacement, and so the reference to "some" may be to Plato and the Platonists. Cf. Plato, *Timaeus* 59A-5; 60B5-C5; 79B-7; 80C5-9. Aristotle's own commitment to these views remains ambiguous.

36 Cf. Aristotle, *Physics* VII, 1, 241b34; *Physics* VIII, 4, 256a3. Although his physics is very different, Plato too utilizes this principle. Cf. Plato, *Timaeus* 57C-58. The full import of this principle for Plato emerges in *Phaedrus* 242B and *Laws* X 894B5–895C. Ross, 598, claims that Aristotle is thinking of a medium; a "medium" may indeed make a motion fast, but it is not the only such cause. Here there is no textual evidence for this view. We shall see an argument about the so-called medium in a moment.

37 I have considered this problem at length in *Aristotle's Physics and Its Medieval Varieties*, 63–84. Hussey, 130, claims that Aristotle requires a physical explanation in terms of bodies. This is false: he requires actuality in contact with potency.

ture is a source of being moved or at rest, and, just as the earlier argument assumes Aristotle's definition of motion, this one assumes his definition of rest. Because the void is undifferentiated, just as it fails to cause elemental motion, it also fails to produce rest when an element is in its natural place.[38]

(2e) Proponents of the void claim that a body can move into the void because void yields. But an undifferentiated void would yield equally everywhere and hence things should move in all directions.[39] But things do not move in all directions; rather, some things (fire) are always moved up, while others (earth) always down.[40] Hence, as undifferentiated (and so yielding equally everywhere), void would fail to cause natural elemental motion.

Throughout both sets of arguments, the "failure" of the void to serve as a cause of motion – indeed, it renders motion impossible – does not rest on the issue of void and place as full or empty of body. It rests solely on the contrast between place as differentiating and void as failing to differentiate the cosmos into up, down, and middle. Although the void is empty in the sense of containing no body, it is also, more importantly, empty in the sense of being indeterminate. Hence, it fails to differentiate the cosmos and in this sense to render the cosmos orderly. Therefore, it fails to cause motion in things. Aristotle's arguments that the void fails to cause motion depend ultimately on his own views of motion and things moved as requiring actuality and a cosmos actually differentiated into "up" and "down." Place as a limit does differentiate the cosmos and so explains the order of nature in regard to "where things are and are moved."

Motion, Void, and τὸ δι' οὗ

Aristotle's final argument against the claim that the void is required for motion is considerably longer and more complicated than its predecessors. Although I shall consider it only within the context of *Physics* IV, 8, this argument has held enormous interest not only for modern readers but also in both the Arabic and the Latin Aristotelian traditions. As with the preceding arguments, its structure is crucial to its meaning.

The argument opens with an assertion: "Again, what is being said [is] clear from these things," i.e., what now follows (215a24). It is subdivided into several parts that are intended to establish the conclusion that the void cannot serve as a special term required by things in motion. Hence, the argument should close with a rejection of the void as required for motion, and indeed Aristotle reaches just this conclusion: the void is asserted

---

38 Hussey, 130, comments, "Once again Aristotle's finitism blights the development of physics."
39 Aristotle, *Physics* IV, 8, 215a22–23; cf. Ross, 589.
40 For Plato's explanation of light and heavy in conjunction with motion up and down – an argument that also rejects the void – cf. *Timaeus* 62C2–63E5.

because its proponents think it will serve as a cause of motion – but in fact
the opposite result follows – the assumption of a void renders motion impossible.[41] Together the opening assertion and this conclusion form the
outer limits of the overall argument.[42] Before proceeding to the internal
divisions of the argument, we must consider the implications of this rejection.

The conclusion that the void fails to serve as a cause of motion is entirely negative and reflects the purpose of the arguments specified in
Physics III, 1: to find the special terms required by motion. The assumption is that a void is required by motion. Therefore, if this assumption
leads to an absurdity, the void and the claim that it serves as a cause of
motion must be rejected. Only one conclusion follows: a void cannot serve
as a cause of motion.

Within this framework, Aristotle asserts that a body may be moved
faster for two reasons, each of which founds a distinct argument that
reaches a contradiction.[43] First, a body may be moved faster or slower depending on that through which it is moved, e.g., water or air; but on the
assumption of a void, a given body will go through both a plenum and
a void in equal time (216a2–3). Second, bodies may be moved faster or
slower because of a greater inclination of either heaviness or lightness;
but in a void "all things will be equally fast" regardless of their heaviness
or lightness (216a20). In each case, assuming a void produces a contradiction, it must be rejected as a cause of motion.

The dividing line between these two arguments is not problematic.
Consideration of the first reason why a thing may move faster, "that
through which," leads unequivocally to a contradiction, and the second
reason, differences among things that are moved (i.e., greater internal
inclination [ῥοπή], either of heaviness or of lightness) produces the second. The main conclusion, i.e., the rejection of the void, then follows.

The first argument, which concerns "that through which" a body is
moved, is the longest and most complex argument against the void as a
proposed cause of motion. Its internal divisions are crucial to its construal
and have been a source of several serious difficulties. Hence they require
consideration.

41 Aristotle, *Physics* IV, 8, 216a21–26. We may note that there is either a typographical er-
    ror or mistranslation – an added negative not found in the Greek – in Hardie and Gaye's
    translation that renders the argument nonsensical. I italicize the mistake: "It is evident
    from what has been said, then, that if there is a void, a result follows which is the very
    opposite of the reason for which those who believe in a void set it up. They think that if
    movement in respect of place is to exist, the void *cannot* exist, separated all by itself."
42 The next line clearly begins a new argument: "but even if we consider it in virtue of it-
    self the so-called vacuum will appear truly vacuous," *Physics* IV, 8, 216a27; Thorp notes
    that this is a rare joke and comments that it may be taken to indicate that Aristotle's ar-
    guments here are not very serious, "Aristotle's *Horror Vacui*," 149–166. I shall argue that
    they could hardly be more serious.
43 Aristotle, *Physics* IV, 8, 215a25–27.

The argument breaks into three parts, and its construal rests on the relation among these parts. In the first part, Aristotle takes body A as moving through B in time C and through D in time E and concludes that as "that through which" something is moved is more incorporeal and more easily divided, the motion will be faster (215a31–215b12). There is no reference to a void, and no contradiction results. The point here is to establish a set of ratios.

In the second part of the argument, Aristotle first asserts that the void has no ratio by which it is exceeded by body, just as there is no ratio of "nothing" [τὸ μηδέν] to a number (215b12–24). The void cannot have a ratio to the full, nor can motion through a void have a ratio to motion through a plenum. Therefore, a body is moved through what is most fine a certain distance in a certain time – but the motion through the whole void exceeds ratio.

The status of this conclusion is problematic: is it a complete argument (either in itself or in conjunction with the first section), or is it another step in a larger argument? It has often been taken as producing an impossible result, i.e., a void implies instantaneous motion, and so constituting a *de facto* rejection of the void.[44] For example, Averroes, who has been enormously influential on this point, divides the text after the second part of the argument and treats the whole as complete at this point.[45]

There are two serious problems here. (1) Aristotle concludes *only* that motion through a void would exceed any ratio. This conclusion might look sufficient for a rejection of the void as a cause of motion, but he does not in fact reject it. (2) The next (third) part of the argument is now problematic. If part two produces a contradiction, then the next section should start a new argument; instead, it reverts back to the first part and develops the ratios posited there in order to generate a contradiction. But connecting the first and third parts leaves the second, "mathematical," part in an anomalous position dividing two parts of a continuous argument.

The third part of the argument provides the key. Aristotle returns to the ratios established in the first part and allows a void to function within them even though he has just shown that it produces an impossible result (215b23–216a7). Only here does he draw the requisite contradiction: body A will traverse magnitude F in an equal time whether F be full or void (216a2). A strong summary sentence repeats the assertion central to the second part of the argument: there is no ratio of void to full (216a9–12).

This strong conclusion clarifies the function of the second part of the argument: when the void functions in a ratio (as its proponents presumably claim, since they posit it as a requirement for motion), it is the only

---

44 This view goes back at least as far as Philoponus. Cf. *In Phys.* IV, 683.8.
45 Averroes's commentary on this argument became known as "Comment 71." It can be found in Grant, *A Source Book in Medieval Science*, 253–262.

possible source of the contradiction that follows in the third step of the argument, i.e., a given body will traverse the same distance in an equal time whether the distance is full or empty (216a3–4). Taken together, the return in the third part to the ratios established in the first, the hypothetical role of a void in the ratios, and the strong rejection of the void with which the argument closes confirm that the formal structure of the argument requires all three parts.

This construal yields a single unified argument beginning with the assertion that "that through which" a body is moved causes it to be moved faster and ending with the assertion "insofar as [those things] differ through which [bodies] are moved, these things result" (215a29). Furthermore, it identifies the assumption of a void as the sole source of the contradiction. The first step establishes a set of ratios. The second shows that theoretically the relation of void to full is like that of nothing to number or point to line; hence a void cannot enter into a ratio, and a body traveling through a void would do so with a motion exceeding any ratio. The third step returns to the ratios established in the first, assuming that a void functions within them even though we know (from the second step) that this assumption is false. A contradiction follows, which must be produced by the assumption of a void.

The Argument, Part I: The Ratio Required by Motion

Aristotle distinguishes two reasons why a body may be moved faster, and he immediately takes up the first:

> For we see the same heaviness, or body, being moved faster due to two causes, *either* because that through which [it is moved] differs, e.g., either through water or earth or through water or air, *or* because that being moved differs, if the other things are the same, on account of the excess of heaviness or of lightness. Thus, that through which [the moved] is carried is a cause when it impedes, most of all being moved against [the moved] and when it rests; again, that which is not easily divided [impedes] more. And the more dense is such. (215a25–31)

Hussey comments that this argument is "hardly intelligible except in the context of an Aristotelian theory of dynamics."[46] Because he claims to be "giving sense to this section," Hussey outlines "Aristotle's argument" prior to any analysis.[47] The argument, he says, rests on the claim that "the speed of a given body through a medium is a function of the nature of the medium."[48] But in Hussey's outline, Aristotle's word "faster" becomes "speed" and "that through which" the body is moved becomes a "medium." Hussey even translates Aristotle's phrase "δι' ὧν" as "media" (216a12).

To some extent Hussey's approach to this argument represents natu-

46 Hussey, 131.   47 Hussey, 131.   48 Hussey, 130.

ral modern assumptions. But these assumptions, as well as the language in which they are expressed, do not represent Aristotle's conception of the project of physics – neither the problems that it addresses nor the solutions to those problems. For example, "dynamics," a post-Galilean conception of physics, treats motion as separable from [moved] things. But Aristotle defines physics as the science of things that are by nature, and he takes up motion explicitly because things that are by nature are moved and at rest *per se* (*Physics* III, 1, 200b12–14). For this reason – and I shall argue this point later – Aristotle's physics is not a stage on the way to modern physics; rather, Aristotle presents a different starting point, nature as intrinsically orderly, and defines the problems of physics differently than do Galileo and Newton. It seems to me that the viability of different starting points and the importance of different problems can be compared – but only after one understands them fully as spelled out in a coherent position.

The language of "speed" and "medium" are, I shall argue, seriously misleading. Indeed, the upcoming argument rests exclusively on differences in the "that through which" a thing is moved as a cause of a body being moved faster. (Differences in a body's inclination to heaviness and lightness form the second argument, after which the main conclusion follows.[49])

The phrase "τὸ δι' οὗ" means literally "that through which" (*id per quod*), and, with one exception, this is its only occurrence in Aristotle.[50] "That through which" a body is moved impedes its motion in several ways, i.e., by moving against the moved or being at rest or being not easily divided, as when it is dense. Both Hussey and Hardie and Gaye translate it as "the medium through which."[51] However, although "medium" in its modern sense is less cumbersome than "that through which," it is not an innocent term. It implies something external and neutral to the moved and so calls to mind passive resistance.[52]

The English word "medium" derives from the Latin *medium,* and a number of Latin commentators, e.g., Albert of Saxony (*Physics* bk. 4, quest. 10,

49 All but one of the preceding arguments is introduced by the word ἔτι; in the overall structure of this argument, the opening distinction founds two separate arguments, which explains why after the ἔτι there are two arguments before the main conclusion rather than one as with the preceding arguments.

50 The exception is in the *Post. Anal.*, where the phrase, referring to a premise in an argument, is best translated as "that through which" as, e.g., in Barnes's translation; cf. Aristotle, *Post. Anal.* I, 25, 86b4.

51 The notion of a medium so permeates the literature on this argument that it is impossible to list all the secondary sources here. Among the more serious and interesting recent work done on the assumption that Aristotle means a medium are Bickness, "Atomic *Isotacheia* in Epicurus," 57–61; and Sedley, "Philoponus' Conception of Space," 142; however, Trifogli avoids the language of "medium" by using the phrase "in something undifferentiated," "Giles of Rome on Natural Motion in the Void," 137.

52 Hussey does not use the word "passive," but seems to take it in this sense; cf. 131–132, 191, especially when he uses the phrase "'inertial' resistance." Ross too seems to think of passive resistance, 590.

fol. 49v col. 2) and Walter Burley (*Physics* fol. 116v), use it in discussing the *id per quod*. But they (and others) discuss the *id per quod* as the middle between the starting and ending points of motion because the Latin *medium* translates the Greek "μέσον," i.e., middle or mean, and is identical in meaning with it. Their arguments explicitly address the problems of resistance and the relation of moved to mover precisely because these problems require solution and are not yet settled so completely as to become part of the word "medium" itself. Hence in their commentaries, "medium" retains its strict Latin meaning – "middle," not "medium."

This sense of "medium" appears in English expressions such as "a happy medium" or a "spiritual medium" (who occupies a middle position between this world and the next). Indeed, "medium" is not used in the sense of "surrounding body," before 1664.[53] In this new sense, it no longer translates μέσον, because it has developed quite different connotations: there is no Greek (or Latin) word that represents the post-1664 English meaning of "medium" precisely because its connotations derive exclusively from early modern physics. Consequently, the English word "medium" evokes the conceptual setting of early modern physics, including, for example, the neutrality of moved to mover, the indeterminacy of matter, and the passive resistance of a surrounding body.[54]

But on Aristotle's view, nature is none of these: it is a principle of order and determinacy, exhibited by the relation of moved to mover. And the fabric of that order is at work in this argument. Place renders the entire cosmos determinate and differentiated in respect to "where," and each element – Aristotle's examples here in *Physics* IV, 8 – is always actively oriented toward or resting in its natural place. The elements are defined as light (fire) and heavy (earth). That is, each element is differentiated by its nature and oriented toward its proper place, i.e., that place toward which it is always carried by nature, if nothing hinders. Because it is moved toward (or rests in) its natural place, each element is actively oriented toward (or rests in) that place. Thus, a body moved through air or water is moved not through an undifferentiated medium, but through an active element exercising its orientation.

Herein lies the force of Aristotle's examples: they are examples of activity.[55] A body may be moved through an element that, being moved to-

---

53 The *O.E.D.* lists "medium" in the sense of that which surrounds, especially associated with air or water, as a "new" meaning not found before 1664; the occurrences listed are exclusively works of physics; cf. entry 4b in the 2d ed.

54 This is not to say that early modern physics had no antecedents in late medieval science – it most certainly did. Cf. Weisheipl, "The Principle *Omne Quod Movetur Ab Alio Movetur* in Medieval Physics," 89–90.

55 It is worth repeating the opening point, which accords perfectly with elemental motion: "Thus that through which [the moved] is carried is a cause when it impedes, most of all being moved against [the moved] and when it rests; again, that which is not easily divided [impedes] more. And the more dense is such." Aristotle, *Physics* IV, 8, 215a28–31.

ward its natural place, is moved in the opposite direction from it and thus impedes its motion; or if an element is resting in its natural place, it thereby resists being pushed out of that place and so impedes a body moved through it. Again, an element that is dense resists being divided. In each case, as "that through which" a body is moved, each element *causes* the motion of the body to be faster or slower; it causes motion not by passive resistance, as would a "medium," but actively, either by being moved toward or resting in its respective natural place. And each element is active because it is determined in respect to up, down, etc.

Aristotle now explains why a body may be moved faster:

A, then, will be moved through B in time C, and through D, which is thinner, in time E (if the length of B is equal to D), in proportion to the hindering body. For let B be water and D air; then by so much as air is thinner and more incorporeal than water, A will be moved through D faster than through B. Let the quickness [τὸ τάχος] have the same ratio to the quickness [τὸ τάχος] then, that air has to water. Therefore if air is twice as thin, the body will traverse B in twice the time that it does D, and the time C will be twice the time E. And always, by so much as that through which [the body is moved] is more incorporeal and less resistant and more easily divided, [the moved] will be carried faster. (215a31–215b12)

Prima facie, the point seems easy. As "that through which" hinders less, the motion of a given body is faster; and Aristotle expresses this relation as a ratio: time C is to time E as the "thinness" of B is to that of D.

But the case is not so easy. One criticism in particular has consistently been raised against it by commentators – the best known of whom is undoubtedly Avempace: Aristotle is wrong in constructing the relation between a body and its resisting medium as a proportion;[56] rather, the resistance provided by an external medium is not a cause of motion so much as a necessary condition for motion; thus, the true relation between the body and the medium is arithmetic: the resistance of the medium should be subtracted from the speed of the moving body, not entered into a ratio.[57] In short, Aristotle has entirely misunderstood the relation between resistance and the speed of a moving body.

This charge is serious. If warranted, the argument is in trouble. But before deciding its truth or falsity, we must consider the concepts "speed," "resistance," and "necessary condition" – to be sure that they are Aristotle's. I have suggested throughout that Aristotle's arguments concerning place and void presuppose his definitions of motion and nature. I shall

---

56 Avempace's original text has been lost, but his criticism of this argument is reproduced by Averroes in his commentary on *Physics* IV; cf. Grant, *A Source Book in Medieval Science*, 253–262. Avempace may well have borrowed the criticism from Philoponus, cf. his *In Phys.* 678.24–684.10. For two modern examples, cf. Ross, 590, and Hussey, 188 ff., esp. 196.

57 Hussey suggests that "Aristotle's laws of proportionality therefore successfully account for the fact that all bodies fall at the same speed, if we abstract from the resistance of the medium," "Aristotle's Mathematical Physics," 237.

now argue that taking them as subordinated makes sense of them, whereas analyzing them apart from these definitions renders them confused.

As a concept, "speed" is associated (in a sense is even identical) with motion thought of quantitatively. Indeed, Hussey rightly calls it a "mathematical quantity."[58] That is, speed is the rate at which a body traverses a certain distance in a given time. Defined as a rate, "speed" bears the conceptual marks of quantitative physics: it is a quantity independent of any particular moving body and is measured by an independent quantified scale.

But Aristotle's definition of motion as an actualization of the potential rests on the relation between mover as actual and moved as potential: motion is neither separate from body nor a quantity; rather, it is a development, or fulfillment, of a capacity in the moved by its mover. Hence, motion can never be conceived independently of mover and moved, i.e., things that are by nature.

Herein lies the problem of motion within Aristotle's determinate orderly world and the physics that studies the world: the motion of the elements – and because all body is composed of the elements, the motion of any body – is explained strictly in terms of the development of a capacity *intrinsic* to the moved by its actuality, i.e., its respective mover. Hence, the identification of a mover (actuality) in contact with the moved accounts for why motion necessarily occurs.[59] In short, within Aristotle's physics the definition of motion turns exclusively on concepts that are incompatible both with quantification and with a concept of motion abstracted from moved things: Aristotle's definition is incompatible with the requirements of speed as a concept.

As a quantitative concept brought to bear on Aristotle's physics, "speed" is seriously misleading. It implicitly redefines motion – making it a quantity separable from moved things – with the result that the identification of a mover is no longer a meaningful explanation. Solutions to the problems involved in motion rest on quantitative measurement and arithmetic manipulation.[60] Hence translating "faster" into "greater speed" redefines the problems of physics – it replaces Aristotle's determinate world with an indeterminate world. For this reason, in Aristotle "faster" does not mean "greater speed."

---

58 For Aristotle as read with this conception of speed, cf. Hussey, 188 ff. Hussey develops this theme further in "Aristotle's Mathematical Physics," 227–230. For a broader view of how modern historians generally have found a notion of speed in this argument, cf. Weisheipl, "The Principle *Omne Quod Movetur Ab Alio Movetur* in Medieval Physics," 83.
59 Cf. H. Lang, *Aristotle's Physics and Its Medieval Varieties*, 63–84.
60 Hussey points out that equipment in Aristotle's time would have been inadequate to measure terminal velocities (131). "Velocity," of course, is different from both "speed" and "faster"; terminal velocity is a special case of velocity, and there is no evidence that Aristotle possessed this concept, let alone tried to measure it.

I have already suggested that "that through which" a thing is moved is not a medium with its overtones of being undifferentiated and passive. Water or air through which a body is moved, *cause* it to be moved faster. Aristotle repeatedly concludes that the void (unlike place) fails to *cause* motion. But what does "cause" mean? The word "cause," αἰτία, originates in the adjective "responsible" and might be characterized as "whatever is responsible for a thing's being (or motion)."[61] Indeed, Hussey translates αἴτιον as "responsible" at 215a29. "Cause" answers the question "why" in its various senses including "of necessity."[62] What is "of necessity" causes a thing as its matter, e.g., walls are necessary to have a house.[63] More importantly, however, cause is associated with the end, that for the sake of which: the house provides the reason why the walls are necessary. And nature is unequivocally included among "causes that act for the sake of something" (*Physics* II, 8, 198b10). Aristotle is explicit that although both "of necessity" and "the reason why" must be included within the domain of physics, the physicist must be primarily concerned with the end that is a cause – the reason why – in the fullest sense (*Physics* II, 9, 200a32–34). Indeed, as the cause of motion in the potential, actuality is complete and is aimed at by the potential; hence, it is *not* a cause in the sense of "necessary condition" but resembles "the end."[64] We must conclude that there is no reason to assume at the outset that "cause" means "necessary condition."[65]

---

61 On the history and development of the notion of cause in Greek philosophy, cf. Allan, "Causality Ancient and Modern," 10–18, and Frede, "The Original Notion of Cause," 125–150.

62 The *locus classicus* for this notion would obviously be *Physics* II, 3–8, where Aristotle discusses the four causes, incidental and essential causes, chance and spontaneity as candidates for the role of causes, that which is unmoved as a cause in nature, and nature itself as a cause. In his commentary on the void as responsible, Hussey suggests that "natural places are *final* causes of movement" (his italics) (128). This assertion is attributed directly to Aristotle by Lettinck, *Aristotle's* Physics *and Its Reception in the Arabic World*, 322, in his summary of *Physics* IV, 8. We may note that both moving causes and final causes produce motion by being actual relative to a potency. Hence, specifying something as actual fully specifies it as a mover – there is no need to designate it as a moving cause or a final cause. Indeed, these causes are not exactly correlative – final causes move as objects of desire and may be either extrinsic or intrinsic, whereas moving causes always move from without; therefore, they may not be exclusive and exhaustive categories. On why place as actual is fully designated as a mover but should not be thought of as either a moving cause or a final cause, cf. H. Lang, *Aristotle's Physics and Its Medieval Varieties*, 83–84. Criticizing this view, Graham, "The Metaphysics of Motion," considers this problem and reaches a different conclusion: "But how can an actuality move anything? . . . The actuality is the goal of a process, but not as a mover, in any obvious sense of 'mover.' Rather, it is what Lang does not wish to admit, a final cause of the process." (179). Graham wholly fails to understand Aristotle's definition of motion as well as why god, the first unmoved mover, and the final cause of the cosmos in *Metaphysics* XII, 7, must be completely actual so as to produce the first motion always.

63 Aristotle, *Physics* II, 9, 200a-10; also *Pos. Anal.* II, 11, 94a20–24; cf. R. Friedman, "Matter and Necessity in *Physics* B 9 200a15–30," 8–11.

64 Cf. Aristotle, *Physics* II, 8, 199a30–32. Cf. Schofield, "Explanatory Projects in *Physics*, 2.3 and 7," 29–40.

65 Ross, 587–588, argues that the Atomists take αἰτία to mean "necessary condition,"

The meaning of the phrase "that through which a thing is moved is a cause" is problematic. At least since Avempace, commentators refer it to the ratio between mover and moved that appears in *Physics* VII, 5.[66] Here Aristotle argues that a ratio always describes the motion of some moved thing by its mover through a certain distance in a given time. "If, then, A is the mover, B the moved, and C the length moved, and D the time, then in an equal time the equal power A will move 1/2 B twice the distance C, and in 1/2 D it will move 1/2 B the whole distance: for thus there will be a proportion" (249b30–250a4). This ratio may seem parallel to that of *Physics* IV, 8. But I have suggested that arguments are subordinated to the problems that they are designed to solve. Hence we must ask: are the problems at stake in these two arguments the same? In *Physics* IV, 8, Aristotle asserts that "that through which" causes a body to be moved faster and describes hindrances. But he mentions neither the mover nor any relation between the mover and the moved; conversely, in *Physics* VII, 5, he discusses the mover/moved relation and never mentions "that through which." Hence, although both arguments assert a ratio, the problems under consideration in them are quite different.

I would suggest that the relation between a moved body and "that through which" reappears in *De Caelo* IV, 6.[67] Concluding his account of heavy and light things, Aristotle asserts that the shape of bodies being moved is not a cause of their being moved upward or downward, but of their being moved faster or slower [θᾶττον ἢ βραδύτερον] (313a15). The problem, he goes on, concerns the relation between a moved body and the "continuous things," air or water, water or earth, through which it is moved (313a16–21). He concludes that the downward motion of a body is due to two things: (1) the body has some strength by virtue of which it is moved downward, and (2) the continuous body through which it is moved has some strength by which it resists being broken up; and therefore, there must be some relation between them (313b16–18). The argument (and the *De Caelo*) conclude with a description of this relation: if the strength of the heavy things for disruption and division exceeds that of the continuous body, the heavy body "will be forced downward faster" (313b18–20).

---

whereas Aristotle answers that it cannot mean "determining cause" and that this ambiguity renders the argument "worthless." Ross is correct in noting that this term need not mean "necessary condition" for Aristotle; but Aristotle often assumes his own meaning, and the argument is not worthless insofar as it constitutes a refutation from his own point of view.

66 Hussey, 131, 189. This identification is very common. For a review of modern historians on this problem, cf. Weisheipl, "The Principle *Omne Quod Movetur Ab Alio Movetur* in Medieval Physics," 83–85. A recent example appears in O'Brien, "Aristotle's Theory of Movement," 61–62.

67 Hussey, 179, mentions this text, but never connects it with the argument of *Physics* IV, 8; it appears in a list of texts under the heading "*One body, motion vs. inertial resistance.*"

There is good reason to identify the argument of *De Caelo* IV, 6, with that of *Physics* IV, 8. (1) The continuous things *are* "that through which" a body is moved. (2) The examples, air, water, and earth, are identical and function in the same way in both arguments. (3) Both arguments account for the same thing, i.e., not that a body is moved, but that it is moved faster (or slower). Thus, the argument of *De Caelo* IV, 6, expands upon the relation asserted in *Physics* IV, 8.

The continuous body through which something is moved does not cause its motion; rather it causes the moved (being moved by its mover), to be moved *faster* or *slower*.[68] "That through which" is one of the elements, e.g., water or air, that by definition must be actively oriented toward or moved toward its natural place or else at rest in that place. If we take two elements, each resting in its natural place and one more rare than the other, then by virtue of being in its place and so resisting being moved out of that place, each element actively hinders a moved body and so causes it to be moved more slowly. The more dense the element, the more natural activity and hence the more it slows the actualization of the moved body by its mover. Thus, by virtue of its nature, "the continuous body through which" a thing is moved causes motion in the moved to be faster or slower.

Now we seem to be talking explicitly about "resistance." Why not think in terms of a resisting medium? The phenomenon is the same, a body being moved through water or air, but it is being explained by different concepts: "that through which" and "active cause" or "media" and "resistance." That "resistance" seems obvious proves nothing more (or less) than that the concepts affirmed by modern physics are native to us. In Aristotle's physics, "that through which" is differentiated and active, exercising its nature, and so causing a thing to be moved faster (or slower). Insofar as "medium" connotes resistance of something undifferentiated, passive, or indifferent, it fails to represent Aristotle's causal account of nature as an intrinsic source of being moved and being at rest.

"That through which" causes actively, extrinsically, and because of its determinate nature, as does a hindrance or impediment – and therefore causes a body to be moved faster or slower. This extrinsic, "hindering," relation can be expressed by a ratio; namely, as "that through which" hinders less, a moved body – being moved by its mover – is moved faster and as "that through which" hinders more, the body is moved more slowly. And such is the force of Aristotle's assertion in *Physics* IV, 8.

But if neither "speed" nor "resistance" nor "necessary condition" are concepts appropriate to Aristotle's argument, what accounts for their use by Aristotle's critics? A pathology of the problem is not difficult. One would conclude that αἰτία [cause] means "necessary condition" only if

---

68 Cf. also *De Caelo* III, 2, 301b17–22.

one first assumes that "that through which" a body is moved is an undifferentiated passive medium. In effect, the criticism so consistently raised against Aristotle's argument on this point starts from the [for Aristotle false] assumption of an undifferentiated passive medium and promptly calls up associated concepts, i.e., speed and passive resistance, as necessary conditions for motion, itself conceived of as quantified. That, finally, is why they are not at work in Aristotle's argument and when brought to bear on it make it (on Avempace's view) wrong and (on Hussey's) "hardly intelligible."[69] Indeed, that *none* of them is appropriate is further evidence that Aristotle's physics is not a proto-mechanistic physics with teleology added on. Rather, it is fully teleological and offers a systematic alternative to quantitative accounts of motion and moved bodies.[70]

The Argument, Part II: Motion, Void, and Ratio

One obvious question concerns what difference it makes to the argument itself to understand "motion," "faster," and "that through which," as I propose. The proof of such claims is in the argument itself:

> But the void is in no way able to have a ratio by which it is exceeded by body, any more than nothing has to number. (For while four exceeds three by one, and two by more than one, and one by even more than that by which it exceeds two, yet with nothing there is no longer any ratio by which four exceeds it. For that which exceeds must be divided into the excess and that which is exceeded, so that four will be the amount of the excess plus nothing.) Therefore, a line does not exceed a point, unless it is composed of points. Similarly, the void cannot bear any ratio to the plenum; so neither can the motion, but if [a thing] is moved a given length through what is finest in such-and-such a time, then through the void it exceeds all ratio. (215b12–22)

The empty cannot enter into a ratio with the full, so that if a thing is moved through what is full but "finest" in a certain time, then through a void it would exceed all ratio. The issue is treated as analogous to mathematical ratios and to the impossibility of a ratio between a point and a line.[71]

What is the "it" that exceeds all ratio? There are several options. Perhaps "speed": in a void, the speed of the moved body would transcend any limit.[72] But, I have already argued, there is no notion of speed here –

---

69 Hussey, 131.

70 The problem of teleology and mechanistic explanations is very difficult and exceeds the bounds of this study. For two classic (and still valuable) treatments, cf. Wieland, "The Problem of Teleology," 141–160, and Carteron, "Does Aristotle have a Mechanics?" 161–174; Charlton, 92–93. More recently, Witt, *Substance and Essence in Aristotle,* 98–100.

71 Cf. Heath, *Mathematics in Aristotle,* 117, on these relations. Heath refers Aristotle's point to Euclid V, Def. 4: "Magnitudes are said to have a ratio to one another which are capable, when multiplied, or excluding one another; for no multiple of zero can exceed 1 or any number." Hussey comments, 132, that "his arithmetical parenthesis serves no clear purpose."

72 Hardie and Gaye reads "it moves through the void with a speed beyond any ratio"

or in Aristotle's physics more generally. Perhaps "length": in a void the length traversed by a body in a given time would exceed any ratio.[73] But as we shall see in the final part of the argument, Aristotle assumes a void equal in length to a given plenum; so the length does not exceed all ratio.

There is another option: motion itself.[74] Aristotle conceives of motion as a relation between mover and moved and that through which a thing is moved causes its motion to be faster or slower. Hence, the relation between them may be expressed as a direct proportion between the moved body and "that through which." Advocates of the void propose it as a "that through which" a body is moved; but the void cannot operate in the ratio because as empty place, its value must be "nothing." Hence motion expressed as a ratio between the moved and a void serving as "that through which" would exceed all ratio: the void cannot serve as "that through which," because it cannot enter into the ratio required by motion conceived of as a relation between mover and moved.

Given this failure, one might reasonably expect Aristotle to reject the void immediately. Instead he proceeds directly to the next step in the argument without comment. But not so his commentators and here we find, historically speaking, one of the most vexed moments of this argument.

Philoponus argues that motion in a void will take *no* time and because all motion requires time, there will be no motion.[75] Perhaps relying on Philoponus, Averroes finds instantaneous motion here. Commenting on the history of this argument in medieval science, Grant concludes that in response to this passage, taken together with Averroes's commentary, it became commonplace to argue (*contra* Averroes) that in a hypothetical void a body would be moved with a finite and temporal motion (Thomas' view, for example).[76] These issues are also reflected in modern scholarship. For example, Ross and Hussey draw the same conclusion as does Averroes, namely, that Aristotle concludes here that motion in a void has no proportion and so, presumably, is impossible.[77] And medievalists too attribute this conclusion directly to Aristotle.[78]

(215b22). Hussey too, 131, asserts "speed" here because in a void "resistance must be put equal to zero, and so there is no finite proportion between the speeds."

73 Apostle translates "it" as the distance traveled.

74 Thus "the moved" is the subject of φέρεται and "motion" of ὑπερβάλλει.

75 Philoponus, *In Phys.* 676.20–23. This passage has been translated by Furley in *Place, Void, and Eternity*, 50. For an analysis of Philoponus's criticism of Aristotle's argument here in *Physics* IV, 8, cf. Wolff, *Fallgesetz und Massebegriff*, 23–27.

76 Thomas, *In Phys.* IV, lect. 12, par. 534; for a discussion of medieval interpretations of this problem, cf. Grant, *Much Ado About Nothing*, 26 ff.

77 Ross, 590. Hussey, 132, commenting directly on this passage, notes that Aristotle is right "through a void there is no terminal velocity: the velocity increases without limit"; in an additional extended note, 192, he refers to this as a *reductio ad absurdum* argument; cf. also Bickness, "Atomic *Isotacheia* in Epicurus," 57.

78 For an example in an otherwise excellent piece of work, cf. Trifogli, "Giles of Rome on Natural Motion in the Void," 137.

There are serious grounds for caution here. Were there a conclusion, then the argument would be complete. But the argument continues, as Aristotle returns without comment to the ratios required by motion and allows the void to operate within them. Why does he fail to draw any conclusion? As we have seen, the arguments against the void set out from a false starting point, a void proposed by his opponents. Consequently, conclusions concerning a void or motion in a void taken independently of this argument do not follow – and Aristotle does not draw them out.

But if no contradiction about the void follows here, then what is the point of this argument? In order to reject the void, when he produces a contradiction, Aristotle must show that the assumption of the void *alone* produces the contradiction. Consequently, he adds a step to the argument to show that the assumption of the void within the ratio required by motion necessarily involves an impossibility. When, in the final step of the argument, a contradiction results, he immediately attributes it to the assumption of a void. This attribution is justified because here he has shown that it cannot function in a ratio.

The Argument, Part III: The Contradiction

The third and final step of the argument now follows. Aristotle returns to the ratios established in step one and assumes that a void functions in the ratio between the moved and "that through which." The ratio, we may recall, posits that if body A is moved through B in time C and through D (equal to B but more rare) in time E, then C is to E as B is to D. For example, as air is more rare than water, so the time required for a body to be moved through air will be shorter than that required for its motion through water.

The argument continues. Let there be a void F equal to B (and D) through which body A is moved in a given time G; time G must be less than time E, that is, the time required to traverse the rare body D; and we would have the ratio "time G is to time E as the void [F] is to the full [D]" (215b22–26). These ratios add nothing new to the argument. Rather, they utilize the relation established in step one: as "that through which" a body is moved becomes more rare, the motion of the body becomes faster; hence, it would be moved even faster in a void than in a rare body.

Aristotle now makes a point that is often found puzzling. If body A is moved through rare plenum D in time E and void F (equal to D), in time G, time G must be less than time E, and in a time equal to G, body A would go through only some part of rare plenum D, let's say H of D (215b26–30). Hussey claims that defining H "is pointless."[79] I would sug-

79 Hussey, 132. Ross, 590, does not mention this definition.

gest, however, that defining H moves the argument forward by clarifying a serious ambiguity.

The assumption of a void as "that through which" may operate in either of two ways. (1) We assume a ratio with "nothing" to operate in the argument along with its impossible result; or (2) we assume a void to operate in the ratio as if [contrary to fact] the result were possible. The comparison of time G to time E along with the assertion that in time G, body A would cover some part (H) of D indicates unambiguously that Aristotle takes the second alternative. That is, he allows a void to operate in the ratio required by motion, not by allowing an impossible ratio, but by assuming that the ratio represents a possible number.[80] Indeed, his identifying H as a number to which, presumably, a value could be assigned tells us unequivocally that Aristotle is not dealing with instantaneous motion.[81] The void is assumed [falsely, as we know from the second step] to operate as if it had a ratio to the full, like any "that through which."

Aristotle now has the situation in hand. Specific ratios obtain between a moved body and that through which it is moved. The less rare that through which is, the less distance a body will be moved in equal time. Conversely, as that through which becomes more rare, a body will be moved faster. Furthermore, and this point prepares the way for the final step of the argument, a body would be moved through void F in finite time, G – the same time in which it would be moved through some part (H) of rare plenum D.

And that step now follows (215b30–216a2). We know that in time G body A will be moved through void F and part H of D. Now suppose another body [another plenum], I, more rare than D in proportion as time G is shorter than E. (A reminder: E is the time required for body A to go through the whole of D.) But body A will be moved through some part H of rare body D in the same time that it will traverse the whole of a very rare body I – and we called that time G. So D:I::E:G. Void F is equal to B and D, whereas I is a *very* rare plenum also equal to B and D; so I must also be equal to void F. Conclusion: void F and rare plenum I (equal to F) will both cause the same body to be moved in the same time (G).

But this is impossible. It is clear then, if there is a time in which it will be moved through any part of the void, then this impossible result will follow: in an equal time, it will be found that both a plenum and a void let something pass through. For there will be some ratio of one body to another as time to time.[82]

---

80 Heath, *Mathematics in Aristotle*, claims that H is introduced "in order to show that the *distances* have a finite ratio to one another . . . ," 118.

81 Cf. Hussey, 132. Grant, *Much Ado About Nothing*, 24.

82 Aristotle, *Physics* IV, 8, 216a4-7: ἀλλ' ἀδύνατον. φανερὸν τοίνυν ὅτι, εἰ ἔστι χρόνος ἐν ᾧ τοῦ κενοῦ ὁτιοῦν οἰσθήσεται, συμβήσεται τοῦτο τὸ ἀδύνατον· ἐν ἴσῳ γὰρ ληφθήσεται πλῆρές τε ὂν διεξιέναι τι καὶ κενόν· ἔσται γάρ τι ἀνάλογον σῶμα ἕτερον πρὸς ἕτερον ὡς χρόνος πρὸς χρόνον. Hussey translates the οἰσθήσεται here as present active, Hardie and

As soon as the void functions in the argument so as to produce a positive ratio, the time assigned to motion in a void will also result for some plenum and a contradiction results. The reason why is clear: the void is not just the "most rare" plenum any more than "nothing" is the smallest number – void is of an entirely different order from plena. But allowing a void to function in the ratio required by motion treats it as if it *were* of the same order as the most rare plenum. As soon as a positive value is allowed, a rare plenum for which this number also results can always be found. (And the contradiction follows if and only if a body were moved through the void in a finite time – the motion cannot be instantaneous.)

The argument begins with the assumption that "that through which" a body is moved causes its motion to be faster (or slower) and concludes that in a plenum and a void, a body would be moved in the same time. Aristotle concludes the argument about "that through which" by identifying the source of the contradiction: there is no ratio of void to full (216a8–11). Thus, the argument closes: the void cannot serve as a term required by motion (216a21–26).

Construing the argument in this way accounts for its various features, including those found problematic on other construals. It may be briefly summarized. This argument refutes the view that the void can serve as a cause of motion by showing that if we assume the void to be a special term required by motion, a contradiction results. Aristotle first asserts that "that through which," e.g., water or air, a body is moved causes a motion to be faster (or slower) by actively impeding it less (or more). The second step of the argument specifies the void as the source of the upcoming contradiction because a void cannot enter into the ratio required by motion and taking it within this ratio would yield an impossible result. In the final step of the argument, a void is assumed to operate, as if (contrary to fact) it produced a rational positive number within the ratio required by motion. A contradiction results: a void and a plenum will let a body pass through in the same time. So the argument begins "that through which" causes a body to be moved faster and concludes that a void and a plenum cause a body to be moved in the same time. Therefore, the void must be rejected because it cannot serve as a cause of motion.

I have suggested that a number of concepts traditionally brought to bear on this argument must be rejected. Aristotle is not discussing a medium in its modern sense, there is no "passive resistance," no use of a "necessary condition" for motion, and "faster" does not mean "greater speed." Rather, the argument turns on his own definitions of motion and nature, and in this sense the rejection of the void is subordinated to these definitions.

Gaye as future active. It is unequivocally future passive. Hussey translates διεξιέναι as "traverse," even though it is a strengthened form of ἐξίημι.

This analysis returns us to a question posed earlier: if neither "speed" nor "passive resistance" nor "necessary condition" are concepts appropriate to Aristotle's argument, what accounts for their widespread use by his commentators? I would suggest that we need not one but two such accounts – one for Aristotle's medieval critics and another for recent treatments of Aristotle. After Aristotle, much of philosophy (including physics) was conducted within commentaries on Plato and Aristotle. But Byzantine, Arabic, and Latin commentaries are not neutral restatements of Aristotle's physics; rather they are aggressive reworkings of this physics to suit ongoing conceptions of work needing to be done within the science of physics. Hence in these commentaries nothing remains untouched from rhetorical structure to concepts to conclusions.[83] I might speculate why: like any set of ideas, philosophy (including physics) is a profoundly coherent enterprise, and it is not possible to change one feature without redefining everything else too. Medieval commentators use a rhetorical structure that is native and concepts to which they feel committed in order to reach conclusions that they believe to be true. In detecting the coherence of their positions we can also detect their relation to Aristotle's own coherent position and so understand the history of ideas in all its multiplicity.

But what of modern readers who claim to be "making sense" of Aristotle with their notions of "speed" and "medium" etc.?[84] The problems involved in such substitutions reveal what is at stake for the modern project of understanding the history of ideas both in Aristotle's argument and in his commentators. Substituting "greater speed" for "faster," and "medium" for "that through which" represents a systematic decision – and so, I would conclude, a systematic mistake. First, it presupposes an atomistic (not to say mechanistic) view of language: one word can replace another without argument or consideration of larger constructive meaning. Indeed, Hussey speaks of "speed" as a "common sense notion" without any acknowledgment that it has become such *because* modern perceptions are deeply informed by quantitative physics.[85] As a result, Aristotle's arguments are translated into an idiom that far from representing his meaning transforms it into a project quite different from his own.

---

83 I have argued this point at some length in H. Lang, *Aristotle's Physics and Its Medieval Varieties*, 161–172.

84 Drabkin, "Notes on the Laws of Motion in Aristotle," 60–84; for a critique of Drabkin, cf. Owen, "Aristotelian Mechanics," 227–245, For more recent versions of Drabkin's thesis, cf. Cohen, "Aristotle on Heat, Cold, and Teleological Explanation," 255–270; Denyer, "Can Physics Be Exact?," 73–83. Hussey himself further develops the ideas discussed here, i.e., those from the notes to his translation, in "Aristotle's Mathematical Physics," 213–242. Even more recently, cf. O'Brien, "Aristotle's Theory of Movement," 64–67. Explicitly opposed to this project, cf. Carteron, "Does Aristotle Have a Mechanics?" 161–174.

85 Hussey, 188.

Even this translation is not the end of the affair. This tradition evaluates Aristotle's arguments (and his physics as a project) in terms of quantitative physics – the model of which usually seems to be Newton.[86] It wishes to make Aristotle speak the language of "speed," "medium," etc., because this is the language of true physics – Newtonian physics – and Aristotle's physics will be truer if expressed in it and brought as close as possible to it in every way.[87] Hence, "to make sense" of Aristotle one must make this the first task.

But the history of ideas (including science and philosophy) is thereby, in its turn, being defined. One assumes a model physics, thought to be true or best, and then translates (and evaluates) Aristotle's physics into the terms and concepts of this model. Thus behind this translation (and evaluation) lies the assumption that there is only one "real" physics, quantitative (Newtonian) physics and the history of philosophy is the history of stages on the road to this achievement. But Aristotle's teleological physics rests on different starting points, defines different problems, and utilizes different concepts than does quantitative physics. Consequently, if we are to understand Aristotle's arguments *and* if we are to understand quantitative physics, we must be prepared to rethink what constitutes physics as a science. (The issue clearly lies beyond this study, but within quantitative physics, Newton has not provided the model for "the best physics" for some time.) In short, we must recognize that the history of philosophy (and science) cannot be as teleological as this view suggests.

I am proposing a different understanding of Aristotle's terms ("faster" is not "speedier"), his concepts (nature is everywhere a source of order), and his physics (knowing things that are by nature), *and,* finally, of the history of philosophy (and physics). In short, the meaning of particular terms or concepts must be determined by the larger context, and the coherence of that larger context must in its turn be determined as a whole presenting logical and conceptual coherence. The initial presumption is always in favor of coherence, and in the example of Aristotle's difficult argument concerning "that through which," the presumption seems justified. In my conclusion, I shall make some suggestions about establishing criteria for comparing and evaluating positions expressed in their full context. Here the prior point is at issue: a full understanding of the history of ideas requires that before comparisons can begin we must un-

---

86 Cf. Hussey, 188, 196. Also, Drabkin, "Notes on the Laws of Motion in Aristotle," 75. Commenting on the text, i.e., *Physics* IV, 8, 215a22, preceding that which we have examined here, Heath, *Mathematics in Aristotle,* 115, says of *Physics* IV, 8, 215a22, that "it constitutes a fair anticipation of Newton's First Law of Motion . . . "; Hussey, 130, makes the same comment.

87 See the methodic remarks of Biagioli, *Galileo Courtier,* 13; also O'Brien, "Aristotle's Theory of Movement," 47–79 (although the model here is Galileo rather than Newton).

derstand each moment of that history, its starting points, problems, and concepts, on their own terms (insofar as is possible).

A Void and the Inclination of Heaviness and Lightness

Strictly speaking, the larger argument is not yet complete. There are two reasons why a body may be moved faster, and the second remains (216a12–14). One body may be moved faster than another because of a greater inclination (ῥοπή) of either heaviness or lightness. Again, the argument turns on ratios. If bodies are alike in other respects, but have greater inclination of heaviness or lightness, they will be moved faster or slower over an equal distance according to the ratio that the magnitudes have to one another (216a14–17). But why do proponents of the void think one body would be moved faster than the other (216a17)? In a plenum, one body must be moved faster than another because the greater body divides that through which it is moved faster by means of its strength, i.e., either its shape or its inclination to go down (earth is heavy) or up (air is light). In a void, all bodies will be moved equally fast [ἰσοταχή], which is impossible.

The conclusion of the two-part argument follows: the void cannot serve as a cause of motion. In fact, if there were a void, the result would be the opposite of that anticipated by its proponents. Rather than serving as a cause of motion, a void would render motion impossible (216a20–23).

The argument based on the inclination of heaviness or lightness of a moving body again reveals Aristotle's view of the world as determinate and the void as indeterminate. The problems associated with inclination are serious, and I shall consider them in the second half of this study.[88] But we can see what Aristotle must presuppose to make this argument work. Fire is by nature light and must be moved upward, if nothing hinders; likewise, earth is heavy and must be moved downward, if nothing hinders. The cause of motion in natural things lies, as we have already seen, in actually "up" and actually "down" – and these are constituted by place, which, as a limit, renders the cosmos determinate. Here we see not the cause of motion, but the cause of its being faster. What causes the motion to be "faster"? A body is moved faster either up or down because it has a greater inclination to lightness or heaviness.

What is inclination? It is, as I shall argue, a body's source of being moved, i.e., its active orientation toward its natural place as determined by the elements of which the body is composed. Thus, the more inclination a body possesses, the more active orientation it possesses and the

---

88 Hussey, 132–133, and 188–189, considers a number of possible interpretations of inclination and concludes that there are good reasons for taking it to mean "weight." I shall consider this issue at length later.

faster it will divide that through which it is moved. But if there were nothing to divide, all orientation would appear to be the same and would be actualized at the same rate. Thus, what we see to be universally the case, i.e., bodies are moved faster as they have a greater inclination to heaviness or lightness, is violated by the assumption of a void. In a void, all things would be moved equally fast, regardless of their inclination.

This argument requires Aristotle's notions of "light," "heavy," and "inclination," all of which will be examined in the second part of this study. Here I wish to make only the limited point required by this argument. "Light" and "heavy" cannot be translated directly into a notion of weight. "Light" means "a power of that which is moved upward," and "heavy" means "a power of that which is moved downward," where upward and downward are defined respectively as away from and toward the center of the cosmos (De Caelo IV, 1). In short, inclination is an intrinsic orientation up (or down) in fire (or earth) as movable – the potency of the moved for its appropriate actuality, i.e., respective natural place. And in this sense, inclination, like the cosmos, is determinate.

In quantitative physics, weight, unlike inclination, is defined as an extrinsic relation between distance and mass, without reference to directional determination such as up or down. Hence, weight is indeterminate in respect to direction. But inclination as determinate in respect to direction is the concept required by this argument, by Aristotle's concepts of motion and nature, and, finally, by Aristotle's physics more generally – the physics of a determinate cosmos.

Ross and Hussey both claim that Aristotle's "contradiction," i.e., in a void all bodies would move equally fast, is now an accepted fact within Newtonian physics.[89] But Newtonian physics posits that in a void bodies of whatever *weight* fall with equal *speed*. That is, the rates at which any two bodies of any weight would traverse a given distance in a given time would always be equal. Aristotle, however, is talking about neither a rate at which motion occurs nor about bodies of different weights. Motion is intrinsic to the moved, and in *Physics* IV, he compares two motions as relatively faster and slower or equal because of an intrinsic inclination in a determinate direction.

This difference between Aristotle's physics and Newton's defines them as different. And it rests on different conceptions of nature and motion. Newton's nature is Christian in important ways, e.g., God's action in it is indispensable, and motion is transferred from one body to another; for Aristotle, natural things have an immediate orientation toward their actuality. His definition of motion requires that it be treated as a mover/moved relation, i.e., on contact with a mover, an ability or potency in the moved is actualized. The most important kind of motion is change of

89 Cf. Ross, 591; Hussey, 192.

place. Motion is not a quantum, such as speed, imposed on the moved, and it cannot be treated independently of this mover/moved relation. Aristotle's physics rests on the definition of nature as that which contains an *intrinsic* principle of motion, whereas Newtonian physics sees matter as inert and acted upon by forces fed into nature *extrinsically* (by God).[90]

Consequently, modern physics does not assert what Aristotle took to be an obvious contradiction. The contradiction reached by Aristotle in *Physics* IV follows only on his definitions of motion and nature, the interlocking concepts at work in these definitions, and the physics that Aristotle develops in support of it. Modern physics does not oppose Aristotle's conclusions so much as replace his conceptions with different ones, including motion, nature, speed and weight. As a result, Newtonian physics – like Newton's universe – is genuinely different from Aristotle's.

Finally, Aristotle's physics requires that everything moved be moved by something, whereas in Newtonian physics things in motion tend to stay in motion and things at rest tend to stay at rest. Aristotle has not been proved wrong; rather, his concept of nature as an intrinsic source of being moved and being at rest has been replaced by a notion of matter as inert. At the same time, the concept of motion has changed from actualization of a potential by its actuality to a quantum of energy imposed on a body from the outside. So the same conclusion follows: Newtonian physics has not proven Aristotle wrong, but has substituted a radically different enterprise under the same name, physics.[91]

This point is important for two reasons: (1) understanding what is at stake in Aristotle's argument and (2) analyzing historical treatments of this argument so as to detect conceptual shifts that have their origins outside of Aristotle. The first understanding is important and desirable in itself – and without it the second is impossible. To compare the two enterprises, we cannot pick through them to find bits and pieces that look alike from a post-Newtonian point of view; rather, we must consider the origins of these differences in their different starting points and problems.

And this understanding is at stake when the rejection of the void as a cause of motion concludes:

Some, then, think that the void must be, separated in virtue of itself, if there is to be motion according to place; this is the same thing as to say that place is

---

90 Cf. Hahn, "Laplace and the Mechanistic Universe," 256–276.
91 Hussey, 130, identifies Aristotle and Newton. I largely cite Hussey throughout because his account is so consistent and clear. He is part of a tradition central to modern studies of Aristotle. Cf. Drabkin, "Notes on the Laws of Motion in Aristotle," 60–84; Heath, *Mathematics in Aristotle,* 115; and Owen, "Aristotle: Method, Physics, and Cosmology," 151–164. For a consideration of temporal order as parallel to "spatial order," which measures Aristotle's account by Newtonian standards and concludes that Aristotle's "temporal theory is by modern standards relatively undeveloped," cf. Corish, "Aristotle on Temporal Order," 68–74.

something separate, and that this is impossible has already been said. (216a23–26; Hussey translation)

The void fails to serve as a cause of motion either when "that through which" or a "greater inclination to heaviness or lightness" causes the motion of a body to be faster. In both cases, the assumption of a void leads to a contradiction. Indeed, the assumption of a void produces the opposite result than that desired by its proponents: they believe that the void will serve as a cause of motion – but in fact, the assumption of a void renders motion impossible.

Final Arguments against the Void

Several further arguments against the void complete Aristotle's analysis and rejection of it in *Physics* IV. Two arguments that there cannot be a void into which a body is moved conclude *Physics* IV, 8; *Physics* IV, 9, returns to the claim that void must be present within the cosmos. I shall consider these briefly.

Even if, Aristotle says, we consider the void in virtue of itself, it will be found to be truly vacuous.[92] He may mean that he will now consider the void simply as empty place rather than as a cause of motion.[93] But even so, he works strictly in terms of his own definitions of motion and place, seeking an answer to the question that opens *Physics* IV: what is "the where" of things that are. The first argument returns to motion, place, and the elements whereas the second concludes that place functions as "the where" of things and so renders the void an unnecessary concept serving no purpose in physics.

The first argument about "the void in virtue of itself" presupposes the doctrine of the elements and place. When a cube is placed into water or air, if the water or air is not compressed, it must be displaced in a quantity equal to the cube. The displaced body must be displaced in the direction appropriate to its nature; fire will go up, earth down, and the middle bodies in both directions (216a33). But if a body were placed into void, displacement toward appropriate natural place would be impossi-

---

92 Aristotle, *Physics* IV, 8, 216a26–27; Thorp, "Aristotle's *Horror Vacui*," 149–150 notes, claiming that modern commentators have "pretty generally shared this view," that here "Aristotle cracks a joke . . . To be sure, this is not a very funny joke; what is interesting about it, though, is that it underlines the general attitude of dismissive flippancy that seems to run through Aristotle's consideration of the void. He seems to refuse to take the hypothesis of the void at all seriously. He never argues *directly* that the void does not or cannot exist, but contents himself with criticizing the arguments that other thinkers had advanced in its favour. And even this criticism seems disorganized and strawmannish – it doesn't really meet these thinkers on their own terms; moreover, it is heavily bound up with Aristotle's peculiar views about the phenomena and the laws of motion." Thorp goes on to argue that Aristotle really does deny the void. I have suggested that these arguments deserve to be taken seriously.
93 Hussey, 133; Apostle, 256.

ble because void is not a body. Thus, the void would seem to interpene-
trate the cube being placed into it in an amount equal to the cube – just
as if the water or air instead of being displaced had penetrated the cube.[94]
But the cube already has just as much magnitude as the void that inter-
penetrates it; and, being a body, the cube must be hot or cold, heavy or
light (216b2–6). This magnitude cannot be separated from the substance, the cube, in which it inheres; but if the magnitude were separated
from the cube and stripped of all its characteristics, it would not only be
itself but would also possess an equal amount of void, i.e., the displaced
void that interpenetrated when the cube was placed into a void. But how
could the magnitude of the cube occupy a place (or void) equal to the
displaced void, since the cube now contains two equal magnitudes, i.e.,
that of the cube and that of the interpenetrating void?[95] If it could, then
two things – the magnitude of the cube and the magnitude of the void –
would occupy the same magnitude, which one magnitude alone, that of
the cube, could (and originally did) occupy. And this argument could be
extended so that an indefinitely large number of things occupy the same
magnitude. Such a view is absurd.

Here Aristotle treats void as a sort of empty magnitude. Because void
is not one of the four elements, it will not be actively oriented, and so displaced, toward its natural place. Hence if a cube with a certain magnitude
were placed into a void, an impossibility would result. Two equal magnitudes, that of the cube and that of the void, would be in the same magnitude, and it would be possible for an indefinitely large number of things
to be in the same place, which is impossible.

The final argument of *Physics* IV, 8, is little more than an addendum
to this argument. The cube, like all bodies, must have volume.[96] If the
volume, bounded by the sides of the cube, does not differ from its place,
why assume yet another place, i.e., the void, over and above the volume
of the cube (216b14–15)? The void, in fact, is an unnecessary and useless
assumption (216b15–16).

---

94 Aristotle, *Physics* IV, 8, 216b-2. The word "διάστημα" appears here again, and Hussey
   translates it as "extension" – a notion that he develops in his commentary on this argument, 133. But there is no need for such a technical translation, and it adds unnecessary associations to Aristotle's argument. Apostle translates it as "interval," and Hardie
   and Gaye as "distance."

95 Aristotle, *Physics* IV, 8, 216b5–11. For a somewhat different construal of this argument,
   cf. Grant, "The Principle of the Impenetrability of Bodies in the History of Concepts of
   Separate Space from the Middle Ages to the Seventeenth Century," 551–571. Grant connects this argument with the notion of place as three-dimensional at *Physics* IV, 1,
   209a5–7, which he takes as a conditional, i.e., "if place or space were three dimensional,
   it would be a body" (551). But as we have seen, place, although not a body itself, is in
   the genus of body and is three-dimensional. The sentence at *Physics* IV, 1, 209a5–7, cannot on any account be taken as a conditional.

96 Aristotle, *Physics* IV, 8, 216b12–13. Hussey, 134, again insists that magnitude or volume
   (ὄγκος) must mean "extension."

For Aristotle a clear conclusion follows: "That there is no separate void, then, from these things is clear" (216b20). The assumption of the void, far from serving as a cause of motion, always produces a contradiction and makes motion impossible. (In the final addendum, the void is shown merely to be an unnecessary assumption.) A separate void must be rejected because it cannot serve as a term without which motion would be impossible.

In *Physics* IV, 9, a final argument against a void returns to a problem taken up earlier, the problem of compression (*Physics* IV, 6, 213b15–18). The earlier argument concerned "local compression," e.g., a jar full of ashes to which water is added. Here the argument concerns the universe as a whole – that it requires compression and expansion if there is to be change at all – and is directed against Xuthus, who was probably a Pythagorean.[97] So its inclusion here continues the rejections of Pythagoreanism and returns to the general problem of the void as a term without which motion would be impossible.

The internal structure of the argument is not difficult.[98] A void is required for things to be compressed and expanded.[99] In an apparent digression, Aristotle next argues against a "nonseparate" void as a cause of upward motion (216b30–217a9). The argument then returns to the problem of compression and expansion, which Aristotle explains using his own definition of matter, the dense and the light (217a20–217b19). A final summation concludes the chapter and the discussion of the void. I turn to these parts in order.

Some think that the rare and the dense prove that there is a void; for if there were no rare and dense, it would not be possible for things to contract and be compressed (216b21–22). If things cannot contract and compress, then either there will be no change at all, or the universe will bulge (as Xuthus said), or alteration must be in exactly equal amounts so that things may be compressed and expanded (216b25–29). This point seems easy: if the cosmos has a finite volume, then any changes must always balance out across the whole, or else the whole will "bulge," i.e., pulsate in and out as it contracts and expands.[100] Void present in the rare allows for the possibility of expansion and contraction.

So Aristotle takes up "the rare" and its relation to the void. The rare cannot be "that having many separate voids" because a separate void is clearly impossible.[101] But the rare may have some void within it. If so, this

97 Cf. Simplicius, *In Phys.* 684.24; Ross, 593.
98 On this structure, cf. Hussey, 134.
99 Aristotle, *Physics* IV, 9, 216b22–29; this argument is repeated at 217a10–19, which Hussey, 134 and 136, takes to be a later intrusion.
100 Cf. Apostle, 257, n. 2.
101 Aristotle, *Physics* IV, 9, 216b30–32; the reference is clearly to the arguments just concluded in *Physics* IV, 8.

intrinsic void would allow motion to occur without a bulge in the universe. This view produces two problematic (for Aristotle) consequences concerning motion.

(1) If one assumes the presence of void in the rare, the void would account not for every movement, but only for upward movement because the rare is light and goes up (216b34–217a). But even this restricted domain is problematic. As we shall see in a moment, the real issue lies in the nature of the rare and the dense. *Either* there is some one matter that is rare or dense depending on the presence or absence of void *or* some other account will be required.[102]

(2) The second consequence of assuming a void within the rare concerns the way in which the void causes motion. It will not cause motion in the same way as does "that through which" a thing is moved. That is, a void would not cause motion to be faster or slower; rather, it would cause as something that moves and so carries continuous things with it – just as when a wineskin is moved upward and carries the wine contained within it, so the void would carry contained things upward by itself being moved upward (217a-3).

The argument turns on a distinction drawn in the account of place: the difference between essential and accidental causes of motion.[103] It reappears in the *De Caelo* in the context of elemental motion. The immediate point here is that even on this account the void serves only as an accidental cause of motion. An accidental cause does not produce motion because of a direct relation of mover to moved, but only by an indirect relation.[104] If what is moved, A, happens to contain or be continuous with something else, B, then B must also be moved whenever A is moved; in this case, A moves B only because B is contained within, or is continuous with, A. Soul, for example, is accidentally moved by body because it is contained within body.[105] And this relation provides the model for Aristotle's analysis of the void as a cause of motion.

The claim that the void causes motion in the rare by containing what is compressed into it implies that the void must be moved and by being moved upward carries what is contained in, or continuous with, it. Thus, the void produces motion not by originating it directly, but only indirectly insofar as when the void is moved, what is continuous with it must also be moved. So the void resembles body that accidentally moves the soul contained in it.

---

102 For a related argument about the dense and the light that also rejects internal vacua, cf. Aristotle, *De Caelo* IV, 2, 308a28–309b29. Hussey, 135, suggests that this argument (and the next) are directed against early Atomists who wanted the void to act as a medium in which motion takes place.

103 Aristotle, *Physics* IV, 4, 212a16–18; 5, 212b10.

104 Cf. Aristotle, *Physics* II and VII; H. Lang, *Aristotle's Physics and Its Medieval Varieties*, 68.

105 Cf. Aristotle, *De Anima* I, 3, 406a3–12.

This argument minimizes first the domain within which the void may serve as a cause of motion and then void's possible role as a cause. It is not a cause of all motion, but only upward motion; it is not an essential cause, but only an accidental cause. Thus even if [as will turn out to be false] the void were a cause of motion, all motion downward and toward the middle would remain unexplained, and we would still require an essential cause of upward motion.

Even this limited status for the void as a cause of motion, however, is impossible. (1) How can there be a motion of void or a place of void (217a4)? For Aristotle, place must be unmoved because it is the limit within which motion occurs; thus, a position that produces the consequence of a motion or place of void is obviously absurd. (2) That into which the void is moved will be "void of void" (217a4). That is, prior to the motion and consequent arrival of void, the "where" into which void is moved cannot even be void.[106] Hence, it will be void of void, which is ridiculous.

(3) Again, how do proponents of the void account for the downward motion of heavy things (217a5)? If a void within the rare were responsible for any motion, it would be only upward motion. Even its proponents cannot show how it produces downward motion.

(4) Furthermore, if void carries the rare upward, it would be moved faster when it is more rare and void; thus, it would be moved fastest if "the rare" were completely void (217a6–7). But the void must be immobile for the same reason that in a void all things would be immobile: the motions cannot be compared (217a7–10). This argument clearly refers to the ratios established in *Physics* IV, 8. The void cannot enter into the ratios required by motion, and any view that implies that it does entails a contradiction.

Here Aristotle clearly assumes his own account of the nature of change and of the rare and the dense to refute the claim that a void causes motion. The point of his argument rests on his view that there is a single matter for contraries and for both large and small bodies (217a21–217b25). Thus, all actuality, e.g., actually heavy and actually light, are produced from what is potentially such. Obviously, we are back to his definition of motion as actualization of a potential – a definition that is incompatible with a quantitative or atomistic view of nature or change.[107] So he concludes: "the largeness and the smallness of the sensible bulk are extended [προσλαβούσης] not because something is added to the matter, but because matter is potential for both."[108]

For Aristotle, place answers the question "where are things that are"

---

106 For a different interpretation, cf. Hussey, 136.
107 Cf. Hussey, 136–137.
108 Aristotle, *Physics* IV, 9, 217b8–10; of course, "are extended" here has nothing to do with extension in the modern, i.e., post-Cartesian, sense.

because it resembles form and so renders the cosmos orderly. The real problem with the void is that it resembles matter – a quasi-material container, or magnitude, into which things can be placed, or "very light stuff," which, when added to the rare, will make it more rare and so able to rise faster. As long as one works within Aristotle's conception of nature as a cause of order, material accounts of motion must always be inadequate. And it is on the basis of just this view that Aristotle rejects the void as a cause of motion. But for now, Aristotle concludes his arguments about the void: "And concerning the empty, how it is and how it is not, let this be the determination" (217b27–29). The next sentence, which opens *Physics* IV, 10, takes up a new topic, time, another term without which motion seems to be impossible.

Conclusion

Beyond the particular arguments of *Physics* IV, 6–9, Aristotle's rejection of the void can be characterized more generally. Because at the outset he conceives of nature as a principle of order and he seeks the specifics of a category of being, "the where" of things that are, he requires a principle that will be determinate rather than indeterminate. His initial commitment to a conception of nature as orderly can spell itself out only in a physics that yields a determinate world. In this regard, the void fails completely: it is a contradictory concept and, when assumed for the sake of argument, not only fails to cause, i.e., account for, motion, but would render motion impossible. The reason why is clear: as empty place the void is indeterminate whereas the cosmos and things in motion within the cosmos require and are caused by a principle of determinacy.

Place defined as the first unmoved limit of what contains must be the principle of order for the category "where." As a limit, place resembles form: the first constitutive principle of the containing body within which are located all things that are by nature or by art. As "the where," place renders the cosmos directional: "up," "down," "left," "right," "front," and "back" are not just relative to us, but are given in nature. They present the order of nature with the result that we can define "where" anything is that is by nature or by art.

As the first unmoved limit of what contains, place accounts for the characteristics that are rightly thought to belong to it and gives us a universal and common term without which motion would be impossible: place is a single unique constitutive principle that accounts for "the where" of all things that are and are moved. Place first surrounds without being part of what is surrounded. It surrounds as the first unmoved limit. As the first limit of the containing body, it is neither less than nor greater than what is contained. Place is separable from what is in place, just as any container is separable from what is contained. Finally, because place as a limit is a determinative principle, every place within the contained is "up" or "down" etc. and each element is carried to its natural place.

The connection between place and the elements is central to Aristotle's account. Everything, both natural and made, is composed of the elements. Hence the rules of motion for the elements will be the rules of motion for all things within the cosmos. And insofar as place, by rendering the cosmos determinant in respect to "the where" is a cause of elemental motion, place is a term without which motion seems to be impossible.

And yet, as we have noted, the relation between place and the elements remains obscure in *Physics* IV and perhaps in the *Physics* as a whole. Place must be a principle, a limit that renders the cosmos determinate in respect to where; the elements, Aristotle argues, in *Physics* VIII, 4, must, like all things, be moved by another. But the mover of the elements – is it place? – remains obscure. If place is the cause of motion of the elements, why it is such is also obscure. If it is not the cause of elemental motion, then what precisely is the relation between place and the elements?

Aristotle's account of place remains incomplete without an answer to these questions. And without a full account of place, his larger project, i.e., accounting for things that are by nature and for the cosmos as a whole, also remains incomplete. Indeed, his view of nature as everywhere a cause of order remains incomplete. Consequently, these questions could hardly be more pressing.

I turn now to the second part of this study, an analysis of the elements and their motions as involving inclination, ῥοπή. Here, I shall argue, we find an account not of common and universal terms, but of proper terms. These terms are not those involved in the container, but in the contained, not the cause without which motion seems to be impossible, but the effect, i.e., things that are by nature and so contain within themselves a source of being moved and being at rest. Here, I shall conclude, lies the full articulation of the relation between place and the elements. With it, we shall achieve a fuller understanding of the determinate world of Aristotle's physics – of nature as always a source of order.

Part II

The Elements

5

Inclination: An Ability to Be Moved

With this account of "the where" – place, not void – I turn to the *De Caelo* and its topics, particularly the elements. As a topic, the elements present a problem: the elements – themselves unquestionably "things that are by nature" – appear in the *Physics* only within the examination of other topics, but are never themselves examined as a topic. A direct examination of them appears in the *De Caelo*. But historically the coherence of the *De Caelo* as a set of *logoi*, the definition of its topic(s), and the relation of its arguments to those of the *Physics* have been problematic. A consideration of the *Physics* and the *De Caelo* as topical investigations solves both problems, substantive and historical.

The Topic of the *De Caelo*

In the *Physics*, Aristotle investigates strictly defined topics.[1] *Physics* II, 1, identifies things that are by nature: animals, their parts, plants, and the elements earth, air, fire, and water (192b9–11). Nature is a principle of "being moved and being at rest in that to which it belongs in virtue of itself" (192b21–22); consequently, Aristotle asserts at *Physics* III, 1, we must know motion if we are to understand nature, and he lists the "common and universal things" without which motion seems to be impossible (200b20–23). The investigation of proper things, he says, will come later because universal things, including the continuous, the infinite, place, void, and time should be investigated first (200b23–25). Investigation of these universal things occupies *Physics* III through VI. *Physics* VII establishes that "everything moved must be moved by something," and *Physics* VIII first establishes that motion in things must be eternal and then resolves the most serious objections to this view.[2]

1 I have argued this case at length in *Aristotle's Physics and Its Medieval Varieties*, 2–13; for *Metaphysics* XII as topic oriented, cf. H. Lang, "The Structure and Subject of *Metaphysics* Λ," 257–280.
2 On these two *logoi*, cf. H. Lang, *Aristotle's Physics and Its Medieval Varieties*, 35–94.

As we have seen, one of the terms required by things in motion is place – "the where" of all things that are. Place is a "universal and common thing" in the sense that it is a single, unique, constitutive principle that renders the cosmos determinate in respect to where and thus defines place for all things, both natural and made, insofar as they are composed of the four elements. (Void is rejected and plays no positive role in Aristotle's physics.) Place immediately raises special problems concerning both the elements and the all, i.e., the relation between place and the elements, as well as how the heaven is or is not in place. Consequently, the elements and the heaven are included among things that are by nature, and *Physics* IV reaches conclusions that bear upon them precisely for this reason. But the immediate problem of *Physics* IV, 1–9, is "the where" of all things that are. An investigation of the elements in and of themselves would in effect substitute the elements for "the where" as the topic of investigation. Because the topic, "the where," strictly defines the arguments here, both the elements and the heavens enter into the argument only insofar as they are relevant to this topic. And no further. Consequently, as specific topics of investigation, they remain unexamined within the accounts of nature, motion, and "the where."

Physics VIII asserts the thesis that motion in things must be eternal. The resolution of an objection against this thesis involves the claim that "everything moved must be moved by something," and the elements appear as a specific case of moved things that require a mover (*Physics* VIII, 4, 255a–255b31). Again, the examination of the elements is defined by the thesis at hand. Thus they are considered only insofar as, like all moved things, they must be moved by something.

In these arguments – and I would suggest that this methodological point is true (with a few exceptions) throughout the corpus – Aristotle does not drop his announced topic in order to take up a subsequent topic, even when the subsequent topic seems pressing.[3] The primary topics of the *Physics* are nature, motion, common and universal things without which motion seems to be impossible, and finally two specialized problems concerning motion in things.

But even if one grants the topical character of Aristotle's arguments and identifies the particular topics of *Physics* II through VIII, the question remains: why do the elements fail to form a topic in the *Physics*? They are universal, insofar as they both comprise all natural things and artifacts and are regularly listed among "things that are by nature" (*Physics* II, 1, 192b10–11, 19–20). And they seem an obvious, even necessary, topic for an investigation of nature and/or common and universal things without which motion would be impossible.

---

3 I have argued this point in respect to *Metaphysics* XII in H. Lang, "The Structure and Subject of *Metaphysics* Λ," 273–275, 280.

The difficulty with the elements as a topic lies in their definition as things that are by nature. Being by nature, each element must contain an intrinsic source of being moved and being at rest. But in *Physics* II, 1, Aristotle specifies the nature of each element as unique: it belongs to each element to be moved to its proper place, e.g., fire by nature is carried upward and earth downward (192b36). We shall see later that being oriented toward its respective natural place is the very nature of each element.[4] Here is the crucial point: the elements share neither a common nature nor a common motion.[5] Consequently, although they are included among things that are by nature and all things within the cosmos are composed of them, there is no one universal element, nor is there a universal nature that is shared by the elements. Hence, an investigation of them cannot be of something common and universal but must in some sense bear on what is unique and proper to each. For this reason, the elements are not included among the common and universal things investigated within the *Physics:* an investigation of the elements must be an investigation of proper things.

Although in *Physics* III Aristotle promises that an investigation of proper things will come later, he never identifies such an investigation.[6] What are "proper things"? Contrasting with "common and universal things," this phrase can mean private, personal, or idiosyncratic; but Aristotle often uses it to mean what pertains properly to the individual as appropriate or specific.[7] In fact, in *Physics* IV, 2, he contrasts "common place" with "proper place" in just this sense (209a32–33); in *Physics* III, 1, it has the same sense: whatever is specific to things in motion rather than what is common and universal for all things in motion.

The *De Caelo* begins with the words "the science of nature" and clearly continues the project of physics.[8] But here Aristotle emphasizes that the science of nature is concerned with bodies, magnitudes, the affections of

---

4 Cf. *Physics* VIII, 4, 255a3–10, where Aristotle contrasts violent with natural motion for each element, i.e., away from or toward the respective natural place of each, and argues that the elements cannot be animate because they are naturally moved in one direction only.

5 Freudenthal, *Aristotle's Theory of Material Substance*, 2, claims that because each element is moved naturally toward its natural place, "any composite substance should instantly disintegrate, with its components flying off upward or downward." He cites the argument of Gill, *Aristotle on Substance*, 166–167. But although he is certainly correct that the elements do not share a common nature, neither this nor subsequent problems that he raises are found in Aristotle. These problems lie beyond this study, but I believe that a topical analysis of the *De Caelo* and the *De Generatione et Corruptione* would "dissolve" them.

6 Aristotle, *Physics* III, 1, 200b24–25; Ashley, "Aristotle's Sluggish Earth, Part I," 10, refers to the *De Caelo* as presenting "special" problems "after the general problems of the *Physics*," but he does not connect this claim to the investigation promised at *Physics* III, 1.

7 Aristotle, *Topics* V, 1, 128b25; *Metaphysics* IV, 2, 1004b11, 15; XIII, 3, 1078a7; *De Anima* I, 1, 402a9, 403a4. This usage dominates the entire discussion of sensation in the *De Anima;* cf. *De Anima* II, 6, 418a11, 24; III, 3, 428b18; 6, 430b29.

8 Aristotle *De Caelo* I, 1, 268a. For a history of the *De Caelo*, the origins of its name, etc., cf. Elders, *Aristotle's Cosmology* , 59–60. Elders, 61, notes the close relation between the subject of the *De Caelo* and Theophrastus's *Physics*. But he claims that in Aristotle's *Physics* "the nature of an elementary body can only insofar be called a source of motion as it relates

them, their motions and the principles that pertain to them as such (*De Caelo* I, 1, 268a-4). He returns to topics, such as the continuous and magnitude, raised in the *Physics*, reviews his earlier conclusions, and develops them further as applied to body.[9] Thus the *De Caelo* presupposes the conclusions of the *Physics* to continue the science of nature, and in this sense its arguments are subordinated to those of the *Physics;* but in the *De Caelo*, Aristotle turns directly to bodies: an investigation of the elements begins almost immediately, as he takes up "the parts according to its [i.e., the whole's] form" (*De Caelo* I, 2, 268b13–14).

Indeed, examination of the natural motion specific to each sublunar element proves that there must be a fifth element, aether, and that there can be but one cosmos. Hence the investigation of the *De Caelo* accounts not for something "common," but for each element according to its kind, i.e., as light or heavy and as moved to its respective proper place. Here is the contrast between the topics of the *Physics* and those of the *De Caelo:* the *Physics* investigates "common and universal" things, which include a reference to body, whereas the *De Caelo* investigates body as such, i.e., "proper things," including the parts of the whole, beginning with the first body and concluding with an account of inclination, the very nature of the sublunar elements. Body is associated with particulars, and here the investigation includes not only the heavenly body, but sublunar bodies, heaviness, lightness and inclination. In short, the *De Caelo* investigates "proper things," including the elements.

Historically, the status of the *De Caelo* has often seemed problematic. Nussbaum takes the opening of the *De Caelo* as dividing the study of nature into three sharply defined branches: "the study of first principles (the *Physics*), the study of bodies and magnitudes (*DC* [*De Caelo*] and *GC* [*De Generatione et Corruptione*]) and the study of things having body and magnitude – e.g. plants and animals . . ."[10] Setting aside her third category, which clearly leads to biology, I am suggesting, first, that the *Physics* and the *De Caelo* are both investigations within the science of physics and, second, that within physics, as Aristotle suggests at *Physics* III, 1, an examination of "common and universal things" comes first (the *Physics*) and that of "proper things" comes later (the *De Caelo*).[11]

---

it to an external mover, so that it may undergo its causal influence. . . . In the *De Caelo* the elements appear to be self-moving bodies. It is perhaps not impossible to make this theory agree with that of the *Physics*. . . . Yet there can be no doubt that the trend of the arguments of the *De Caelo* as such is different," 30. Solmsen, *Aristotle's System of the Physical World*, 128–129, sees the *Physics* and *De Caelo* as contrasting with one another – the *Physics* presenting place as surrounding and the *De Caelo* presenting a "theory of natural places in its cosmological context." I shall argue against these views.

9 Aristotle, *De Caelo* I, 1, 268a7 ff; cf. *Physics* VI.

10 Nussbaum, 109; cf. *De Caelo* I, 1, 268a1–6, which Nussbaum cites.

11 Although the issue lies beyond this study, I believe that the *De Generatione et Corruptione* continues the work of the *De Caelo*.

In his extended introduction to the *De Caelo,* Moraux raises a number of problems concerning the object and structure of the *De Caelo* as a treatise. These problems are part of the tradition of this text, and he reviews various positions concerning them. The primary question reflects on the coherence of the *De Caelo* (and ultimately of Aristotle's cosmos) as a whole: why should a book "concerning the heavens" include a study of the four elements?[12] And Moraux concludes that we should abandon the project of trying to define "in a word" the proper object of the four *logoi* that we know as the *De Caelo.*[13]

His conclusion requires two assumptions: (1) that the *De Caelo* possesses the heavens as its topic and (2) that the elements constitute a different topic. Moraux gives extensive historical evidence for these assumptions – they go back at least to the Byzantine commentators. But the Byzantine tradition has its own vested interests and its assumptions often represent these interests rather than Aristotle's arguments.

(1) The opening line declares that the *De Caelo* concerns nature, and Aristotle defines the problem at hand as the parts of the cosmos according to the form of the whole. The first part is aether, the first body. A proof of the necessity and nature of this part, which appears first, rests on an examination of the character of the four elements, earth, air, fire, and water as parts moved within the cosmos. As we shall see, the elements play an important role not only in the proof of the necessity of aether but also in the proof that there can be only one heaven (*De Caelo* II, 12). Hence the division of the *De Caelo* into two parts and two separate topics, the heaven and the four elements, is not borne out by the arguments themselves.

Taking the *De Caelo* as the investigation of "proper things" solves the traditional problem of the coherence of the *De Caelo.* The four *logoi* of the *De Caelo* resemble the *Physics,* namely, they form a series of closely related topical investigations, included within physics because they concern things that are by nature and so contain a principle of being moved and being at rest. Just as the arguments occupying *Physics* III-VI are related by virtue of considering common and universal things without which motion seems to be impossible, so the arguments of *De Caelo* I-IV concern body and magnitude, proper things, and are united by considering the parts according to the form of the whole, taking the most important part first.

(2) The problem at stake here is not only textual; it reflects a problem with the traditional understanding of Aristotle's cosmos and hence nature as a principle of order. Moraux asks why the elements should be considered in a book entitled "concerning the heavens" because he assumes that the heaven and the elements are different topics – sufficiently different that it is unclear why they should be included within a single set of *logoi.* Again, this view is not immediately reflected in the *De Caelo,* if we

12 Moraux, vii-xxviii.   13 Moraux, xxvii.

consider its topical character. The *De Caelo* concerns the parts of the whole, and, although they differ in their status as parts, the fifth element and the four sublunar elements are the same insofar as all are parts. All operate within the whole: aether and the four sublunar elements are alike in being by nature and so being moved. The fifth element, aether, is moved eternally with a circular locomotion, whereas the other elements experience limited rectilinear motion – and this is a serious difference between them. But it does not override the fact that as moved *all* the elements are construed according to the same causal model: Aristotle's definition of motion as the actualization of the potential insofar as it is such. Indeed, for this reason they can be examined as a group even though each is unique and specific of its kind.

In recent scholarship, the *De Caelo* has received considerably less attention than has the *Physics*. This situation may reflect the view that its topics or arguments are somehow more "physical" or less "philosophical" than those of the *Physics*.[14] Graham, who claims to give an overall interpretation of the entire Aristotelian corpus (including the "physical works"), refers to the *Physics* more than seventy times, often including discussions of it in his main text; but he refers to the *De Caelo* only eight times, mentioning it only once in his main text.[15] Scholars who do discuss the *De Caelo*, often either focus on the absence of any reference to the unmoved mover – with an apparent implication that the first heaven is self-moved – or stress that the theory of homocentric spheres that appears in *De Caelo* II, 12, "is essentially mechanical";[16] hence, they conclude, there is a problem with the relation of this account of motion and moved things with teleological accounts found elsewhere in Aristotle.[17]

I shall address these issues later, but wish here to make two methodological points. (1) Before the absence of any reference to the unmoved mover can be taken as significant in any way, it must be shown that this mover is relevant to the topic being investigated and/or the arguments themselves. (2) To show that Aristotle has a mechanical account of heavenly motion, it is inadequate merely to point to a "theory of homocentric spheres"; one must also show that the causal relations among the spheres are mechanistic.

---

14 Although he does not characterize the *De Caelo* as a whole – indeed, he refers to it only five times – Freudenthal speaks of "an essential postulate of Aristotle's physics that each element has a natural movement toward its natural place"; surely he would not characterize the *De Caelo* as a mathematical treatise nor the account of the elements as a mathematical deduction. Freudenthal, *Aristotle's Theory of Material Substance*, 2.

15 Graham, *Aristotle's Two Systems*, 15, 236, 348, 350–351.

16 For an account that takes the mechanistic movement of the heaven for granted and tries to place it within a context of proving god, cf. Gomez-Pin, *Ordre et substance*, 103–104.

17 Cf. Easterling, "Homocentric Spheres in *De Caelo*," 138–153, esp. 142; Kosman, "Aristotle's Prime Mover," 135–153; Judson, "Heavenly Motion and the Unmoved Mover," 155–171; Guthrie, xi–xxxvi, is a standard guide here.

I shall argue that the topics of the *De Caelo* do not require a causal reference to a first mover and that Aristotle's arguments are not mechanistic but teleological. Like the arguments of *Physics* IV, they are best understood as requiring, and resting upon, intrinsic relations. Hence, nothing follows from the absence of any reference to a first mover, and there is no contradiction between the "mechanistic account" of elemental motion (including aether) in the *De Caelo* and "teleological accounts" found in other texts, primarily the argument for the unmoved mover in *Metaphysics* XII, 7.

An examination of the elements requires an account of ῥοπή, which I shall argue is nothing other than the very nature of each element. I translate ῥοπή as "inclination"; it is often translated as "impetus," which is quite misleading.[18] Impetus derives from the Latin term meaning a violent impulse and connotes vehemence or passion; but ῥοπή is exclusively associated with natural as opposed to violent motion. It derives from the verb ῥέπω, which is used almost exclusively in the present and imperfect tenses and means to be directed toward or to sink the balance of the scale.[19] "Inclination" best captures this sense; for example, if we say that the famous tower at Pisa inclines at such and such an angle, we mean that it is directed downward at that angle continuously, but neither impetuously nor impulsively.

The problems with ῥοπή do not stop with its translation. Zimmerman claims, for example, that for Aristotle it means "tendency," which he identifies with weight, and he concludes that "Aristotle had balked at making weight a mover."[20] I shall argue that "inclination" does not mean weight and there is no evidence of any "balk." Indeed, as I shall conclude, the account of the elements and inclination in the *De Caelo* is in full agreement with the account of nature and things that are by nature in *Physics* II, 1. Inclination is nothing other than an ability to be moved toward or to rest in its proper place: the nature of each element. Because the elements are simple and constitute all natural things as well as artifacts, inclination lies at the heart of Aristotle's natural world and hence his physics.

In the *De Caelo*, we find that "inclination" occurs seven times.[21] The first occurrence reports a view attributed to Empedocles, but rejected without comment by Aristotle, and I shall not consider it.[22] The remaining six

---

18 So, e.g., Stocks translates ῥοπή as "impetus" at 301a22, 24, and 307b33, although he uses "tendency" at 284a25 and 305a25, whereas at 297a28 and b7 he translates it as "impulse."
19 For an example of "inclination" as continuous action, cf. Pindar, *Odes* 8.23.
20 Zimmerman, "Philoponus' Impetus Theory in the Arabic Tradition," 121; the issue of "weight" has a long history; cf. O'Brien, "Aristotle's Theory of Movement," 47–57.
21 These texts are *De Caelo* II, 1, 284a25; 14, 297a28, b7, 10, 14; III, 2, 301a22, 24; 6, 305a25; and IV, 1, 307b33.
22 At *De Caelo* II, 1, 284a25, Aristotle refers to Empedocles, who explains (Aristotle says) that "the world, by being whirled round, received a movement quick enough to overpower its own downward inclination . . . ," although this passage is not generally treated as a

occurrences appear in four separate arguments, which I shall examine in the order of their appearance. In these arguments, a clear account of inclination as the intrinsic principle of motion in each of the four elements emerges. At the same time, a view of the cosmos, i.e., "the whole" (τὸ πᾶν), also emerges. The *De Caelo* concludes with a return to place as established in *Physics* IV and a specific reference to the argument concerning elemental motion in *Physics* VIII, 4.

Given Aristotle's definition of motion as the actualization of the potential *qua* potential, an account of potency and actuality is essential to an account of the elements, their respective motions, and inclination. Hence in the analysis of inclination in *De Caelo* IV, I shall turn to the investigation of potency and actuality in *Metaphysics* IX. This account, I shall conclude, forms a coherent pattern with the arguments of both the *Physics* and the *De Caelo*.

In *Metaphysics* IX, 8, Aristotle reaches a remarkable conclusion:

The imperishables are imitated by those involving change, such as earth and fire. For these too are ever active; for they have motion in virtue of themselves and in themselves. [1050b28–30: μιμεῖται δὲ τὰ ἄφθαρτα καὶ τὰ ἐν μεταβολῇ ὄντα, οἷον γῆ καὶ πῦρ. καὶ γὰρ ταῦτα ἀεὶ ἐνεργεῖ· καθ' αὑτὰ γὰρ καὶ ἐν αὑτοῖς ἔχει τὴν κίνησιν.]

I shall argue that this conclusion, reached in the context of its topic, i.e., potency and actuality, also applies to the investigation of the elements in the *De Caelo*. On both accounts, the cosmos is unified because its parts go together "by nature" with the result that the cosmos as a whole presents order as its primary feature and this order is teleological, i.e., rests on an immediate active intrinsic orientation of what is potential toward its proper actuality.

The elements, I shall conclude, are "ever active" because each is by nature nothing other than inclination: an orientation toward respective proper place or resistance against being moved out of (or away from) that place. Such place is constituted by place as a limit because it renders the cosmos determinate in respect to "where." Inclination is an intrinsic ability in each element to be moved toward its proper place as found in nature; hence inclination too is an immediate constitutive principle of all natural things or artifacts made from natural things. Therefore, an intrinsic relation obtains between the elements and the structure of the cosmos in respect to the category "where." Herein lies both the unity of the

---

legitimate fragment; Diels and Kranz, *Die Fragmente der Vorsokratiker*, mention this reference but do not treat it as a fragment; Empedocles A 30 Nach. (I, 499, 6). This text is not discussed at all, e.g., in Wright, *Empedocles: The Extant Fragments*. The word ῥοπή does seem to appear in Anaxagoras's cosmology, and we might speculate that Aristotle himself is not altogether original in his use of this word, although the evidence is certainly very slim. Cf. Diels and Kranz, *Die Fragmente der Vorsokratikers*, Anaxagoras B 10 (II, 37, 10).

topics examined in the *De Caelo,* the *Physics,* and *Metaphysics* IX and the order of the cosmos constituted by nature.

Physics II, 1, begins an investigation of nature and when, at the end of the *De Caelo,* the account of the motion of each element toward its natural place is complete, a full view of nature emerges. Nature is a principle of order for a determinate cosmos, bounded by place, as the determining principle of "where" things are, and having earth as its motionless center. The stars, constituted by aether, exhibit eternal circular locomotion, and the four sublunar elements, which imitate the heavens, are potentially up, down, or in the middle in relation to their proper actuality, which is the up, the down, or the middle of the cosmos as constituted by place. The elements – inclination being their intrinsic principle of being moved toward or at rest in their respective natural places – are "by nature," and all other moved things are constituted by them. Hence by establishing first place as "the where," something universal and common without which motion in things would be impossible, and then inclination as the proper nature of each element, Aristotle provides an account of nature and things that are by nature.[23]

If one anticipates an explicit systematic conjunction of Aristotle's *logoi,* then it appears that he fails to provide it. But if one recognizes first their topical character and then the relation among the topics, i.e., they are ordered from "common and universal things" to "proper things," then the logic of the larger argument appears: the proper presupposes (and in this sense is subordinated to) the more common and universal. Consequently, although as topical the investigation of "proper things" is separate, nevertheless it is not independent of that of "common and universal things."

Aether: The Fifth Element, the First Body

As we have seen, the *De Caelo* begins by affirming that the investigation concerns nature, i.e., the science that for the most part concerns bodies and magnitudes, their affections and motions and the principles, however many, of this sort of substance.[24] As in the *Physics,* the announced topic defines the questions to be addressed, i.e., the parts of the all according to the whole.[25] And the highest part comes first: there must be a fifth element.

---

23 Aristotle, *Physics* IV, 4, 211a11; cf. also *De Anima* I, 1, 402b25.
24 Aristotle, *De Caelo* I, 1, 268a-4. For a history of how the subject of the *De Caelo* was understood by both Neoplatonists and Byzantine commentators, cf. Moraux, vii ff., and Elders, *Aristotle's Cosmology,* 59–60. For a brief survey of modern treatments of the *De Caelo,* including efforts to date it, cf. Bos, *On the Elements,* 5–22. Bos develops his own genetic account of the *De Caelo* and its relation to the *Physics.*
25 Aristotle, *De Caelo* I, 2, 268b13–15: περὶ δὲ τῶν κατ' εἶδος αὐτοῦ μορίων νῦν λέγωμεν ἀρχὴν ποιησάμενοι τήνδε. On the problem of the translation of the phrase τῶν κατ' εἶδος αὐτοῦ μορίων, cf. Matthen and Hankinson, "Aristotle's Universe," 434, n. 10, who argue

Matthen and Hankinson suggest that this argument constitutes "a derivation of the theory of the elements from first principles."[26] Although an exact meaning of "first principles" is unclear in their account, this argument clearly rests on several important points established in the *Physics*. (1) "For we say all bodies and magnitudes are moved in virtue of themselves according to place; for we say that nature is a source of motion for them."[27] (2) All motion insofar as it is according to place, which we call locomotion, is either straight or circular or a combination of these; for these two are the only simple motions.[28] (3) A thing is moved according to nature when it is moved toward and remains in a place without force; conversely, if it is carried toward and/or held in a place by force, the change of place is contrary to nature.[29] So the motion (or rest) of each element is natural when it is moved toward (or remains in) its natural place and is contrary to nature when it is moved (or held) away from that place. Thus fire is moved naturally upward toward the periphery (and air too, although air is a middle element and is moved upward only relatively), while earth is moved naturally downward (and water too, but again as a middle element). The present investigation of the elements puts these points to work.

Aristotle first argues that there must be a fifth element and this element constitutes the first containing body, of which place is the limit. The necessity of a fifth element follows as a direct consequence of the nature of "simple bodies," i.e., the elements, which "have a source of motion according to nature, such as fire and earth, and their forms and those things akin to them (*De Caelo* I, 2, 268b27–29)." The fact of simple bodies, Aris-

---

persuasively that traditional translations such as "formally distinct parts," which presumably mean "parts distinct from one another in form," must be set aside in favor of "the natural reading [which] is 'parts of the totality in virtue of the form of the totality.'"

26 Matthen and Hankinson, "Aristotle's Universe," 421.

27 Aristotle, *De Caelo* I, 2, 268b14–16; cf. also *Physics* II, 1, 192b13–15, 20–23; III, 1, 200b12–13. Elders makes a remarkable claim in respect to this line: "By 'nature' Aristotle means the specific nature. Yet it would seem that we must understand this as the specific nature which is dependent on place in the cosmos, and thus depends on cosmic order," *Aristotle's Cosmology*, 84. He then distinguishes this formulation from that of *Physics* II, 1, 192b21–23. It is difficult to see how he can derive the first claim from this line, although I shall argue that in a certain sense it is true. I shall also argue that Aristotle's formulation here is perfectly consistent with that of the *Physics*. Elders also claims that this definition of nature is specifically anti-Platonic; I have argued this for the account of motion in *Physics* VII, 1, in H. Lang, *Aristotle's Physics and Its Medieval Varieties*, 35–62.

28 Aristotle, *De Caelo* I, 2, 268b17–18; cf. *Physics* VIII, 8, 261b29–30.

29 Aristotle, *De Caelo* I, 8, 276a22–24: ἅπαντα γὰρ καὶ μένει καὶ κινεῖται καὶ κατὰ φύσιν καὶ βίᾳ. κατὰ φύσιν μέν, ἐν ᾧ μένει μὴ βίᾳ καὶ φέρεται, καὶ εἰς ὃν φέρεται, καὶ μένει· Compare this line with *Physics* VIII, 4, 254b20–22: καὶ τῶν ὑπ' ἄλλου κινουμένων τὰ μὲν φύσει κινεῖται τὰ δὲ παρὰ φύσιν, παρὰ φύσιν μὲν οἷον τὰ γεηρὰ ἄνω καὶ τὸ πῦρ κάτω. The two lines, the divisions into natural and constrained motion, and even the examples that follow, i.e., fire and earth, are virtually identical – and the verb in *Physics* VIII, 4, is unequivocally passive.

totle argues, implies simple motions, and if the motion of a simple body is simple and if a simple motion is of a simple body, then there must be some simple body that by nature is carried in a circular motion according to its nature (268b29–269a). Although it may be moved by force with some other motion, this is impossible according to nature if there is one natural motion for each of the simple bodies (269a7–9). So, for example, if the motion is upward, the body is fire or air, and if downward, water or earth, and likewise the motion that is natural to one element will be unnatural to the others (269a17–18; 32–269b). From the relation between the elements and their motions, the necessity of a fifth element follows directly: circular locomotion is a distinct natural motion; every simple motion belongs to a unique simple body; therefore, circular locomotion must belong to a simple body as its unique natural motion.

This account is not complete, however. The necessity of a fifth element raises further problems about the nature of the element itself. Because these problems concern aether, they concern a part of the cosmos, and so Aristotle takes them up directly. As we shall now see, the fifth element possesses a special status within the cosmos: it is the first body and more divine than the others.

In *Physics* IV, 4, when Aristotle defines place as the first limit of the containing body, he says virtually nothing about that body. But an account of it is required, if the account of place is to be fully coherent. Here in the *De Caelo*, in the investigation of the parts of the all, he provides the necessary account: the fifth element, aether, is the first body, i.e., the body that possesses place as its immediate limit. Hence, the account of aether is an account of the first body referred to in the definition of place in *Physics* IV.

Aristotle first argues that the fifth element is neither heavy nor light, and he briefly defines these terms: "Let heavy then be what is moved naturally toward the middle, while light is what is moved away from the middle" (*De Caelo* I, 3, 269b23–24). The heaviest is what most of all is carried downward and the lightest upward (269b24–26). And what is light or heavy can be such relative to other things, e.g., air is lighter than water and water lighter than earth (269b26–29). So, Aristotle concludes, the fifth body is neither heavy nor light and thus is moved neither away from nor towards the middle (269b29–32).

In short, the primary characteristic of the four sublunar elements, i.e., being heavy or light, is by definition nothing other than the ability to be moved each to its respective proper place upward or downward. Hence, the determination of the cosmos as up, down, and middle, the natural motion of each element, and the definition of an element as light or heavy are intimately related. Furthermore, Aristotle uses this relation to conclude that the fifth element is moved neither up nor down *because* it is neither heavy nor light. Consequently, whereas the other elements are moved *within* the cosmos, each to its respective proper place, aether is not moved within the cosmos, and in this sense aether is both unmoved and is never "out of place." As we shall soon see, the account of each element as possessing a single unique inclination and as exhibiting its respective natural motion within the cosmos is an account of the relation between the directional determination of the cosmos and the characteristics of the elements as heavy or light.

Before turning to the sublunar elements, we must consider the account of aether, the fifth element, as it relates to the account of the containing body referred to in the definition of place. As place is the first limit of the containing body, aether is the first body. And Aristotle gives a full account of it. Aether is prior to the other elements and more divine (*De Caelo* I, 2, 269a30–32). Circular motion is natural to this body, and this motion is able to be continuous and eternal.[30] Consequently, aether is different and separate from the things around us, having a more honorable nature (269b14–17). And the perfections of the first body follow: it is neither heavy nor light; it is neither generated nor destructible; it neither grows nor diminishes nor alters in any way; and, perhaps most importantly, there is no contrary to its motion, i.e., circular locomotion (*De Caelo* I, 3–4). Circular locomotion is the first, highest motion, and it must be both without a contrary and unique. Were we to suppose a contrary, it would be purposeless; but god and nature make nothing "in vain."[31]

A modern impulse toward systematization is required to combine topics that Aristotle distinguishes as belonging to different accounts. If we follow this impulse, the consistency between the two accounts readily appears. Place and aether, the first limit and the first containing body, share important characteristics. Both are first and both are separate. Place is separate in the sense of independent, and it is apart in definition, although it can only actually be as the limit of the containing body; aether, which as body must be limited and presumably possesses place as its limit, is separate from the four bodies that surround us *both* in the sense that it constitutes the heaven whereas they constitute the sublunar world *and* in the sense that the first body is apart from change except for the highest motion, circular locomotion. Aether, exhibiting only the first and highest motion, is bounded and so completed by what is immovable, i.e., place.

Most importantly, the priority of the first body is a priority that it possesses by nature. The complete, i.e., a circle and hence circular locomotion, is by nature prior to the incomplete, i.e., a line (*De Caelo* I, 2, 269a19–20). Circular locomotion is natural to the first body and, whereas what is unnatural passes away most quickly, what is natural is able to be continuous and eternal (269b2–16). Place is a principle of order rendering the

---

30 Aristotle, *De Caelo* I, 2, 269b7; this text agrees with *Physics* VIII, 8.
31 Aristotle, *De Caelo* I, 4, 271a33: ὁ δὲ θεὸς καὶ ἡ φύσις οὐδὲν μάτην ποιοῦσιν.

cosmos determinate in respect to "the where," and the first body exhibits the highest degree of order – eternal, continuous, circular locomotion. As the power of place is something remarkable, so the body of which place is the limit is the most divine of all bodies. In short, the first body is perfectly fitted to the first limit.

But the account of aether as the necessary fifth element has been the source of considerable debate beyond the problem of its nature and relation to place. Examining the history of the idea of impressed infinite power, Sorabji raises the most serious issues in the debate:

> There is another reason for wondering why infinite power should be needed for everlasting celestial rotation. In an earlier treatment of celestial motion in *On the Heavens*, we find no integral and undisputed reference to the role of God. Instead, Aristotle combines two ideas, that the heavens are made of an indestructible fifth element which can undergo no change but motion and that circular motion is natural to that element. If it can undergo no other change, what reason could there be for its natural motion to cease? . . . It looks, then, as if in his earlier work Aristotle allowed for a motion that was eternal not because of any infinite power, but because of immunity to further change in something to which circular motion was natural.[32]

The last issue raised by Sorabji evokes a conclusion about the chronology of the *De Caelo* based on the absence of a reference to god – namely, that it is an "earlier" work.[33] Beyond chronology lie different substantive conclusions concerning god and self-motion in *De Caelo* I. Kosman argues that the motion of the heaven must be a form of self-motion; he ultimately seems to conclude that the unmoved mover of *Metaphysics* XII and *Physics* VIII is the soul of the first heaven in the *De Caelo*.[34] Judson too argues that "the idea that the sphere is made of stuff that naturally moves in a circle is not incompatible with the idea that the outermost sphere is moved by its soul."[35] Cherniss argues that "if the natural motion of the fifth essence is circular translation and if this sphere has no potentiality implying the contraries of rest and motion, the influence of the prime

---

32 Sorabji, "Infinite Power Impressed," 197. Elders, *Aristotle's Cosmology*, 95, commenting on *De Caelo* I, 3, makes the point absolutely explicit: "In the *De Caelo* . . . Aristotle generally only speaks about bodies which have the power to move themselves." Sorabji raises this issue because it has a long history in a variety of forms. On this history, cf. also Sharples, "The Unmoved Mover and the Motion of the Heavens in Alexander of Aphrodisias," 62–66.

33 Jaeger, *Aristotle*, 300 ff.; Guthrie, xv-xxxvi; "The Development of Aristotle's Theology," 162–171; Ross, 94–102; Bos, *On the Elements*, 89–91.

34 Kosman, "Aristotle's Prime Mover," 138–158. Kosman claims, 139, to propose a new reading; however, a form of this thesis (that the mover of the *Physics* is soul) was proposed some years ago, first by de Corte, "La Causalité du premier moteur dans la philosophie aristotélicienne," 105–146, and Paulus, "La Théorie du premier moteur chez Aristote," 259–294 and 394–424; these arguments are criticized and rejected by Pegis, "St. Thomas and the Coherence of the Aristotelian Theology," 67–117.

35 Judson, "Heavenly Motion and the Unmoved Mover," 157.

mover upon it is gratuitous."[36] The view that this argument in the *De Caelo* implies that the heavens are self-sufficient and without need of an unmoved mover is rejected by Nussbaum, although she offers no argument and concedes that "no one could deny that there is far less attention given to the unmoved mover, far more to the natural motion of the spheres, than in *Physics* VIII, *Metaphysics* XII or the *MA* [*De Motu animalium*]."[37] Gomez-Pin connects the proof for aether to the notion that aether has divine characteristics, which he ultimately relates to the proof of a first mover in *Physics* VIII, 6.[38]

Not by accident is the argument of *De Caelo* I, 2 regularly related to those of *Physics* VIII and *Metaphysics* XII. *Physics* VIII explicitly argues that "everything moved is moved by something," including the elements, and both *Physics* VIII and *Metaphysics* XII argue that the first motion of the world requires a first unmoved mover.[39] But recognition of the topical character of Aristotle's *logoi* provides a general pathology of the problems: because these texts constitute different investigations – *Metaphysics* XII investigates substance, *Physics* VIII the eternity of motion in things, and the *De Caelo* the parts of the all – they present different interests and different arguments. These differences can be clearly defined.

(1) The purpose of the argument here in *De Caelo* I, 2, is to show that there is some body separate from the other four and of a higher nature. The argument rests on the relation between the elements and their proper motions. Mover/moved relations are not mentioned because they are not at issue in this argument.

(2) In *Physics* VIII, 4, Aristotle takes up mover/moved relations, arguing that "everything moved must be moved by something"; he examines both animals and the elements, natural as well as unnatural, or violent, motion (254b14–15). After considering all the permutations of these categories, he concludes that in each case "everything moved must be moved by something" (256a2).

Here in *De Caelo* I, 2, an examination of the elements as parts of the all leads to the conclusion that aether exhibits a natural, eternal, and continuous motion. Because mover/moved relations do not constitute the topic here, this conclusion remains unspecified as to whether it is uncaused, self-moved, or moved by another. But if we combine this argument with that of *Physics* VIII, 4, we must conclude that it cannot be uncaused because everything that is moved must be moved by something, although we would still not know whether it is self-moved or moved by another. In *Physics* VIII, 6, and *Metaphysics* XII, 7, Aristotle argues that the fact of nat-

36 Cherniss, *Aristotle's Criticism of Presocratic Philosophy*, 183.
37 Nussbaum, 131.
38 Gomez-Pin, *Ordre et substance*, 103.
39 Aristotle, *Physics* VIII, 6, 260a15–18; 10, 267a21–267b25; *Metaphysics* XII, 6, 1072a15–18; 7, 1072a23–24; 1073a4–9.

ural locomotion implies that there must be a first mover that produces this motion eternally and continuously; and if we combine these arguments with that of *De Caelo* I, 2, then we are entitled to draw the obvious, if unstated, conclusion: aether requires an unmoved mover. This conclusion does not appear in *De Caelo* I, 2, because here Aristotle intends to establish the necessity and nature of the fifth element, not a mover/moved relation.

(3) Why does Aristotle "fail" to mention god in this argument? The answer is already clear. The topic here concerns the nature of the whole according to its parts. Hence we see the part, the fifth element, and the motion natural to it. The cause of that motion is a different topic and a different problem, i.e., the problem of an effect/cause relation and the cause required or implied by the effect. That god is omitted here does not mean that Aristotle denies god (or the necessity of a mover) in this argument and is in no way either a failure or an omission. Rather, given the topic at hand, the first mover is irrelevant to the argument. There is not "less attention" given to the unmoved mover here than in *Physics* VIII or *Metaphysics* XII – the unmoved mover does not appear in *De Caelo* I, 2, at all.[40]

Of course, in a general way, Aristotle does mention god in this argument. His account for the reasons why the first of bodies is eternal, unaging, unalterable, and without increase or decrease, both confirms and is confirmed by phenomena; Aristotle argues that all who believe in the gods, Greek and barbarian alike, give the highest place to what is divine "because the immortal, is joined to the immortal."[41] That is, immortal body would naturally be conjoined to what is immortal. And indeed, the problem of that conjunction – a problem taken up in *Physics* VIII and *Metaphysics* XII – yields an account of the relation of the first heaven to the unmoved mover, god.[42]

(4) Kosman seems to identify the limit of the heaven with the "Prime Mover."[43] But place is not mentioned in the account of aether any more than god is. And for the same reason: the topic here is the whole according to its parts, and the first part, aether, comes first. Its necessity and nature form the immediate topic of the account. But if god and place are equally absent from the account, what justifies the rejection of a conjunction of the arguments about aether with those concerning god in

---

40 I speculate that the insistence of Aristotle's commentators on the importance of the absence of god from this argument reflects interests that originate in a Judeo-Christian culture.

41 Aristotle, *De Caelo* I, 3, 270b-9, esp. 8–9; for further argument concerning whether there is evidence for the first unmoved mover in the *De Caelo*, cf. Easterling, "Homocentric Spheres in *De Caelo*," 138–153.

42 Such is the argument at *Physics* VIII, 6, cf. esp. 260a17–19, and *Metaphysics* XII, 6, cf. esp. 1072a10.

43 Kosman, "Aristotle's Prime Mover," 149–150.

favor of a systematic conjunction of this argument with that of *Physics* IV, 4, concerning place? First, the account of god in *Metaphysics* XII requires that god be completely separate from the world and act as an object of desire, whereas the limit of the world cannot be completely separate and would not, presumably, act as a final cause.[44] Consequently, Kosman's argument requires that Aristotle either "changed his mind" or reworked his theory. In short, the arguments of *Metaphysics* XII and *De Caelo* I do not immediately go together and so require additional extrinsic hypotheses.

Second, place is defined as the limit of the containing body, whereas aether is defined as the first body. Indeed, Aristotle concludes his argument about aether with an account of the name of "the first body" – it derives from the fact that it "runs always" (*De Caelo* I, 3, 270b20–24). Place and aether are treated as different topics, and they can be combined only speculatively. I combine them as a "universal and common term" taken together with a special term because this speculation rests directly on the conclusions of the two arguments without the requirement of further hypotheses.

Indeed, the argument about god can also be speculatively introduced here, as long as the topical character of Aristotle's arguments is respected. As the first mover, god produces motion not in the first body, but in the first heaven.[45] Aether is not the first heaven without further qualification; rather, aether is the fifth element and the first body, the first part of the cosmos. Body requires limit, and the limit of the first body is place, which resembles form. Together body and limit, aether and place, presumably form the heaven. Thus god by moving the heaven also, albeit indirectly, moves aether because it is part of the heaven – a problem that may be referred to the discussion of parts in a whole in *Physics* IV.

The Nature of the Cosmos

De Caelo I continues with several arguments against the possibility of an infinite body. Their results are largely negative, and I shall not consider them. Two final problems about the nature of the world remain. The first concerns how many worlds there are, and Aristotle argues that there cannot be more than one. The elements and their motions play a crucial role in these arguments. The second problem, whether or not the world is ungenerated and indestructible, completes *De Caelo* I. As in *Physics* VIII, Aristotle argues that the world can be neither generated nor destroyed, and even the fact that there is but one world provides evidence for this view.

Inclination – however its meaning is finally interpreted – is central to

---

44 Kosman, "Aristotle's Prime Mover," 151–153.
45 Aristotle, *Physics* VIII, 6, 259b30–260a17; 10, 267a21–267b8; *Metaphysics* XII, 1, 1069a30–32; 6, 1071b11–12; 7, 1072a21–26; 1072b14–31; 8, 1074a31–37; 10, 1075a12–24.

elemental motion. Consequently, insofar as the arguments that there is only one heaven complete the account of place and utilize Aristotle's account of elemental motion, they also set the stage for the problem of inclination. I shall consider these arguments insofar as they present an account of elemental motion to conclude that there is only one heaven. Finally, I shall suggest that these points allow us to anticipate the meaning and role played by inclination within Aristotle's account of elemental motion and his physics more generally. Against these conclusions, we can then turn to the first appearance of "inclination" in *De Caelo* II, 14.

The first argument is that there can be only one world. Because the cosmos is differentiated into "up" and "down" and the simple motion of each element must be either away from, toward, or about the middle, the natural motion of each element must be one.[46] Furthermore, fire, for example, or whatever takes the name "fire," would in all cosmoi have the same power that fire has in our world, namely, the power to be moved upward.[47] So fire everywhere is by nature moved upward, i.e., away from the middle, and likewise earth is everywhere by nature moved downward, i.e., toward the middle.[48]

Because the motions of the elements are the same, the elements too must also be everywhere the same (*De Caelo* I, 8, 276b10–11). If there were another world, its elements would be one and the same as those in our world; therefore, because all elements are the same and are moved toward their respective natural places, each element would be moved to its proper place in our world, e.g., earth would be moved to the center of our world (276b11–12). But such a situation is impossible because from the point of view of its own world, earth would be moved upward, i.e., away from its center, just as earth from our world would be moved upward if moved toward the center of another cosmos (276b12–17). In short, the assumption of more than one cosmos entails the denial of the identical natures of the elements and/or the oneness of their respective motions throughout the different cosmoi (276b19–20). But this assumption is intolerable, and Aristotle immediately concludes: there cannot be more than one world (276b21).

The crux of the argument rests on the determinateness *both* of the cosmos in respect to "up" and "down" *and* of each element and its respective natural motion in relation to the cosmos. Each element must be one in nature and hence must always be moved with one motion, e.g., up for fire, down for earth. That the elements and their motions are determined in respect to place, i.e., "where" as up and down, explains why the heaven

46 Cf. Aristotle, *Physics* IV, 4, 211a2–6, 212a25–29; *De Caelo* I, 2, 268b21–24; I, 4; 8, 276a30.
47 Aristotle, *De Caelo* I, 8, 276a30–276b4; for an analysis of linguistic difficulties associated with this line, cf. Verdenius, "Critical and Exegetical Notes on *De Caelo*," 272.
48 Aristotle, *De Caelo* I, 8, 276b4–9; cf. the account of nature in *Physics* II, 1, 192b35–193a2; cf. *Physics* VIII, 4 255a2–5; 255b13–31.

must be one. But the oneness of the heavens follows if and only if the cosmos is determinate in respect to direction and each element is always moved to its own place. The cosmos is rendered determinate in respect to direction by place because it orders the cosmos in respect to "where" and elemental motion is orderly in respect to place.

The second argument that the heavens must be one pursues the claim that each element and its respective natural motion are formally the same. The elements might be thought to differ insofar as they are more or less distant from their proper places – but such a view is unreasonable (276b22–23). Distance from respective proper place can make no difference to the elements; the form of any element is the same regardless of the distance between the element and its proper place (276b23–25). Therefore, each element must be one.

The problem with differentiating (and so defining) each element by means of distance from proper place is that such distance is extrinsic to the element whereas a thing's nature constitutes it intrinsically. The elements are by nature, and such things are defined by form with a reference to matter. Hence distance cannot constitute the nature of the elements; rather, the principle of motion for each element must be identified as its intrinsic nature. This point is crucial to the argument and must be explained satisfactorily if the account of the elements – and the cosmos as composed of them – is to be coherent.

That each element must have some motion is clear because the elements are moved.[49] Can their motions be due to constraint? If so, then they could be moved anywhere, e.g., to various centers of various cosmoi. But their motions cannot be due to constraint. Natural motion is prior to motion by force because motion by force is defined as contrary to natural motion and so presupposes it; indeed, that which is not moved by nature cannot be moved at all by force.[50]

Although unnatural motion is defined in relation to natural motion, natural motion is never defined in relation to any other kind of motion; rather, it is defined in respect to place, and the defining place must be one in form and number too, such as place that is middle or place that is outermost (*De Caelo* I, 8, 276b29–32). Furthermore, place cannot be one in form, but many in number, e.g., different cosmoi, because when things that are formal in their nature are identical, they must be one in number.[51]

Although Aristotle does not refer explicitly to *Physics* IV, his account of

49 Aristotle, *De Caelo* I, 8, 276b26–27; cf. also *De Caelo* II, 6, 288a27–28.
50 Aristotle, *De Caelo* I, 8, 276b28–29. For natural motion as prior to motion by force, cf. *Physics* IV, 8, 215a-5.
51 Aristotle, *De Caelo* I, 8, 276b32–277a9; cf. *Physics* IV, 4, 211b10–13. By implication, what has no natural motion must be unmoved, which accords with Aristotle's account of god in *Metaphysics* XII.

place as the principle rendering the cosmos determinate in respect to "up," "down," and "middle" is presupposed by this argument. And place as a boundary is a container and so resembles not matter but form. When individuals are identical in form but different in number, e.g., different members of the same species, their difference in number can be due only to matter. In the absence of matter, identity of form immediately implies identity of number.[52] The natural motion of each element indicates such oneness of form; it thus implies that the heavens too must be one in both form and number.[53] In short, because each element is one, its natural motion, which is defined solely by reference to place, must also be one. Hence the heaven and place too must be one in form and number.

The third (and last) argument that there can be only one heaven also depends upon the relation between the heavens and place. A consideration of other kinds of motion makes it "clear that there is some 'to which' earth and fire are moved naturally" (277a12–13). What is moved always changes from something into something: "that from which" and "that to which" differ in respect to form, and the change is always limited.[54] So, for example, a person changes qualitatively from sickness to health, and a thing may increase from something small to something great (277a16–17).

Locomotion is no different from alteration or increase: it must have both that from which and that to which a thing is carried by nature, and they differ in form (277a18–19). So fire and earth are moved not to infinity but to opposite points, namely, up and down, and these points serve as the limits of their respective motions (277a20–23). Furthermore, that earth is moved faster as it nears the center and fire as it nears the periphery – the respective natural place of each – shows that local movement both cannot be continued to infinity and must be natural (277a27–277b9). If motion were to become progressively faster and continue to infinity, then the quickness [ταχυτής] as well as the heaviness and lightness would also be infinite (277a29–31). The impossibility of this view is so obvious that it goes without saying.

The opposite is also true: constrained motion always becomes slower as it is carried away from its source. And the places of constrained and natural motion contrast: the place from which a body is moved by force is the place to which it is moved without force, i.e., by nature (277b6–8). Hence, because fire and earth are moved faster as the one is moved upward and the other downward, their motions cannot occur by constraint – and cannot occur by the action of another body either – but must occur by nature.[55] Again, the conclusion is obvious: the heavens must be one –

---

52 This is, of course, the argument that the first mover must be one in definition and number at *Metaphysics* XII, 8, 1074a35–37.

53 Aristotle, *De Caelo* I, 8, 277a10–12; cf. also 9, 278a19–20.

54 Aristotle, *De Caelo* I, 8, 277a14–16; cf. also *De Caelo* I, 4, 271a3–5.

55 Cf. Aristotle, *De Caelo* I, 8, 277b1.

the unique one that provides the starting point and the end point for all natural motion of the elements.

Like its predecessors, this argument unambiguously reveals three major points presupposed by Aristotle's entire account of the elements, their motions and their respective natural places. Indeed, these points show why this argument is subordinated to the account of "where" in *Physics* IV. (1) Place and the elements are irrevocably locked together, and in some sense place functions as actuality for the elements. By rendering the cosmos determinate in respect to "where," place constitutes the "to which" for each of the four elements within the cosmos. Thus, place renders the cosmos determinate in respect to "up," "down," etc.; each element is aimed at, moved toward, and rests in its proper natural place by nature, if nothing hinders. And this argument clarifies further the ambiguity remaining at the end of the account of place in *Physics* IV, 5: the heaven, i.e., the first containing body, must be one in number.

(2) Aristotle conceives of the cosmos as bounded by place, which also must be one in form and number. "Where" things are within the cosmos is thus determinate neither as part of the matter of the cosmos nor merely accidentally; rather, place determines the cosmos in respect to up, down, and middle because it is one in form and number. And place, together with the natural immediate orientation of each element toward its natural place, accounts for the motion of the elements (and hence of all body) *and* the regularity of that motion, e.g., fire by nature is moved always upward and earth downward. This relation underlies the cause/effect relation operating throughout these arguments.

Finally, (3) because the cosmos is one and each element is everywhere identical in nature, the subsequent account of elemental motion and inclination will be universal even though each element is distinct in kind. Hence by definition this account will explain all motion everywhere even though it bears on "proper things," i.e., the four elements, each a unique kind. Aristotle's physics establishes a first determining principle, universal and common for all things, i.e., place, and it also investigates the four elements. Consequently, this physics both achieves the highest standard of science, first and universal principles, and accounts for the world itself in all its multiplicity. Aristotle's physics solves the difficulties associated with these topics, reveals the causes of these difficulties, and shows why characteristics rightly thought to belong to things do in fact belong to them (*Physics* IV, 4, 211a7–11).

Matthen and Hankinson raise a difficult issue concerning the *De Caelo*. They insist that "[t]he line of thought that we find presented in *De Caelo* is meant to be explanatory" and that the elements are being considered as parts of a whole.[56] These points lead them to claim that Aristotle's sci-

---

56 Matthen and Hankinson, "Aristotle's Universe," 419, 421.

ence is "anti-reductionistic" in the sense that "parts are ontologically and causally subordinate to wholes. . . . In Aristotle's system, parts that maintain the nature of the whole are typically treated as deriving their nature from the whole, and as posterior to it in being (however that is to be understood)."[57] A remarkable conclusion follows:

> Returning now to the elements, we have seen that Aristotle defines them in terms of the sphere that constitutes the totality. The performance of their characteristic activities thus presupposes the existence of that whole – they move towards the centre and the periphery of *this sphere*. So they are ontologically posterior to the whole: they cannot pre-exist the sphere, because without the sphere they would lack their defining innate tendencies. Thus the doctrine of natural places is how Aristotle's notions about the explanatory priority of the whole come in the end to be bashed out in his cosmology.[58]

The first two points of Matthen and Hankinson here are in a sense true. The *De Caelo* is explanatory and considers the elements as parts of the whole in the sense that it solves a variety of specific problems associated with the heavens and its parts.[59] But their subsequent view and conclusion fail to recognize that the cosmos as a whole has itself been rendered determinate by place as the first containing limit. Thus, the whole is not itself a first principle of which place is an application; rather the whole is that which has been caused in respect to "the where" by place, which as a cause resembles form. Consequently – and I shall elaborate later on this point – the account of the motion of the elements and their proper natural places is not about the priority of a whole; rather, it is a causal account in which a first principle, itself common and universal, renders the whole determinate, with the result that parts and the determinate whole are ordered to one another as potency and actuality. So, for example, fire is not subordinated to a whole or totality, but is actively oriented toward its respective determinate natural place, up, or the outermost circumference of the heaven. And the circumference of the heavens is "up" because as the first unmoved limit place has rendered the whole determinate in respect of "where" and this determination is expressed within the cosmos as "up," "down," "left," "right," "front," and "back".

The final problem of *De Caelo* I (10–12) concerns the question of whether the heavens are generated or destructible (*De Caelo* I, 10, 279b4–6). And Aristotle concludes unambiguously: it is impossible that the heavens were at any time generated or will at any time be destroyed.[60] Place is the first limit of this ungenerated and indestructible containing body,

---

57 Matthen and Hankinson, "Aristotle's Universe," 426.
58 Matthen and Hankinson, "Aristotle's Universe," 430. (Italics in original.)
59 McCue emphasizes that the questions posed by Aristotle in *De Caelo* I and the first half of II were the traditional questions of his predecessors and are neither original with him nor peculiar to his account, "Scientific Procedure in Aristotle's *De Caelo*," 11–12.
60 Aristotle, *De Caelo* I, 12, 281a29–281b2, 283b17–21; cf. *Physics* VIII, 1, 252a5–6; 252b5–7.

and so place too must not only be one in number, but also ungenerated and indestructible.

The Elements and ῥοπή: Why the Earth Must Be a Sphere

De Caelo II, 1, opens with a leisurely summation of the view that the heavens neither came into being nor admit of destruction.[61] Aristotle refers to the ancients and the old [but false] tale of Atlas supporting the world (284a12, 19). And *De Caelo* II, 2, clearly begins a new set of problems and argues *contra* the Pythagoreans concerning the structure of the heavens, namely, that the heaven possesses right and left, top and bottom, front and back – rather than merely right and left as they claim.[62] Both the eternity of motion in things and the inherently directional structure of the heaven recall the *Physics*. Even Aristotle's language in *De Caelo* II strongly recalls the account of place in *Physics* IV, 4: the heaven is a limit and surrounds – it is the containing body, and here we possess the account of it.[63]

De Caelo II, 3–6, concerns problems that presuppose the account of the heaven but are more specialized. *De Caelo* II, 7–12, takes up problems concerning both the stars, e.g., their composition, about which there is little direct evidence, and their motion, order, and shape.[64] The solution of some of these is referred to the astronomers – a sign of their specialization.[65] I omit consideration of these because they do not bear on place and the elements as reflecting the order of nature.

In *De Caelo* II, 13, Aristotle begins his account of the situation and shape of the earth by criticizing alternate accounts, such as those of Pythagoras, Plato, Xenophanes of Colophon, and Thales. He then extends his criticisms to include Anaximenes, Anaxagoras, Democritus, and finally Empedocles.[66] The extent of Aristotle's disagreement with his opponents concerns not only the parts, e.g., earth and its movement, but the very nature of the all (294b30–32). Some decisions, Aristotle points out, must be made from the beginning, for example, that bodies having no natural movement cannot have constrained movement, and in his own account these have been made and may now be used as starting

---

61 The shift in style here has often been noticed. For a good summary of this view, cf. Bos, "*Manteia* in Aristotle, *de Caelo* II 1," 29.

62 This point too appears in *Physics* IV, 1, 208b14–23, and 4, 211 a3–5. Against the Pythagoreans, cf. *De Caelo* II, 2, 285a6–11. Cf. also *De Caelo* I, 4, 271a26–27. For an interesting account of the history of the notions "up," "down," "left," "right," "front," and "back," cf. Lloyd, "Right and Left in Greek Philosophy," esp. 41 for the *De Caelo*.

63 Aristotle, *De Caelo* II, 1, 284a3–11; cf. *Physics* IV, 4, 211b10–12; 5, 212b5–8; on the eternity of motion in things, cf. *Physics* VIII, 1, and the identification of that motion as the first circular locomotion at *Physics* VIII, 7–10.

64 Aristotle, *De Caelo* II, 7, 290a17–24; 12, 291b24–28; cf. *Parts of Animals* I, 5, 644b23–645a.

65 Aristotle, *De Caelo* II, 11, 291a29–31; 12, 291b22–23.

66 Aristotle, *De Caelo* II, 13, 294b14, 295a30. On the success of these *vis à vis* his opponents, cf. Furley, *The Greek Cosmologists*, 23–24, 198–200.

points (294b32–295a2). In short, both Aristotle's criticisms and his own constructive account of the shape and situation of the earth rest on the assumption of what he has already established, including his account of nature, of place, and of the elements.

Aristotle next (*De Caelo* II, 14) argues that the earth is at rest and that its middle coincides with the middle of the cosmos. The middle of the earth (and of the cosmos too) is that to which every part of earth is carried and where it rests; likewise fire is carried into the opposite place, "the periphery of the place containing the middle" (296b7–24).[67] The conclusion that now follows sets the stage for the appearance of inclination (and elemental motion) by returning to the question of what the elements and their motions are by nature:

> For if [earth] by nature is always carried naturally toward the middle, just as it appears, and fire away from the middle again toward the extremity, then it is impossible for any part whatever of this [i.e., earth] to be held away from the middle except forcibly. For one motion belongs to what is one and a simple motion to what is simple but not the opposites [i.e., opposite motions do not belong to what is one and simple]. And motion from the middle is contrary to motion toward the middle. If then it is impossible for any part [of earth] to be held from the middle, it is clear that for the whole it is even more impossible. For [the place] into which the part is by nature carried, the whole also by nature [would be carried there]. Therefore, if it is impossible [for the whole] to be moved except by some stronger power, then it is necessary for it to remain at the middle.[68]

The argument defines each element in terms of its natural motion, by virtue of which it is oriented toward its respective proper place – earth toward the middle, fire the periphery, of the cosmos. Because each element is one and simple, each is moved by nature with only one motion. Hence earth must rest at the middle of the cosmos.

Having established that the earth is immobile and at the middle of the cosmos, Aristotle now uses his account of earth's motion toward the middle to show that the earth must be spherical (*De Caelo* II, 14, 297a6–8). The motion of the element is explained first by the term "ῥοπή," inclination, and then "βάρος," heaviness. Proper consideration of these terms accounts for the motion of earth toward the middle and shows that the earth must be spherical.

The argument itself is not difficult. Imagine, as some [falsely] do, that

---

67 Aristotle, *De Caelo* II, 14, 296b14, is quoted: πρὸς τὸ ἔσχατον φέρεται τοῦ περιέχοντος τόπου τὸ μέσον.

68 Aristotle, *De Caelo* II, 14, 296b27–297a2. In the final sentence of this passage, Stocks abandons all relation to the grammatical structure of the sentence rather than use the passive voice. His translation reads: "Since, then, it would require a force greater than itself to move it, it must needs stay at the centre." Moraux translates: "Dès lors, puisqu'elle ne peut être mue, faute d'une force supérieure, elle doit nécessairement demeurer au milieu."

the earth at one time came into being.[69] Earth, in every part, has heaviness until it reaches the middle (297a9). And the arrangement of the different sized parts would not produce a bulge, but the parts would compress together until the center is reached.[70] This downward locomotion is not, as some would have it [again falsely], due to constraint; rather, it occurs because what has heaviness is carried by nature toward the middle.[71] Because all earth is actively oriented toward the middle, all earth will be moved downward as far as possible. Consequently, parts of earth will work out so that the extremity is everywhere equidistant from the middle: such is the definition of a sphere, and so the shape of the earth will be by definition spherical (297a23–24).

Two objections to the claim that the earth must be a sphere might be raised. (1) What if the parts of earth being carried downward are not alike but are of different sizes: would the larger piece produce a bulge, or would the earth still be spherical? It would still be spherical – such possible differences would make no real difference (297a25–27). And the reason why there would be no difference lies with ῥοπή and βάρος.

If the imagined pieces of earth were of different sizes, they would necessarily have different heaviness and different inclinations. In this case:

> For the greater must always drive on the lesser before it until the middle, both having inclination for the middle, namely the more heavy drives on the less heavy until this [i.e., the middle]. [297a27–30: τὸ γὰρ πλεῖον ἀεὶ τὸ πρὸ αὐτοῦ ἔλαττον προωθεῖν ἀναγκαῖον μέχρι τοῦ μέσου τὴν ῥοπὴν ἐχόντων ἀμφοῖν, καὶ τοῦ βαρυτέρου προωθοῦντος μέχρι τούτου τὸ ἔλαττον βάρος.][72]

How does inclination account for the motion of the pieces that results in a necessarily spherical earth? The argument is not easy.

Aristotle explicitly sets out from his account of elemental motion. Hence, however we interpret this argument, one condition must be met. Inclination and heaviness here presuppose *natural* motion, i.e., the motion of earth *toward* the center. Because extrinsic force explains only constrained motion, as when a body is moved *away* from its natural place,

---

69 Aristotle, *De Caelo* II, 14, 297a12–14. Aristotle, of course, argues for the eternity of time and motion, cf. *Physics* VIII, 1.

70 Aristotle, *De Caelo* II, 14, 297a10–11; the notion that the cosmos might "bulge" occurs with the same word at *Physics* IV, 9, 216b25.

71 Aristotle, *De Caelo* II, 14, 297a12–17. Stocks's translation is hopelessly unliteral and introduces modern concepts: "The process should be conceived by supposing the earth to come into being in the way that some of the natural philosophers describe. Only they attribute the downward movement to constraint, and it is better to keep to the truth and say that the reason of this motion is that a thing which possesses weight is naturally endowed with a centripetal movement." Moraux is also unliteral: "Mieux vaut partir d'une base vraie et expliquer ce mouvement par la tendance naturelle qu'ont les choses pesantes à se porter vers le centre."

72 Again, Stocks's translation is unliteral and interpretive: "For the greater quantity, finding a lesser in front of it, must necessarily drive it on, both having an impulse whose goal is the centre, and the greater weight driving the lesser forward till this goal is reached."

extrinsic force cannot be at work here. Either implicitly or explicitly, the natural motion of earth toward the middle must presuppose the causal structure that accounts for all natural motion. Natural motion rests on a thing's nature, and nature is an intrinsic ability to be moved by appropriate actuality.

This condition returns us to the definition of earth and the identity between earth's definition, its intrinsic ability to be moved, and its actuality: earth is by definition heavy or downward, i.e., it possesses an intrinsic ability to be moved downward (where it rests); the cosmos, as Aristotle reaffirms in his rejection of the Pythagorean view, is fully determinate, possessing front and back, right and left, up and down, and its actuality defines "down" as opposed to "up," which is at the periphery. The determination of the element earth to be moved downward (and to rest there), and the actuality, i.e., the middle of the earth (and the cosmos) that defines down, are one and the same.

And the determination of earth for the middle provides the key to understanding this argument. Both pieces of earth, "the greater" and "the lesser," possess inclination for the middle. This inclination cannot be other than the very nature of earth: the ability to be carried by nature toward the middle that each piece possesses to the extent that, being large or small, it is determined to be heavy. Both pieces possess [natural] inclination to be carried toward the middle by virtue of being earth. Etymologically, the notion of inclination here implies orientation that is continuous. Being bigger, the heavier piece has greater inclination; consequently, the less heavy piece has two motions: (1) that resulting from the actualization of its intrinsic ability to be moved downward, its inclination, and (2) that resulting from the greater inclination of the heavier piece of earth by which it is driven downward.[73] (Aristotle assumes that from the point of view of the middle, the heavier piece is behind the less heavy piece and so pushes it.)

In effect, the less heavy piece of earth is moved both intrinsically by its inclination (its own principle of motion) *and* extrinsically by the heavier piece of earth pushing from behind because the heavier piece has greater inclination.[74] For this reason, all the pieces of earth incline toward the middle and are pushed and compressed together as tightly as possible; consequently, the pieces sort themselves out until the shape of the earth must be spherical. In short, even if (contrary to fact) we imagine the earth to be generated from different sized pieces of earth, its shape would nevertheless necessarily be spherical because the pieces would be inclined as

---

73 It may be well to recall here the one appearance of ῥοπή in the *Physics*, namely, IV, 8, 216a14–15: "bodies which have a greater inclination either of heaviness or of lightness, if they are alike in other respects, move faster over an equal space".

74 On pushing as one of the four ways in which one thing moves another, cf. *Physics* VII, 2, 243a15–243b6.

much as possible toward the center until everything was compressed and equidistant from that center.

Clearly, there is a puzzle here: what relation obtains between inclination as an intrinsic principle of motion and the extrinsic relation between the heavier piece as it pushes the less heavy? Again, the account must rest on Aristotle's conception of natural, rather than violent, motion: both pieces of earth experience natural motion because, insofar as it is heavy, each is moved toward the middle, i.e., the natural place for both pieces. Natural motion is strictly defined in terms of natural place, and here that definition is met: both pieces of earth are being moved downward, i.e., toward their natural place.

But the motion of the less heavy piece is a compound of its own intrinsic motion (actualization of its own inclination downward) and an added motion, that due to the heavier piece pushing it extrinsically from behind. These two motions combine to move the less heavy piece downward more quickly than it would be moved by the actualization of its own inclination alone. This relation lies at the heart of the solution to the question "would the earth bulge or still be spherical if it were generated from different sized pieces of earth moving downward?"

This is what makes the case seem odd: nature (and natural motion) is by definition an *intrinsic* principle of being moved; motion by force is by definition motion away from a thing's natural place and requires an *extrinsic* mover.[75] But the case proposed here presents the less heavy piece of earth being moved both *intrinsically* by its own inclination and *extrinsically* by the heavier piece. Hence, it might appear to be moved naturally by its inclination but violently, or unnaturally, by the heavier piece that moves from the outside.

But although violent motion requires an extrinsic mover, motion produced extrinsically *need* not be violent. The *sine qua non* of violent motion is that it be *away* from a thing's natural place. And this condition is clearly not met here. Even though the smaller piece of earth is moved extrinsically by the larger, the motion must nevertheless be natural because it is downward. The less heavy piece of earth is moved naturally because it experiences only one motion – motion toward the middle, i.e., natural motion.

In effect, a single natural motion is produced by two causes. So we must ask about the role of each cause. Both the larger and the smaller pieces of earth have inclination for the middle, and this inclination is the ability of each piece to undergo motion.[76] Furthermore, this inclination, which is greater as the piece of earth is heavier, is the intrinsic nature, or ability to be moved, in each piece. And each piece is moved by virtue of

75 This case is taken up explicitly at *De Caelo* III, 2, 301b17–30.
76 Again, cf. *Physics* VIII, 4, 255b30–31.

this ability. Consequently, the heavier piece of earth does not cause the less heavy piece to be moved absolutely; its own inclination toward its natural place constitutes its natural ability to be moved and is the primary source of its being moved naturally. Proof of this point is simple: the less heavy piece would go down in the absence of a heavier piece behind it. However, because the heavier piece has more inclination, it is moved faster than the less heavy piece; because it is moved faster, the heavier "drives on" the less heavy before it.

The solution to the puzzle of this argument lies here: the heavier piece of earth does not cause the less heavy to be moved, but causes it to be moved *faster*. Furthermore, we know why the larger piece "drives on" the smaller: because both are naturally oriented toward the middle, but the larger possesses greater inclination by virtue of being heavier. "Inclination" expresses the natural orientation of the element toward its natural place and so constitutes its very nature. Hence, it must be an intrinsic principle of motion, specifically of being moved downward, i.e., toward the middle. The actuality for this principle is the middle, i.e., "down." By implication, all pieces of earth, both small and large, go down as far as is possible until each is moved as close to the middle as possible; and the heavier pushes the less heavy before it with the result that both go as quickly as possible, not stopping until they are as close as possible to the middle. The same conclusion follows as before: parts of the earth will work out so that the extremity is everywhere equidistant from the middle; therefore, by definition the shape of the earth will be spherical.

This argument at once reveals Aristotle's determinate world, how it operates by virtue of being determinate and the kind of account that such a view produces. Aristotle presupposes that each element, in this case earth, is always determinate, i.e., any piece of any size contains an intrinsic principle of motion oriented toward its natural place (for earth, this is the middle). The account of earth's motion rests squarely on the identification of this principle. In a determinant world, cause and effect go together "naturally" with the result that (in the absence of hindrance) an identification of the principles of determination specifies the reason why the motion takes place.

The middle is the principle, in the sense of the "to which" the element earth is moved. It is actually down and so serves as the actuality for what is potentially down, i.e., earth when it is out of its natural place. Both ῥοπή and βάρος serve as the intrinsic principles within the element earth. ῥοπή, inclination, specifies the principle of determinateness whereby earth is oriented downward, and βάρος, heaviness, specifies the intrinsic principle whereby a larger piece of earth is moved faster than a smaller toward its natural place.

We may pause to note that Aristotle's teleology lies here, properly speaking. The relation between each element and its proper place is such that in the absence of hindrance no third cause is needed to combine them. No third cause is needed because the relation between them is one of an immediate active orientation in the moved, the element, for its actuality, to which it is moved – its natural place. And Aristotle's teleology is, properly speaking, nothing other than this immediate intrinsic relation of moved to mover.

What of the extrinsic relation between the heavier and the less heavy pieces of earth? A post-Newtonian instinct may jump in anticipation of a mechanistic possibility here, but it simply cannot be found or even anticipated. The argument is constructed around a concept of natural motion, and given the definition of natural motion, this extrinsic relation has little explanatory status. That is, it does not explain the motion as such of the less heavy piece, but only why that motion is faster, i.e., faster than it would have been if produced solely by its own intrinsic ability to be moved. The proper cause of natural motion remains squarely centered on the intrinsic orientation of earth toward the determinate middle of the cosmos.

Indeed, the extrinsic relation between the two pieces plays no role in Aristotle's initial argument that elemental motion shows that the earth must be spherical. The relation between the heavier and less heavy pieces of earth is introduced into the argument only through an objection – the search for a configuration of earth that might challenge the view that the earth is spherical. And Aristotle's solution to that objection turns on the same principles as does his initial explanation: the intrinsic ability of earth to be moved downward. The extrinsic relation between the two pieces is an accident of the configuration posed by the objection. And this relation plays only an accidental role in Aristotle's resolution of the objection. That is, it does not explain the motion of the less heavy piece, but why the motion is faster.

Before proceeding to the second objection to the view that the earth must be spherical, we may note that on this construal the argument here in the *De Caelo* bears a close resemblance to that of *Physics* IV, 8, concerning "that through which" a body is moved. There too, as I have argued, the extrinsic relation between the moved body and that through which it is moved does not cause the motion of the body, but causes it to be "faster." Again, both partners in the relation must be conceived as expressing an intrinsic nature and the extrinsic relation between them plays only a marginal role in the account of motion.

(2) The second objection to the claim that the earth must be a sphere remains, and inclination reappears as Aristotle replies by expanding his account of earth's motion toward the center. If a piece of earth many times greater than the whole earth were added to one hemisphere, would the middle of the earth and the middle of the whole no longer be the same (*De Caelo* II, 14, 297b2–3)?

This then is the difficulty. But it is not hard to understand if we look further and work through how we require a heavy [body], of whatever magnitude, be carried to the middle. It is clear that it is not [in place] when the edge touches the center, but it is necessary that the greater [piece of earth] prevail until its middle possesses the middle of it [i.e., the whole]. For it has the inclination for this. It makes no difference if that which is spoken of is a clod, i.e., a piece of any chance [size], or the whole earth; for the result asserted is not because of a smaller or [greater] magnitude, but according as each has inclination [ῥοπή] toward the middle. Therefore, whether a whole or just a part is carried thence, it must be carried until that [i.e., the middle], likewise from every side it takes the middle, the lesser being driven on by the greater by means of forward push of the inclination. Therefore, either it [i.e., the earth] was generated, [in which case] it must have been generated in this way so that it is clear that the generation of it was spherical, or being ungenerated it is such [i.e., spherical], remaining always, being in the same way as if it had been generated in respect to a first coming-to-be.[77]

Here the inclination of earth, either the whole or any part, is an inclination for the middle, such that the middle of the earth cannot rest until its middle is identical with the middle of the whole. The motion of earth toward the middle must occur by nature (rather than by force) and always occurs if nothing hinders. Furthermore, as the concluding lines indicate, once it is at the middle, whether always or because generated, earth remains there without constraint. That is, it rests by nature in that place toward which it is moved by nature.

As in the first objection, inclination constitutes the very nature of earth, such that "more" earth means greater inclination to be moved. In the absence of hindrance, this inclination is actualized not when the outer edge of the earth touches the middle of the whole, but when the middle of the whole coincides with the middle of the earth. The middle of the whole must, therefore, function as the actuality of the earth, i.e., that toward which it is oriented.[78] Only so will earth be moved until its middle is one with the middle of the whole. And the earth then rests there, i.e., expresses the activity proper to earth, remaining at the middle, either from its generation or from eternity. Consequently, its shape must by nature be spherical.[79]

The relation between the inclination of earth and the middle of the cosmos exemplifies that defined by the relation of potency and actuality as mover and moved. That is, in the absence of hindrance, what is potential cannot fail to be actualized by its proper actuality until actualization

77 Aristotle, *De Caelo* II, 14, 297b1–17; on this text, cf. Verdenius, "Critical and Exegetical Notes on *De Caelo*," 280–281. We may note that Stocks translates ῥοπή as "impulse" throughout this argument.

78 Cf. *Physics* VIII, 10, 267b6–7, where Aristotle identifies the middle and the periphery as the principles of the heaven.

79 A few lines later (*De Caelo* II, 14, 297b21–23) Aristotle emphasizes that the earth's motion and resulting spherical shape must be "by nature."

of the potential is complete; when actualization is complete what was potential has become identical with its actuality, and the activity that is the
fullest expression of this actuality immediately ensues. In short, the inclination of earth must be moved by the middle of the cosmos until the
middle of the one is identical with the middle of the other, whence earth
rests. Thus, inclination is continuous – either the active orientation of
earth toward the middle or the activity of resting once earth is at the middle. And this relation is so strong that it serves as the basis of a proof that
the earth must be spherical.

In the context of rather different interests, Nussbaum makes an important point for this argument: "his [Aristotle's] arguments for its [the
earth's] immobility depend far more on his doctrine of natural motions
than on his theories about heavenly motion."[80] This argument establishes
the structure of the cosmos, i.e., earth is at its center, is immobile, and is
spherical, by utilizing an account of the natural motion of the elements
rather than some form of cosmic or celestial motion. Consequently, the
argument provides further evidence that neither the cosmos nor the *De
Caelo* are divided into celestial and sublunar parts; rather, the natural motion of earth toward the center serves as the basis of a proof of an essential feature of the structure of the cosmos. And at the same time, the argument provides a partial account of the natural motion of the elements
with earth as the example.

In this argument, inclination may tentatively be identified as the very
nature of the elements: the source or cause of being moved and remaining at rest in that to which it belongs essentially and not accidentally. Thus, earth is moved whenever it is out of its natural place and remains at rest when it is in place. However, the elements remain in many
ways unexplained, and there are several further occurrences of the term
"inclination." They must be considered before a final judgment can be
reached.

80 Nussbaum, 129; she refers to *De Caelo* II, 3, 268a12–21.

6

Inclination As Heaviness and Lightness

Having answered the most serious criticisms of his account, Aristotle concludes his arguments that the earth must be a sphere – and so concludes *De Caelo* II – with the particulars of its size. *De Caelo* III, 1, opens with a brief summary of what has been discussed thus far concerning the first heaven and its parts, the stars carried in the heaven, and the composition and nature of these things, including, finally, that they are ungenerated and incorruptible (298a24–27). Having dealt with the first (and highest) element, aether, Aristotle turns to the other elements.[1]

These elements – each of which possesses its own specific nature – are "by nature." Natural things are *either* substances *or* operations and affections of substance. "Substances" refers to the elements – aether, air, fire, water, and earth – and things composed of them, such as the heaven, as a whole and its parts, as well as plants and animals and their parts; "operations and affections" include movements of each of these according to their proper power as well as their alterations and transformations into one another.[2] Obviously, Aristotle concludes, the study of nature is for the most part concerned with bodies because all natural substances are either bodies or come to be after bodies, and all involve magnitude.[3] The investigation must also include generation and destruction, as "operations and affections" of the elements and all things composed of them (298b8–11). The elements, their proper source of motion, and their generation (and destruction) form the topics of *De Caelo* III.

Aristotle's emphasis on the link between physics as a science and body as its object may reflect the fifth-century tradition within which he defines

1 Aristotle, *De Caelo* III, 1, 298b6–8. Aristotle speaks of two elements, not four. Simplicius, *In Phys.* 555.6–12, suggests that he refers generally to heavy (earth and water) and light (fire and air). In their respective translations, Stocks and Guthrie agree with him, as does Elders, *Aristotle's Cosmology*, 271.
2 Aristotle *De Caelo* III, 1, 298a27–298b; cf. *Physics* II, 1, 192b10, which also lists things that are by nature but does not include the heavens.
3 Aristotle, *De Caelo* III, 1, 298b–4; cf. *Physics* II, 2, 194a12–15; 194b10–14; and *De Caelo* I, 1, 268a-6.

his project against that of Plato and other thinkers.[4] Physics concerns natural substances, i.e., bodies composed of the four elements. Hence, the account of the four elements is the account of "proper things," i.e., the constituent parts of all natural substances, i.e., bodies, and as such falls squarely within the science of physics.

The opening of De Caelo III, 1, shows a strong continuity with De Caelo II. The "higher" element, aether, and the more important part of the cosmos composed of it, the heaven, together with the larger structure of the cosmos, e.g., earth is at the center, have been examined. The sublunar elements remain and, as we shall see, their examination presupposes the conclusions already established not only in De Caelo I and II, but also in the Physics.

By refuting his most serious opponents (De Caelo III, 1), Aristotle clears the way for his own account of body as heavy or light, which begins with an account of the natural motions of the elements (De Caelo III, 2). Hence we must begin with his objections to using mathematics to generate bodies. His own account of body develops his account of the elements, of their natural motions as directed toward natural place, of inclination as the nature of the elements, and of nature as a principle of order.

Aristotle first criticizes his predecessors, presumably Plato and the Pythagoreans, who use mathematics to generate bodies out of planes and to resolve them again into planes (298b35–299a2). This procedure requires a false view of mathematical objects, namely, that solids be constructed from planes, planes from lines, and lines from points and so is bad mathematics.[5] It is also bad physics: bodies that are heavy or light cannot be composed of mathematical parts, points (and hence lines, surfaces, and solids too), which are neither heavy nor light (299a26–300a14). Likewise, number or mathematical monads, even in combination with one another, can never make bodies (300a16–19).

In effect, this argument spells out the primacy of body as the proper object of physics and confirms its primary nonderivative characteristic: all body must be either heavy or light. Insofar as physics identifies body, i.e.,

4 Cf. Kahn, "La Physique d'Aristote et la tradition grecque de la philosophie naturelle," 45–52. Commenting on the parallel text of Physics II, 2, De Groot, "Philoponus on Separating the Three-Dimensional in Optics," 160–161, also sees Aristotle's emphasis on the inclusion of body within physics as antimathematical and anti-Platonic; however, she insists that "Aristotle ruled out the separability even in thought of physical forms from matter." That the proper objects of physics must retain a reference to matter (or body) in their definitions means neither that such forms are "physical forms" nor that they are not separable even in thought. On this issue, cf., e.g., Mansion, "Tò σιμόν et la définition physique," 124–132; Grene "About the Division of the Sciences," 10; Berti, "Les Méthodes d'argumentation et de démonstration dans la Physique (apories, phénomènes, principes)," 56.
5 Aristotle, De Caelo III, 1, 299a6–15; for Aristotle's views on points, lines, etc., cf. Physics VI, 1, 231a21–231b9. In Plato, cf. Timaeus 56B. For an analysis of just this problem and various Hellenistic responses to Aristotle, cf. Konstan, "Points, Lines, and Infinity," 1–32; for a critique of Konstan, cf. H. Lang, "Points, Lines, and Infinity: Response to Konstan," 33–43.

the elements, as its proper subject-matter, being heavy or light constitutes a starting point for it. As we shall see, inclination is identified with
being heavy or light; thus it is a starting point for physics as the science
of body.

The central question of *De Caelo* III, 2, is whether the elements are generated. Aristotle concludes that, although they may be generated from
one another, absolute generation is impossible.[6] The argument takes him
directly to a broader account of nature as always a cause or source of order. Again he substantiates his own view of nature and natural motion by
criticizing alternate views.

The conclusion that the elements cannot be generated absolutely follows, for Aristotle, from an analysis of natural motion. First, he argues that
some natural motion must belong to each of the "simple bodies," i.e., the
elements (300a20). This argument entails a direct account of why order
must be prior to disorder, or more precisely, why to be natural *is* to be orderly. Second, the elements must possess a natural motion, and this necessity implies that they "must have an inclination in respect to heavy and
light" (301a22–23). Taken together, these two arguments present an account of natural motion.

This account of natural motion establishes the meaning of inclination
and the view that the cosmos must be a determinant and orderly whole
because nature is an intrinsic source of motion. Hence it reflects back to
the definition of nature in *Physics* II, 1. But in it, Aristotle introduces the
role of force during natural motion. Thus to complete the account, he
must explain the relation between natural motion and force, which is an
extrinsic source of motion. Force, Aristotle argues, can contribute to natural motion and is the sole cause of violent motion. He concludes that it
is now clear "that everything is either heavy or light and how unnatural
motion occurs" (*De Caelo* III, 2, 301b31–34). His own view, i.e., that the
elements are generated out of one another, appears in *De Caelo* III, 6,
along with another appearance of inclination as a principle associated
with being heavy or light and whose absence implies immobility. This
point (along with criticisms of his predecessors) completes the account
of the elements and their motions in *De Caelo* III.

The Natural Motion of the Elements

The larger purpose of the arguments in *De Caelo* III, 2, is to prove that absolute generation of the elements is impossible. These arguments constitute the first step of the proof: there must be some natural motion for
each of the natural elements (301a20–22). And this account of natural

6 "Are the elements generated" is raised as "the first question" at *De Caelo* III, 1, 298b11,
and concluded at III, 2, 301b31–32.

motion entails that the order of the heavens must always have been as it is because, being orderly, nature could not be otherwise.[7]

Each of the four elements must have "some motion" (and rest) that is by nature, as can easily be shown.[8] Aristotle first reiterates relations established earlier: each of the elements must have one natural motion and hence one natural rest (*De Caelo* III, 2, 300a20–27). Furthermore, it has already been established that there is a body, earth, that rests naturally at the center. If rest in this place is natural, then motion toward it must also be natural; and if a thing rests in a place by force (or contrary to nature), then it must have been carried there by force (or contrary to nature).[9]

Now another argument that each element has a natural motion is straightforward. If we assume that a body is at rest by constraint, i.e., by force applied from the outside (and contrary to nature), then we must ask what produces the constraint. It must be either a body at rest in its natural place (it hinders the constrained body because, being at rest naturally, it resists being moved out of that place), or a body that, like the first, is at rest by constraint. In the former case, we reach rest that is natural to a body; because motion toward the place in which a thing rests naturally must be natural, we reach a necessary natural motion. In the latter case, we reach a body at rest by constraint and must ask again, what hinders it? And again the answers must be either a body at rest naturally (and thereby entailing a natural motion) or at rest by constraint. Consequently, there must either be an infinite regress of bodies at rest by constraint, which is obviously impossible, or there must be a body at rest naturally, which immediately entails a natural motion toward the place in which the body rests naturally (300a28–300b).

This argument clearly presupposes the definition of nature in *Physics* II, 1, as well as the relation between place, motion, and nature established in *Physics* IV. But the topic of the argument is the elements. We saw in Part I of this study that motion, on Aristotle's definition, must be the motion of something; there is no motion apart from things, and motion is not transferred from one body to another. We also saw that place, which is a limit, renders the cosmos formally determinate in respect to "where." This account yields the elements as moved within a determinate cosmos. For this reason, it is both an investigation of "proper things" and subordinated to the investigations of "common and universal things" in the *Physics*.

By nature each element possesses an *intrinsic* principle of being moved toward (or at rest in) its natural place; by contrast, for that element mo-

---

7 Aristotle connects motion and the very disposition of the cosmos immediately before identifying nature and order: Aristotle, *De Caelo* III, 2, 300b26.
8 Throughout this passage Aristotle uses parallel constructions for "by nature [φύσει]" or "according to nature [κατὰ φύσιν] and "by force [βίᾳ]" or "outside of nature [παρὰ φύσιν]"; cf. also *Physics* VIII, 4, 254b13–14; 255a21–23.
9 Aristotle, *De Caelo* III, 2, 300a28–29.

tion (or rest) by constraint implies an *extrinsic* source of being moved away from (or held out of) natural place. Motion toward (or rest in) a thing's natural place is natural for it and motion (or rest) away from that place can only be by force and contrary to nature for it. Because motion by force is by definition motion of each element away from its natural place, such motion presupposes such a place. Indeed, because a respective place is defined as natural for each element, these places define the various motions of each element as natural or violent. Finally, an element's motion away from a natural place can proceed in a variety of directions, but only one motion can be toward it; therefore, for each element there can be many motions that are contrary to nature, but only one according to nature. And *that* motion is defined by the place toward which it is oriented, e.g., downward, or toward the middle, for earth.

Aristotle's strategy is to identify rest that is natural for each element because (given his earlier definitions) natural rest in a place entails natural motion toward that place. Assuming that a body is at rest by constraint leads *either* to a body at rest naturally, and so a natural motion toward that place, *or* to an infinite regress of bodies at rest by constraint, which is impossible. Therefore, we must reach a body at rest naturally and so a natural motion; by definition, this motion must be toward the natural place in which the body rests. Thus, the cosmos is constituted first and primarily by a determinate relation between body and place: all natural body is moved by nature toward the place in which it rests by nature.

The claim that each element has only *one* natural motion is implicit in the earlier argument that all things that are of the same kind have the same natural place: earth always goes down, or toward the center. Where there is the same natural place, there can be many motions away from it but only one motion toward it. And so for earth there can be only one natural motion, i.e., motion downward toward the center. In short, nature implies order, and the elements express this order by the fact that each is moved by nature with one motion only, i.e., the motion toward its respective natural place.

Before proceeding, we may consider an objection recently raised against Aristotle's view and appropriately named "the direction problem."[10] Even though a stone, for example, possesses an intrinsic source of motion, namely, its nature, "[i]t is still reasonable to wonder how it is that the internal source of motion determines the appropriate direction."[11] A "direction principle" for Aristotle could easily be defined: "a simple body can undergo natural motion . . . only if the immediate neighborhood of

10 Kronz, "Aristotle, the Direction Problem, and the Structure of the Sublunar Realm," 247–257.
11 Kronz, "Aristotle, the Direction Problem, and the Structure of the Sublunar Realm," 247.

the body, the body's place, is inhomogeneous in such a way that the directions 'up' and 'down' are unambiguously indicated."[12]

The "direction problem" raises several important issues. (1) Direction is not something added on or extrinsic to the motion of an element. Natural motion is by definition toward proper natural place. (2) A body's proper place is not its "immediate neighborhood," but "the where" natural to it. And obviously a body can be out of its proper place, e.g., earth can be up. (3) Place within the cosmos should not be thought of as "inhomogeneous" – a term connoting material parts that are formally indeterminate – but as immediately determinate in a formal sense, i.e., rendered determinate in respect to direction by place as the first limit of the containing body. Therefore, natural place is never "neutral" to direction, and there is no "direction problem" in Aristotle's physics or his cosmos.

Implications for the cosmos as Aristotle conceives it emerge in his criticisms of his predecessors. According to Aristotle, Empedocles, Leucippus, Democritus, and Plato think of motion as infinite or indeterminate. Empedocles, for example, argues (Aristotle tells us) that the earth is held in place by a vortex; presumably, the vortex holds by constraint rather than by nature; consequently, were the constraint removed, the earth would be moved naturally. But where would it be moved to? On Aristotle's view, it would stop in its natural place where its rest would be natural rather than by constraint; hence, the motion toward that place would also be natural (*De Caelo* III, 2, 300b-8).

Leucippus and Democritus argue that the primary bodies are always being moved in the void, i.e., the infinite.[13] But (as we saw in *Physics* IV, 8) a void (or the infinite) is undifferentiated and indeterminate; therefore, it cannot provide a natural place for each of the elements, and in a void motion would continue forever (unless something more powerful were to impede it) because there is no natural place in which a thing rests (215a19–22). Hence a void, or the infinite, fails to account for natural motion; because any constrained motion presupposes a natural motion that it contravenes, it is perfectly fair (from Aristotle's point of view) to ask Leucippus and Democritus what motions are natural to these bodies (*De Caelo* III, 2, 300b11–14). And the implication is that there are none. Hence, they fail to account for natural motion; furthermore, if all motion were by constraint, an infinite regress would occur – which is impossible (300b14–16).

The same point follows for Plato's *Timaeus* where the claim appears that before the world was made, the elements moved without order.[14] In

---

12 Kronz, "Aristotle, the Direction Problem, and the Structure of the Sublunar Realm," 251.
13 Aristotle, *De Caelo* III, 2, 300b8–10; cf. also *Metaphysics* I, 4, 985b4, which is Diels and Kranz, *Die Fragmente der Vorsokratiker*, 67A6.
14 Aristotle, *De Caelo* III, 2, 300b16–18; cf. Plato, *Timaeus* 52E-53.

Aristotle's terms, these movements must have been either by constraint or by nature. If by constraint, then we are back to an infinite regress; if by nature, then it can be shown that these elements presuppose an orderly world, i.e., Aristotle's world (300b18–21). The first heaven produces motion by virtue of its own natural motion, and the other bodies, moving without constraint, would come to rest in their natural places and so would produce the order that they now have, with heavy bodies toward the center and light bodies away from it.[15] In this sense, Plato's account of the generation of the world from disorderly motion already presupposes an order that (on Aristotle's view) constitutes what the cosmos is by nature. And his final criticism makes this point explicit.

Empedocles, Aristotle tells us, argues that natural bodies, such as bones and flesh, can result from chance combinations of "bodies," i.e., unordered moved things and their resulting mixtures.[16] Such a view implies that disorder is prior to order (and so natural) and, conversely, that order is posterior to disorder (and so unnatural).[17] Consequently, order would at best resemble art and at worst chance.

Aristotle rejects "Empedocles' view" with an argument for the primacy of order concluding unambiguously: of things that are according to nature, none comes about by chance.[18] His argument rests on the question of whether there is one or an infinite number of locomotions. If proponents of the view that disorder is prior to order make moved things infinite in the infinite, then *either* the mover must be one, and so they must be moved with a single locomotion and will not be moved in a disorderly way; *or* the movers must be infinite and so the locomotions too would be infinite (300b31–301a). But it is impossible that the movers are infinite, and if the locomotions are limited, then obviously there will be some order (301a-2).

The strategy of these arguments uncovers Aristotle's own presupposition: an orderly cosmos. To show the priority of order, one need not show that all things have a single motion or resting point (natural place), but only that locomotions are limited rather than infinite. Indeed, it is false that all things are carried to the same "goal" – only those that are like in kind (301a2–4). So there may be as many different locomotions as there are kinds; if things that are like in kind are moved to the same place, order must be prior to disorder. And this is precisely Aristotle's position.

---

15 Aristotle, *De Caelo* III, 2, 300b21–25. At line 21, I read ἑαυτό on the basis of arguments given by Guthrie, xxxi, n. 6, and Verdenius, "Critical and Exegetical Notes on *De Caelo*," 282.

16 Aristotle, *De Caelo* III, 2, 300b25–29; cf. *Physics* II, 8, 198b16–32.

17 Aristotle, *De Caelo* III, 2, 301a9–11. For an argument that Aristotle seriously misunderstands Empedocles here, cf. Kirk, Raven, and Schofield, *The Presocratic Philosophers*, 296–302.

18 Aristotle, *De Caelo* III, 2, 301a11; for a refutation of the view that chance is among the causes of things, cf. *Physics* II, 4–6.

Indeed, we see here why the elements are "proper things" and not "common and universal."

Thus we find the central point concerning the elements: things that are like in kind are all moved to the same place because it is their very nature, i.e., what each is, to do so. Earth, for example, is moved downward not by constraint, but by nature: it possesses an intrinsic principle of being moved downward (toward the middle); because this principle constitutes earth's nature, it constitutes its definition too. Hence earth's natural motion must be the same for all earth (assuming nothing hinders it) and is a direct consequence of its determinate nature. Furthermore, "disorderly" means contrary to nature, because the proper order of sensible things *is* their nature (301a4–6). For earth, and hence things made of earth, this nature, e.g., the ability to be moved downward or to rest once down must be expressed whenever possible (301a7–9).

The opposite view, i.e., that disorder is prior to order, entails the impossibility that disorderly motion is infinite (301a6–7). Furthermore, it makes disorder natural and order unnatural – whereas in truth, of things that are by nature none comes to be by chance (301a9–11). Hence nature is always and everywhere identified with order.[19]

Anaxagoras bears witness to this view and so starts his cosmography from unmoved things.[20] Even other thinkers must put things together before they try to produce motion and separation: it is unreasonable (perhaps impossible) to make the generation of natural things from things that at the start are moving and separated out (301a13–15). So Empedocles, who intends to make disorder prior, must first have the world ruled by love, a necessary principle of unity and combination, before constructing the heaven out of separated bodies.[21]

In effect, three choices are possible: (1) Like Anaxagoras, one can admit from the beginning that order is prior to disorder and that nature must be identified with order; or (2) like Democritus and Leucippus, one can involve oneself in the contradiction of an infinite regress; or (3) one can try (but fail), like Plato and Empedocles, to make disorder prior and natural while in fact importing order into the world through other principles. The latter two views must be rejected: nature must be identified with order rather than chance. Consequently, each of the elements must have one motion that is by nature and that is always ordered to the same goal – its natural place.

Thus Aristotle completes his proof that some one motion must by na-

19 Cf. Aristotle, *Physics* VIII, 1, 252a12; *Gen. of Anim.* III, 10, 760a31.
20 Aristotle, *De Caelo* III, 2, 301a11–13; Aristotle does not expand the reference to Anaxagoras. Cf. Simplicus, *In Phys.* 300.31.
21 Aristotle, *De Caelo* III, 2, 301a15–20; cf. Aristotle's criticisms of Empedocles at *Physics* II, 8, 198b33 ff. At 199b15, Aristotle claims that this view destroys nature and things that are by nature.

ture belong to each of the four elements. It sets out from the definitions of "by nature" as motion toward (or rest in) natural place and "by force" as motion away from (or rest by constraint out of) natural place. Using this definition, he establishes the requisite natural motion and then criticizes views that deny such motion, explicitly or implicitly. Here the relation between place and moved things is unambiguous: each element possesses a natural motion because each is by its very nature oriented toward its natural place. The implication of this relation is the identification of nature with order and what is orderly.

All Body Must Be Either Heavy or Light

Although Aristotle claims that all body must be either light or heavy, the reason why it is such is not clear. Consequently, the relation between the order of nature, being light or heavy, and being moved is also unclear. And Aristotle turns to this problem in the second set of arguments in *De Caelo* III, 2. He intends to establish a necessary relation between motion and the "fact" that all body must be either heavy or light. And this relation ultimately leads to his desired conclusion: not all things are generated, and nothing is generated absolutely (301b30–33).

Two arguments now follow, each designed to show that all body must be either heavy or light – a conclusion that in turn supports the larger proof that the elements are not generated absolutely. Inclination appears as the principle of motion connected with heavy and light. The first argument takes up natural motion and the second constrained, or unnatural, motion. Because all motion must be either natural or unnatural, these two arguments, taken together, show that as moved, all body must be either heavy or light. Here, as earlier, inclination constitutes the nature and principle of motion in things that are by nature, specifically, the elements and all things composed of them:

That, then, there is some natural motion, which is produced neither by force nor contrary to nature, for each of the bodies [i.e., the elements] is clear from these [i.e., the preceding] arguments. That there must be some bodies having inclination in respect to being heavy and light is clear from these [i.e., the following] arguments. For we say that they must be moved; but unless the moved has by nature an inclination, it is impossible that it be moved either toward the middle or away from the middle. [301a22–26: ὅτι μὲν τοίνυν ἐστὶ φυσική τις κίνησις ἑκάστου τῶν σωμάτων, ἣν οὐ βίᾳ κινοῦνται οὐδὲ παρὰ φύσιν, φανερὸν ἐκ τούτων· ὅτι δ' ἔνια ἔχειν ἀναγκαῖον ῥοπὴν βάρους καὶ κουφότητος, ἐκ τῶνδε δῆλον. κινεῖσθαι μὲν γάρ φαμεν ἀναγκαῖον εἶναι· εἰ δὲ μὴ ἕξει φύσει ῥοπὴν τὸ κινούμενον, ἀδύνατον κινεῖσθαι ἢ πρὸς τὸ μέσον ἢ ἀπὸ τοῦ μέσου.]

As we shall now see, although inclination is specified differently for the different elements, it always acts as a principle of being moved by nature in respect to being heavy or light and so explains natural motion.

The relation of this argument to the preceding account of natural motion is telling. Earlier, Aristotle identified the principle of natural motion in the elements, i.e., what makes them able to be moved: inclination that constitutes the very nature of each. He has shown (on his own view) that each of the elements must have one natural motion. Now, as earlier, he identifies the principle of motion in bodies as inclination but specifies it further: inclination is ability to be moved expressed as being heavy or light. That is, it is the principle by which what is heavy is actively determined downward – while what is light is actively determined upward. Indeed, being heavy or light is nothing other than inclination, present in all body according to its kind.[22]

A *reductio ad absurdum* argument, resembling an earlier rejection of the void, shows that there can be no such thing as a body that is neither heavy nor light.[23] Suppose two bodies, A, which is not heavy, and B, which is heavy; body A will be moved the distance CD whereas body B in the same time will be moved a greater distance, CE (*De Caelo* III, 2, 301a26–29). If one divides the heavy body, B, one will produce a body that is heavy (but less heavy than B) and will also cover the distance CD; thus, Aristotle generates the requisite contradiction: a heavy body and a body that is not heavy will cover the same distance in the same time, which is impossible (301a29–301b). And the same argument works in the case of lightness (301b). After giving another argument for the same point, Aristotle draws the conclusion following from both: every body must be either heavy or light (301b17–18).

The key step in the argument (one that is denied in modern physics) lies in the claim that in the same time, a heavy body, B, will be moved a greater distance than will a body, A, which is not heavy. Aristotle assumes here that B is moved naturally, i.e., toward its natural place; as we have seen, when an element is moved toward its natural place, its inclination is being actualized. Thus B will be moved a greater distance than A because being heavy is B's inclination: its intrinsic principle of being moved, its active orientation toward the center, its very nature.[24]

This claim is best read in conjunction with Aristotle's concept of nature and the identification of nature as an intrinsic principle of being moved and being at rest. By virtue of what it is by nature, B is determined to be heavy and hence to be moved downward. In a determinate cosmos, what is properly determined – such determination constituting at once a thing's unique nature and its ability to be moved – must be actualized faster than what fails to be so determined. What is not heavy, A, has no

---

22 Although he expresses the point somewhat differently (and refers only to the closing lines of *De Caelo* III, i.e., 8, 307b20 ff.), I believe this view is in agreement with case argued by Morrow, "Qualitative Change in Aristotle's *Physics*," 154–167.
23 Cf. Aristotle, *Physics* IV, 8, 215bff.
24 We should again recall the one appearance of "inclination" at *Physics* IV, 8, 216a14–15.

such inclination, i.e., no active determination to be moved downward; hence it can only be moved downward by constraint.

On the one hand, if A has some other determination and hence some other natural motion (since each of the elements has one natural motion), then its motion must become slower as it moves downward because downward is away from its natural place. On the other hand, if we assume (contrary to fact) that A possesses no determination, it still must be slower than B because it fails to be heavy, i.e., possesses no intrinsic orientation downward, and so would be moved more slowly.

In a determinate cosmos, i.e., a cosmos constituted by an intrinsic relation between a thing and its natural place, because B is heavy, it must be moved a greater distance than A, which is not heavy, in an equal time. Consequently, within such a cosmos, Aristotle's contradiction must result. (We might note that before modern physics could deny Aristotle's claim about the motions of bodies A and B, his determinate world had to be replaced by an indeterminate, i.e., directionless, world.)[25]

The same argument can be made for what is light. Aristotle does not spell out the argument, but it is easy to see what it must be. To be light is not mere absence of being heavy; rather, it is a different determination: natural orientation upward. Therefore, being light is an inclination, an intrinsic ability to be moved upward within a cosmos determined in respect to up and down. Hence, in the same time, a light body, C, will be moved a greater distance than another, D, which is not light. Another body that is light but less light than C can always be found that will be moved the same distance as D with the result that two bodies, one light and the other not, will be moved the same distance in the same time, which is impossible.

This argument is designed to prove that all bodies must be either heavy or light.[26] And the contradiction reaches just this conclusion if we think of the cosmos as orderly, i.e., made determinate in respect of "where" by place. Within such a cosmos, the assumption of a body that is not heavy (or light), i.e., that it is indeterminate, must produce a contradiction. Hence on the assumption of a determinate cosmos, all bodies must be either heavy or light.

I wish to emphasize two points before turning to the second argument that all bodies must be either heavy or light. (1) Aristotle does not argue directly *either* for his account of natural place and the sense in which it renders the cosmos determinate *or* for the identification of inclination with heaviness and/or lightness. As I have argued, natural place is a formal

---

25 I have argued that such a replacement can be found at least as early as Duns Scotus, *Aristotle's Physics and Its Medieval Varieties*, 173–187.

26 This point emerges clearly again at the summation of the argument: "from these things it is clear that all things are either heavy or light and how there are motions contrary nature," *De Caelo* III, 2 301b31–32.

principle rendering the cosmos determinate, and the nature of the elements is identified with inclination as an ability to be moved. Consequently, the further identification of inclination with heaviness and/or lightness does not add anything new to the argument; it merely specifies more clearly both the principle of determination in things – the principle that has already been established as inclination – and the relation between inclination as an intrinsic principle of motion and proper natural place in a determinate cosmos. In relation to place, inclination expresses itself by being heavy or light – orientation down or up.

(2) The sharp conceptual difference between heaviness and weight reappears in this argument. "Weight," a post-Copernican concept, is defined as an extrinsic relation between mass and distance that is nondirectional and bears no relation to problems concerning speed.[27] So, for example, in a vacuum a penny and a feather fall (whichever direction that may be) at the same rate. But as principles, Aristotle's "heavy" and "light" presuppose a determinate cosmos in order to act as intrinsic determinations, the one as an inclination downward, the other as an inclination upward. Furthermore, these inclinations are at once a thing's nature and its principle of motion, i.e., its ability to be moved. Hence, they are immediately and directly related both to motion and to how "fast" the moved thing is carried.

The importance of this issue can be seen in Elders's treatment of this text. He claims that Aristotle provides a "discussion of weight" in which the two bodies serve as "concrete examples."[28] But this is not a discussion – it is an argument in the standard *reductio* form. Elders suggests that the possibility reached in this example is "presumably base[d] . . . on his observation of falling bodies" and concludes that "Aristotle's argument only succeeds in ascertaining the relation between size and speed of moving bodies."[29] But Elders's suggestion is impossible. Size plays an instrumental role in producing the contradiction and is not an object of the argument. Finally, as I suggested earlier, "speed" does not appear in this argument. The question of what is at stake in this conceptual difference is important, and I shall return to it.

Whereas this argument rests on natural motion, a second argument, which considers motion by constraint, reaches the same conclusion, namely, that every body must be either heavy or light. (The ultimate conclusion that the elements could not have been generated absolutely will then follow.) A body that is neither heavy nor light would necessarily be

---

27 For an account of Aristotle's position in terms of weight as Galileo conceives of it, cf. O'Brien, "Aristotle's Theory of Movement," 51–57. Stocks translates the opening of the *reductio ad absurdum* argument at *De Caelo* III, 2, 301a26–29: "Suppose a body A without weight and a body B endowed with weight. Suppose the weightless body . . ." But "weightless" is not the same as "not heavy."

28 Elders, *Aristotle's Cosmology*, 287.   29 Elders, *Aristotle's Cosmology*, 288.

moved by constraint (since by definition it would have no inclination, i.e., no intrinsic principle of natural motion actively oriented toward natural place); consequently, having no natural place in which it rests by nature, the moved thing would be moved infinitely (*De Caelo* III, 2, 301b-4). And the mover that produces motion by constraint has some power by which the moved is carried; something smaller and lighter will be moved further by the same power than will something larger and heavier.[30]

Let us take A, which is lighter – here lighter cannot mean oriented upward but must mean "not as heavy" – to be moved ΓE and B, which is heavy, to be moved ΓΔ in the same time (301b6–7). ΓΔ is obviously smaller than ΓE because the smaller and lighter A can be moved by constraint further by the same power in the same time than can B, which is heavy. We could divide off from B some part so as to form a proportion, B is to "this part" as ΓΔ is to ΓE; this part would be carried across ΓE in the same time that B as a whole is carried across ΓΔ because the quickness of the lighter body will be in proportion to the quickness of the greater as the greater body is to the lighter (301b8–13). Because the less heavy a body is, the faster it is moved by constraint, the quickness of the moved body will be in inverse proportion to its heaviness. Consequently, a body that is not heavy, i.e., A, and one that is heavy, i.e., the part taken from B, would be carried the same distance in the same time – which is impossible (301b13–14).

Furthermore, an indeterminate body, being neither heavy nor light, would be carried further than any that is heavy (however small we take it to be); in an indefinite time, such a body would be moved infinitely, in contrast to any body that is heavy or light (301b14–16). Although it is not spelled out, the point seems clear. A body that is determined toward a natural place becomes harder to move as it is moved by constraint away from that place. Consequently, if we assume that the power moving a heavy or light body by constraint remains the same, the body moved by constraint would eventually reach a point sufficiently far from its natural place that this power would no longer be sufficient to move it. And this point holds for any determinate body, whether heavy or light. But an indeterminate body would have no natural place and so could be moved infinitely by the same power. Therefore, the motion of a determinate body, unlike that of an indeterminate body, would never be infinite. Hence, as both arguments show, every body must be either heavy or light (301b17–18).

A comparison of the two arguments, one based on unnatural motion and one based on natural motion, shows that both tell the same story: the

---

30 Aristotle, *De Caelo* III, 2, 301b4–5; we must note an ambiguity in the word "light," or "lighter." "Light" can mean either the determination to go upwards – its stronger and more proper meaning within Aristotle's physics – or the mere absence of the determination of heaviness, a derivative meaning for which there is no other convenient word. Here A is lighter than B in the second sense, i.e., it does not possess the determination of heaviness, i.e. to go downward, and in fact A is indeterminate.

cosmos must be determinate, and within a determinate cosmos all bodies must be either heavy or light. As determinate, the cosmos is orderly,
and natural motion must be prior to unnatural. Here, the first argument
begins with natural motion and concludes that all bodies must possess inclination in respect to being heavy or light; the second argument begins
with unnatural motion in order to reach the same conclusion. Therefore,
Aristotle concludes, every body must be either heavy or light (301b16–
17). And because every body must be heavy or light, i.e., possess inclination, we can understand nature and natural motion as this intrinsic
principle at once constituting a thing's nature and dynamically orienting
it toward its natural place, down for earth, up for fire, and toward their
respective places in the middle for air and water.

Force: The Extrinsic Source of Motion

Aristotle next takes up the problem of force [δύναμις]. The sequence and
structure of his ideas here reveal an important feature of his physics as
the science of things that are by nature. The primary problem is whether
the elements can be generated; the solution rests on an account of motion that focuses primarily on natural motion. Natural motion in things
leads to an argument concerning inclination in respect to being heavy or
light, and this argument concludes that all body must be either heavy or
light. The account of inclination rests on the definition of nature as an
intrinsic source of motion. In its turn, this intrinsic source of motion contrasts with an extrinsic source of motion, called force, which may operate
during natural motion and is the sole cause of violent motion. Hence violent motion enters the account by its association with force, which entered the argument in contrast to nature.[31] An account of force explains
first how it may contribute to natural motion and then how it is exclusively responsible for violent motion. The main conclusion – the elements cannot have been generated absolutely but may be generated from
one other – then follows.

   Because physics is the science of things that are by nature and because
violent motion lies outside nature, violent motion by definition lies outside the domain of physics, properly speaking. Aristotle often treats natural and violent motion as parallel and so may be thought of as dividing
motion into these two kinds and accounting for first one and then the
other. So O'Brien claims: "[m]ovement by force is thus so to speak the
mirror image of natural movement."[32] But in an important sense this
claim is misleading. Aristotle does distinguish between them and treat

---

31 Elders claims that the argument about force has only a loose relation to what has preceded, *Aristotle's Cosmology*, 288.
32 O'Brien, "Aristotle's Theory of Movement," 77.

them as parallel in the sense that for both it is "equally" true that a mover
is required to produce motion in the moved; nevertheless, natural and
violent motion are not of equal status within his physics. The dominant
interest of a physics of things that are by nature must be natural motion;
violent motion (and any account of it) is ancillary at best. We see the evidence of this interest here in *De Caelo* III, 2: natural and violent motion
are not contrasted directly with one another. Rather, the contrast is between an intrinsic source of motion, i.e., nature, and an extrinsic source
of motion (when a thing is moved away from its natural place), i.e., force.
Violent motion appears only in association with force as an extrinsic
mover.

The argument is complex and difficult. Contrasting nature with force,
Aristotle defines force and then explains how force operates during first
natural and then violent motion. Unlike nature, force is a source of motion in something extrinsic to the moved (or in itself *qua* other) (301b17–
18). And force plays a role, albeit a different role, during both natural
and violent motion. Natural motion, for example, a stone being thrown
downward, will be made faster by an external force, but unnatural motion will be due to force alone (301b19–22). In both cases, air is required
as an "instrument" [ὄργανον] because it is by nature both heavy and light
(301b22–23). "Insofar as it is light, it will make the locomotion upward,
whenever it is pushed and takes the source [of motion] from the power
[the original mover, i.e., the thrower of a stone], and, again *qua* heavy it
will make downward motion" (301b23–25). In either case, i.e., in either
upward or downward motion, the original force gives the source of motion by hanging the moved body in the air (301b26–28). For this reason
an object moved by force continues to be moved, even though the mover
does not follow it, and this is why violent motion requires air. Furthermore, the natural motion of each thing is helped on its own course in the
same way (301b29–30). And, Aristotle concludes, it is clear both that
everything must be either heavy or light and how unnatural motion occurs (301b31–32).

A number of points in this argument require explanation. Whereas nature is an intrinsic source of motion, force [δύναμις] is a source of motion in another, i.e., extrinsic to the moved. The contrast between nature
and force is clearly between an intrinsic and an extrinsic source of motion. Extrinsic force makes natural motion faster, e.g., when a stone is
thrown downward, and in violent motion such force is the sole cause of
the motion. In both cases, (1) air is required as an instrument, and (2) the
original force, e.g., the thrower of a stone, hangs the moved body in the
air. These points, I shall now argue, can be fully explained in terms of
Aristotle's physics as an account of things that are by nature, given that
nature is a principle of order and natural motion is prior to and presupposed by violent motion. (A reminder: by definition, natural motion is

motion toward a thing's natural place, and violent motion[s] is away from a thing's natural place.)

The case of natural motion is relatively easy. The cause of natural motion is proper actuality, i.e., place (downward in the case of a stone), toward which the intrinsic principle of motion, i.e., inclination (e.g., heaviness in the case of a stone), is actively oriented. Hence, extrinsic force is not the cause of the motion *per se,* but only makes it faster. A fuller argument for this view is, as we have just seen, spelled out in *De Caelo* II, 14.

During both violent motion and natural motion, air is the "instrument," Aristotle says, because it is by nature both heavy and light. What does "instrument" mean here, and what is the explanatory force of air's nature as both heavy and light? Again, the notion of nature is easy to define: as a middle element, air is sometimes naturally inclined downward, i.e., is heavy, and at other times is naturally inclined upward, i.e., is light.

But the word "instrument" [ὄργανον] is puzzling. It does not appear elsewhere in the *Physics* or the *De Caelo.* We can best understand it if we see why this argument presents a special, if not unique, problem for Aristotle's account of motion as a mover/moved relation – a relation requiring that "everything moved must be moved by something." (In post-Copernican physics this entire problem disappears because motion is not divided into "natural" and "unnatural" – it is not divided this way because the cosmos is not conceived of as immediately orderly – and motion is not explained by the identification of a mover such that "everything moved must be moved by something" because now "things in motion tend to stay in motion and things at rest tend to stay at rest.") Aristotle virtually always distinguishes between natural and unnatural motion. But here they are alike in two important ways, (1) force may be present during either, and (2) both require that "air" act as an "instrument." During both natural and violent motion, e.g., after a stone has been thrown, the air is the extrinsic mover (we do not yet know how); hence, we might expect a single word, e.g., "mover" or "power" to characterize the air in both cases. But "mover" describes the air only during violent motion; during natural motion air is not the mover, but only makes natural motion (itself produced by a mover, who must be actuality but is not identified here) faster. "Power" or "force" also fail to describe the air properly speaking; "power" or "force" describe not the air itself, but what the air exerts on the stone during both natural and unnatural motion. Furthermore, air exerts force during both natural and violent motion, but the role of this force differs in the two kinds of motion: it makes natural motion faster but is the sole cause of violent motion. Hence, "instrument" is best thought of as a general word describing air as the source of force in both its roles, i.e., causing violent motion and making natural motion (caused by actuality and the intrinsic relation of potency to actuality) faster. And air plays these roles because of its nature.

Air is by nature both light and heavy. Like water, air is a middle element and so by nature is moved by its proper place – the place toward which it is actively oriented – either upward, insofar as it is light, or downward, insofar as it is heavy. Insofar as air is light, it is moved upward naturally and, being moved upward by its natural place, can move another, e.g., a stone thrown upward into it; insofar as air is heavy, it is moved downward and so can hurry the downward motion of something heavy, e.g., a stone thrown downward or a stone thrown upward but now returning downward.

Here we reach the heart of the affair. Aristotle says that the air, being pushed, produces upward or downward locomotion whenever it takes the source of motion from the original thrower of the stone. The thrower gives the source of motion "by hanging the moved body in the air." It is easy to see that violent motion requires air as an external mover. But how do these relations work together to explain upward or downward motion of a stone after it is thrown? I shall now argue that they are best understood in light of the arguments of the *Physics*. Indeed, in *Physics* VIII, 10, we find an argument that in the context of its own topic resembles this one.

As with Aristotle's other *logoi*, the topic and structure of *Physics* VIII is important. Its principle thesis is that motion in things must be eternal.[33] After a formal argument for this thesis, Aristotle raises and briefly resolves three objections to it, but claims that these require further consideration (*Physics,* VIII, 2). Two objections collapse into a single question: why are some things in the world at one time at rest and at another in motion (*Physics* VIII, 3, 253a22–24)? And its solution completes the argument: some things are moved by a single eternal unvarying mover and so are moved always, whereas other things are moved by something itself changing and so they too change (*Physics* VIII, 6, 260a14–16).

But a third objection remains: if motion in things is to be eternal, it must be shown first that some motion must be capable of being continuously and secondly what this motion is.[34] This motion, he concludes, must be circular locomotion (*Physics* VIII, 9, 265b12–15). But this motion is not fully explained until we reach its first cause, namely a first unmoved mover.[35]

To establish the cause of the first continuous motion, circular locomotion, Aristotle first establishes that an infinite force cannot reside in a finite magnitude nor can a finite force reside in an infinite magnitude (*Physics* VIII, 10, 266b25–27). Ultimately, he will argue that the first motion must be the motion of a single magnitude and must be produced by a single unmoved mover; otherwise, there will be not a single continuous

33 Aristotle, *Physics* VIII, 1, 252b5–7. Cf. H. Lang, *Aristotle's Physics and Its Medieval Varieties,* pp. 63–66.
34 Aristotle, *Physics* VIII, 7, 260a21–25; cf. H. Lang, *Aristotle's Physics and Its Medieval Varieties,* 88–89.
35 Aristotle, *Physics* VIII, 9, 266a6–8; cf. 1, 252a32–254b4.

motion, but a consecutive series of separate motions produced by separate movers (267a22–24). But before proceeding to this view, he raises a special problem, a problem that even though it appears in a different context, clearly relates to the problem at issue in *De Caelo* III.

If everything moved must be moved by something, then how can things no longer in contact with their movers, e.g., a stone thrown into the air, be moved (266b28–30)? It is impossible to say that the mover moves the air, which being moved, moves the stone, because this claim just postpones the problem; that is, when the mover stops moving, the air stops being moved and so cannot move the stone – indeed, this view implies that all things are moved and cease being moved simultaneously, which is clearly false (266b30–267a2). Rather, "we must say that the first mover makes, for example, air or water or some other such thing, which by nature moves and is moved, move [the stone]" (267a2–5). Hence, when the thrower throws the stone, the air – which itself is able to be both mover and moved – becomes its mover. Because the air is now a mover, it moves what is consecutive with it, i.e., the stone, and passes the stone along to the air next to it, which becomes the mover, etc. (267a2–8). The power of the mover (the air) becomes progressively less, and the motion ceases when there is not sufficient power to make the next member of the series a mover, but only a moved (267a8–10). At this point, the mover, the moved, and the whole motion stops (267a10–12).

Now Aristotle draws a conclusion that is strictly limited by the topical interest of the argument: this motion, i.e., the motion of the stone, is not continuous because it is not produced by a single mover; rather, it is produced by a series of consecutive movers, first the thrower, then the immediate air, and then the air consecutive with that, and so on. Because such motion requires a series of movers, it must take place in water or air and is called by some "antiperistasis" (267a15–17); because it is caused by a series of movers this motion cannot be the primary motion. And there is no account of the stone's motion beyond this conclusion, which addresses the topic at hand.

Two important points follow here. (1) This argument is designed strictly to show that the apparently continuous motion of a stone thrown upward is in fact a series of consecutive motions produced by a series of movers.[36] Considered as an explanation of projectile motion (as it commonly is), the account seems incomplete, if not an outright failure.[37] This

---

36 Machamer, "Aristotle on Natural Place and Natural Motion," 384, reads this argument in light of the claim that "[t]he point of *Physics* VIII is to relate all motion in the cosmos to the motion of the Unmoved Mover as the efficient cause." I have argued against this view in *Aristotle's Physics and Its Medieval Varieties*, 63–94.

37 For an outstanding example of how this argument may be read as an account of projectile motion as well as of the problems raised by such a reading, cf. Wolff, "Philoponus and the Rise of Preclassical Dynamics," 89–90.

"failure," however, originates not in "Aristotle's explanation of projectile motion," but in the economy and efficiency with which the argument operates within and is defined by its larger topic. Given his purpose here, Aristotle discusses *only* those features of "projectile motion" necessary for his conclusion: "projectile motion" cannot be the first motion.

(2) Altogether, the language of "projectile motion" may be somewhat misleading here.[38] The phrase "projectile motion" is associated with problems concerning the dynamics of moving bodies, and these problems are not found in Aristotle's physics because they represent and require concepts entirely foreign to both his science and his determinate world.[39] Finally, insofar as Aristotle's larger purpose may be said to raise the problem of projectile motion – the motion of a stone after it leaves the hand of a thrower – its explanation operates as do all his explanations of motion, i.e., in terms of the relation obtaining between a mover and a moved thing. In this sense, Aristotle affirms the principle "everything moved must be moved by something." Given this principle, the identification of a mover fully accounts for motion, and the force of this argument is to identify consecutive movers that account for the motion(s) of the stone.[40]

If the argument about the mover of the stone after it leaves the hand of the thrower is obscure, or even incomplete, the problems lie here. First, "air" and "water" are identified as able both to be moved and to move [another]. But this claim remains unexplained. Second, after rejecting the view that the original mover moves the air, Aristotle claims that the original mover makes the air (or water) move the stone. This claim too is obscure.[41] In both cases, the problems lie specifically with the identification of a mover – and this is Aristotle's problem, and so we may surely demand that he solve it.

I would suggest that Aristotle provides the requisite solution here in the *De Caelo*. Assuming that *De Caelo* III, 2, is consistent with *Physics* VIII, 10, we can reject the view that the original thrower moves the air that moves the stone and identify air as what moves and is moved. Air is moved naturally by its proper place, and, because it is moved naturally, air is able to move a stone after it has left the hand of the thrower. The key point is the role of natural motion: once the stone is placed into the air, the air is a mover not because of the stone's original thrower (the view rejected in *Physics*

---

38 Wolff speaks of "The Aristotelian theory of projectile motion . . . ," "Philoponus and the Rise of Preclassical Dynamics," 85, n. 3.
39 Although the issue goes beyond this study, I would suggest that the origin of the problem of "projectile motion" lies in late medieval physics and its criticism of Aristotle. For an older but still excellent summary of the issues, cf. Gilson, *History of Christian Philosophy in the Middle Ages*, 513–516.
40 Cf. H. Lang, *Aristotle's Physics and Its Medieval Varieties*, 84.
41 Heath, *Mathematics in Aristotle*, 156, cites Simplicius as helpful on this entire argument, because it is obscure on exactly this point.

VIII, 10) but because of its nature as a middle element, which is moved upward or downward. When a stone is placed into it, the air, by virtue of being moved by its natural place, moves the stone upward, which is by definition violent motion for a stone, or makes its downward (natural) motion faster. Because the air is moved by nature, its power either to move the stone or to make its motion faster does not depend on the original thrower and so does not cease when the thrower stops producing motion.

But then how does the original mover make the air move the stone? Whether the stone is thrown upward or downward, the thrower "hangs it up in," i.e., puts it into contact with, the air. The thrower places the stone into the air at which point the air becomes the mover. That is, the stone is "handed over" from the first mover, the original thrower, to that which by nature moves and is moved, the air. The air moves the stone as any container moves what is contained in it. Within the assumptions of his physics, Aristotle explains motion by identifying a mover, and this account is entirely consistent with his physics: it identifies first the thrower and then the air as consecutive movers during violent motion. Indeed, for this reason, he asserts, were there not some body such as air, violent motion would not be possible (*De Caelo* III, 2, 301b29–30).

And the argument concludes. An object moved by force continues to be moved even though the original mover does not follow it. If there were not some body, e.g., air or water, such motion could not occur.[42] The reason why is obvious: insofar as air is by nature both light and heavy, it serves as the requisite mover for the moved, such as a stone, once the stone is "in" the air. The same case can be made concerning how air helps a thing on its course during natural motion. Once the thrower "hangs," that is, places, a stone into the air by throwing it downward, the air immediately makes the natural motion faster by exerting extrinsic power on the moved by virtue of being naturally heavy (and so oriented and moved downward). In effect, this account resembles that of the lighter/smaller piece of earth being pushed by the heavier/larger piece behind it: actuality (downward or the middle) is the proper mover, and the extrinsic mover makes the motion produced by this mover "faster."

We may speculate that this account is consistent with, if it does not actually presuppose, the arguments of the *Physics*. A thing is contained in the air from which it is separate but by which it is surrounded (*Physics* IV, 4, 211a25ff). And a thing is moved accidentally if it is moved by virtue of the fact that that in which it is contained is moved; for example, water is moved accidentally whenever the glass in which it is contained is moved.[43] In short, a thing that is both moved and a container is itself the

---

42 Although it lies beyond the bounds of this study, we may note that, strictly speaking, water or air would not be necessary for natural motion. I believe this point accords with the arguments of *Physics* IV concerning place and void.

43 Aristotle, *Physics* VIII, 4, 254b8–9; cf. *De Anima* I, 3, 406a6–7.

mover of that which it contains. Because air is moved upward and downward by nature and contains the stone (which by being thrown has been
"hung up" in the air), the air is both moved and, in the case of violent
motion, is the mover exclusively responsible for the motion or, in the case
of natural motion, makes the motion – itself produced by proper actuality – faster.

Indeed, this argument is fully consistent in its most important features
with Aristotle's definition of physics as the science of things that are by
nature. It not only agrees with but further develops both his assertion that
violent motion presupposes natural motion and his denial that a void is
required for motion (in fact, a void would render motion impossible). Violent motion in this account presupposes natural motion because a stone
cannot be moved upward except by being contained in what is moved upward by nature, i.e., air. And, of course, in a void there would be no mover
and hence no motion. Finally, as we saw earlier, given Aristotle's potency/
act relation, the identification of a mover fully explains motion and in
this argument two movers are identified, first the thrower and then the
air – itself being moved by its actuality – and Aristotle clearly thinks that
he has explained violent motion.

Conclusion: The Elements Cannot Be Generated Absolutely

"From these things it is clear that everything is either light or heavy and
how motions contrary to nature occur" (*De Caelo* III, 2, 301b30–31). And
the next words turn to the problem of the larger argument: if everything
were generated, then the place in which things are now would earlier
have had to be a void in which there was no body (302a1–3). Hence, although it is possible for the elements to be generated out of one another,
absolute generation is impossible (302a4–5). For such generation requires a separate void (302a8–9).

The requirement of a separate void completes the argument that absolute generation of the elements is impossible. First, as has been shown
in *Physics* IV, the assumption of a separate void renders motion impossible. Place is the term required by motion. As we have just seen here, all
body must have one natural motion, i.e., toward its natural place, and this
motion shows why the cosmos is orderly rather than disorderly. In short,
the requirement of a void is another way of making disorder or indeterminacy prior to order and determinateness. Hence it must be rejected:
the elements cannot have been generated absolutely.

The remainder of *De Caelo* III develops Aristotle's own view (along with
accounts of his predecessors) that the elements are generated from one
another. Having concluded that everything cannot have been generated
absolutely (unless we assume a separate void, which is false), he raises the
question of what bodies are subject to generation and why. To answer it,

he defines the elements: the elements are those bodies into which other bodies are divided, and they may belong to these bodies either actually or potentially, but are themselves indivisible into further parts different in form (302a10–11; 16–18).

The reason why there are elements is not difficult. Every body has its proper motion, some motions are simple, and these must be of simple bodies (302b5–9). But how many simple elements are there? The elements cannot be infinite in number (III, 4), but there must be more than one (III, 5). Hence we can conclude that there must be many, but a limited number (*De Caelo* III, 5, 304b21–22).

In *De Caelo* III, 6, Aristotle argues that the elements are not eternal, but are generated from each other. They cannot be eternal because "we see fire and earth and each of the simple bodies being destroyed" (304b26–27). In short, the elements of bodies must be generated and corrupted (305a14–32). And the generation of the elements returns us to their natures. That is, what is generated is body, i.e., what is heavy or light, and so must possess inclination. Because the elements must be subject to generation and corruption, they must be generated either from something immaterial or from a body and if from a body, either from one another or from something else (305a14–16). The account that generates them from something immaterial requires a separate void, which is impossible (305a16–17); furthermore, the elements cannot be generated from some body other than themselves because if they were there would of necessity be some body separate from and prior to them (305a22–24). But what would such body be?

But if on the one hand this [body] will have heaviness or lightness, it will be some one of the elements but on the other hand [if it] has no inclination, it will be unmoved, i.e., mathematical. And therefore it will not be in place. For [a place] in which [a thing] rests, into this it is also able to be moved. And if by force, then contrary to nature, and if not by force, then according to nature. If, therefore, it will be in place, namely somewhere, then it will be one of the elements. And if not in place, there will be nothing [generated] from it. For that which is generated and that out of which it is generated must be together. Since, then, [the elements] can be generated neither from something incorporeal, nor from some other body, it remains that they are generated from one another (*De Caelo* III, 6, 305a24–32).

Aristotle has argued in *De Caelo* III, 2, that every body has a natural motion, i.e., motion toward its natural place; furthermore, there must be certain bodies whose inclination is "in respect to being either heavy or light" (300a20–301a20; 301a21–301b30). Here in *De Caelo* III, 6, the elements are identified first with being heavy or light and then with inclination. That is, if a "body" has no inclination, then it will be unmoved and mathematical – it will not be a body; if a body does have such inclination, then it must be one of the elements. And each of the elements must be either

light of heavy. Therefore, there is no body without inclination in respect
to being light or heavy, and any body that has such inclination must be
one of the elements. Because the elements are those things out of which
all things in the world, both natural and made, are composed, then all
things in the world must possess inclination in respect to being light or
heavy.

As we have seen, inclination, again identified with being light or heavy,
is the intrinsic determination of a thing to be moved. And Aristotle calls
on this point when he asserts that what has no inclination is unmoved,
i.e., mathematical (or like the first mover). But a thing that is moved is
moved either toward or away from its natural place, i.e., it is moved either
naturally or violently. Hence, all body is movable and must be in place,
whereas all body that is in place is either an element or composed of the
elements. The elements have heaviness or lightness respectively as their
very natures, and anything heavy or light must be an element.

Investigation of the elements exhibits the deepest order of the cosmos.
The four "terms" of elemental motion turn out to be identical – (1) being
heavy or light, (2) possessing inclination as the very nature of the elements, (3) being moved, and (4) being in place. If something is heavy or
light, then it has inclination in that respect, is movable, and so must be
one of the four elements. The elements are among things that are and so
must be "somewhere." Conversely, absence of inclination entails being
unmoved and indicates that a thing is not in place, i.e., not within the
cosmos as defined by place. Because all movable things within the cosmos
are composed of the four elements whereas nothing outside the cosmos
is so composed, we again arrive at inclination as the very nature of the elements and so of all movable things.

The inclination of each element is an ability to be moved toward its respective natural place and to remain at rest in that place once motion is
complete. "Heaviness" or "lightness" are further names given to the respective inclinations of earth or fire perhaps because such names emphasize the active orientation of the moved toward the actualities that
serve as their movers. Thus everything moved is moved by another, the
elements are moved by natural place, and inclination is the nature of the
elements as oriented toward this actuality or as exercising natural activity, i.e., resting, once they have arrived in their proper place. In short, the
order of nature is not produced by an extrinsic cause but is (1) the very
fabric of the cosmos and (2) immediately teleological as an immediate
active orientation of the moved for its actuality that embraces all things
without exception insofar as they are made of the elements.

These claims are not new. We have already seen the close connection
between movers, moved things, and motion. We have also seen the relation between the elements as moved and place as their actuality as well as
the identification of inclination as the nature of the elements, i.e., as an

intrinsic ability to be moved toward each element's proper place. However, although Aristotle has identified inclination with light and heavy and has used this identification to specify a mover for the violent motion of the elements, he has not yet explained light and heavy and so has not yet explained inclination. The elements are moved by virtue of being light or heavy, and they rest in their respective natural places; but their actualization and/or activities remain unexplained. Given the radical identification of inclination with the elements – and hence with all body – as "heavy" and "light," the need for such an account is pressing. It appears in *De Caelo* IV, and I turn to it now.

7

Inclination: The Natures and Activities of the Elements

The conclusion that the elements are generated from one another raises the question of how this generation occurs. This question – indeed, the issue of the generation of the elements generally – presupposes a prior problem: what differentiates the elements? This problem is serious for two reasons. First, whatever differentiates each element must be generated when the element is generated. Second, whatever generates each element renders it unique and so makes the element be what it is according to its definition. Hence an account of what differentiates each element is central to the nexus of topics concerning the generation of the elements, their natures and motions. This "prior problem" is solved in the remainder of *De Caelo* III and IV, and the final account of the generation of the elements appears in the *De Generatione et Corruptione*.

Aristotle first criticizes his predecessors (*De Caelo* III, 7 and 8). The followers of Empedocles and Democritus explain the generation of the elements as an excretion of what is already there; this view reduces generation to an illusion – as if it requires a vessel rather than matter (*De Caelo* III, 7, 305b-5). And Aristotle quickly shows that it entails that an infinite body is contained in a finite body – which is impossible (305b20–25).

On other accounts, the elements change into one another, by means of shape or by resolution into planes (305b26). These alternatives are related because shapes (e.g., triangles, pyramids) are composed of planes, and the elements too are composed of some prior thing. Thus both require that the elements be composed of parts – and Aristotle produces a contradiction based on the relation between parts and whole: an element identified with a shape must always be divisible; but a division of a pyramid or sphere would leave a remainder that is not a pyramid or sphere and hence a part of an element will not be the element (305b30–306b1).

The same objection applies to the claim that shape causes the generation of the elements – the theory proposed by Plato at *Timaeus* 51A (306b5ff). Most importantly, the simple bodies, especially water and air, seem to receive their shape from the place in which they are (306b9–15).

Therefore, although Aristotle does not elaborate, shape is not primary, but is a consequence of "where" the element is. Hence it cannot cause the generation of the element.

The problem "what differentiates each element?" occupies the remainder of the *De Caelo*. The answer will be what renders each element unique and thereby constitutes its specific defining characteristic. Furthermore, if each element possesses a unique characteristic, when the elements are generated from one another, these characteristics too must be generated. Whatever characterizes, and so differentiates, each element must be proper to it. Therefore, in addition to preparing the way for an account of their generation, a solution to the problem of what differentiates the elements provides the account promised in *Physics* III, 1.

After rejecting Plato's account of the shape of the elements, Aristotle reveals his own strategy, which informs the remainder of the *De Caelo*.

For since the most important differences among bodies are those according to their affections, and functions and powers (for we say of each of the elements [that it has] both functions and affections and powers), first let us speak of what each of these is, thus investigating these things, we take up the differences of each in relation to each. [*De Caelo* III, 8, 307b19–24: ἐπεὶ δὲ κυριώταται διαφοραὶ τῶν σωμάτων αἵ τε κατὰ τὰ πάθη καὶ τὰ ἔργα καὶ τὰς δυνάμεις (ἑκάστου γὰρ εἶναί φαμεν τῶν φύσει καὶ ἔργα καὶ πάθη καὶ δυνάμεις), πρῶτον ἂν εἴη περὶ τούτων λεκτέον, ὅπως θεωρήσαντες ταῦτα λάβωμεν τὰς ἑκάστου πρὸς ἕκαστον διαφοράς.]

De Caelo IV investigates these differences. Here inclination appears within an account of elemental motion – an account that presents both the determinate world of Aristotle's physics and place as its primary principle of determination.

De Caelo IV seems to present a new starting point, reiterating both what is to be studied and that it belongs to the investigation of those things involving motion. We must investigate, Aristotle says, "concerning heavy and light, both what each is and what their nature is and on account of what cause they have such powers" (307b28–30). What renders a thing unique constitutes its specific difference, hence its nature, and hence, at least in part, the object of its definition. Consequently, this investigation concerns what renders each element and its nature unique – and what causes it to have its power. In short, *De Caelo* IV takes up the problem defined at the close of *De Caelo* III.

This investigation belongs to physics because "we call things heavy and light because they are able to be moved naturally in a certain way," and physics is the science of things that are by nature and that contain in themselves "some source or cause of being moved and being at rest."[1] And the account begins with a striking claim:

1 Aristotle, *De Caelo* IV, 1, 307b30–33; cf. *Physics* II, 1, 192b21–23; the verb form κινεῖσθαι here in *De Caelo* IV, 1, is identical to that of *Physics* II, 1. We may note that Stocks trans-

A name is not given for the activities of them [the light and the heavy], unless someone might think "inclination" to be such. Because the investigation of natural things concerns motion and these things have in themselves something, such as a spark, of motion, everyone makes use of their powers although without definition, indeed except a few. [*De Caelo* IV, 1, 307b32–308a4: (ταῖς δὲ ἐνεργείαις ὀνόματ' αὐτῶν οὐ κεῖται, πλὴν εἴ τις οἴοιτο τὴν ῥοπὴν εἶναι τοιοῦτον.) διὰ δὲ τὸ τὴν φυσικὴν μὲν εἶναι πραγματείαν περὶ κινήσεως, ταῦτα δ' ἔχειν ἐν ἑαυτοῖς οἷον ζώπυρ' ἄττα κινήσεως, πάντες μὲν χρῶνται ταῖς δυνάμεσιν αὐτῶν, πλὴν οὐ διωρίκασί γε, πλὴν ὀλίγων.]

Therefore, Aristotle concludes, it is necessary to look at his predecessors before proceeding to his own view (308a4–7).

Here we have the last appearance of "inclination" in the *De Caelo* and in Aristotle's physics. Some take "inclination" as a name for the activities of the elements as light and heavy; all physicists wish to use the power of the elements within their accounts of nature, although only a few define this power. Ultimately, Aristotle concludes (as usual) that his view is best, and in *De Caelo* IV, 3, he provides an account of the elements and of inclination itself. But given his earlier account of being light or heavy, there is a serious problem with identifying inclination with the activities of what is heavy or light. This problem lies at the heart of Aristotle's account of the elements and his physics more generally.

As we have seen, what is heavy or light possesses a power, δύναμις, to be moved naturally. "Power" generally indicates what is potential, which in turn is associated with an ability to be moved.[2] In *De Caelo* III "inclination" was clearly identified with being heavy or light and so with potency, i.e., ability to be moved; indeed, the heavy and the light are identified as "able to be moved naturally" (*De Caelo* IV, 1, 307b31–32).

Now inclination is proposed as a name for the activity of the elements, and activity is an expression of actuality. Can inclination be both potency and actuality for the elements? No. When potency is actualized by its proper mover, it terminates in actuality, which is itself often expressed as activity. So, for example, whereas the potency of each element is to be actively oriented and so moved toward its proper place, the activity of each element is to rest in its proper place because it is now actually, not potentially, in that place. A thing cannot be both potential and actual (or fully active) in the same respect at the same time without breaking the law of noncontradiction (*Physics* III, 1, 201a19–21). This problem, I submit, is fully solved by Aristotle's account of the nature, motion, and activity of the elements here in *De Caelo* IV.

lates it as passive: "we call things heavy and light because they have the power of being moved naturally in a certain way."

2 Cf. Aristotle, *Metaphysics* IX, 1, where it can also mean ability to move; however, that it is primarily associated with ability to be moved and matter comes out quite clearly at 8, 1049b19 ff.

The Account of the Elements

Being heavy or light is intimately connected both with inclination and with being moved. Here Aristotle fully specifies these terms and the relations entailed by them. He first points out that "heavy" and "light" have two meanings. (1) they can be used without qualification, and (2) they can be used in relation to one another (*De Caelo* IV, 1, 308a7–8).

(1) "For some things are always moved by nature away from the middle and others always towards the middle . . . And of these, what is moved away from the middle I say to be moved 'upward' and 'downward' what [is moved] toward the middle."[3] Indeed, to think that there is no "up" or "down" given in the heaven is absurd: "but we call the extremity of the all 'up', which indeed is in the position 'up' and by nature primary and since there is both an extremity and a middle of the heaven, it is clear that there will be an up and a down (just as many say, although inadequately)" (308a17–24). Here is the definition of absolutely light and absolutely heavy. The former is moved upward, or toward the extremity, and the latter is moved downward, or toward the middle (308a29–31). The cosmos is determinate (because place is its boundary) in respect to up, down, etc.; when the terms "heavy" and "light" are used without qualification, they refer not to one another, but to the cosmos as determinate in respect to place. In this sense, "light" and "heavy" lock together with the cosmos as "up" and "down," and, as we shall see, this unqualified sense constitutes the primary meaning of "light" and "heavy." This argument clearly presupposes the earlier arguments of the *De Caelo* as well as those of the *Physics*.

(2) The heavy and the light can also be defined relative to one another – the only sense recognized by Aristotle's predecessors (308a9–13). For example, of things that are heavy, wood and bronze may be called "lighter" and "heavier," respectively (308a7–9). One thing is lighter than another when both are heavy and have equal bulk, but one (the heavier) is moved downward faster than the other (the lighter).[4] "Lighter" here means "less

---

3 Aristotle, *De Caelo* IV, 1, 308a14–17: τὰ μὲν γὰρ ἀεὶ πέφυκεν ἀπὸ τοῦ μέσου φέρεσθαι, τὰ δ' ἀεὶ πρὸς τὸ μέσον. τούτων δὲ τὸ μὲν ἀπὸ τοῦ μέσου φερόμενον ἄνω λέγω φέρεσθαι, κάτω δὲ τὸ πρὸς τὸ μέσον. We may note that φέρεσθαι here is the same form as κινεῖσθαι, which appears at 307b32; but whereas Stocks translates κινεῖσθαι as passive: "they have the power of being moved," he treats φέρεσθαι as active or avoids the issue by treating it as a noun: "There are things whose constant nature it is to move away from the centre, while others move constantly toward the centre; and of these movements that which is away from the centre I call upward movement and that which is towards it I call downward movement." Moraux reads: "De par leur nature propre, certaines chose se portent invariablement loin du centre et d'autres se dirigent invariablement vers lui. De ce qui s'éloigne du centre, je dis qu'il se porte vers le haut et de ce qui gagne le centre, je dis qu'il se porte vers le bas."

4 Aristotle, *De Caelo* IV, 1, 308a31–33: πρὸς ἄλλο δὲ κοῦφον καὶ κουφότερον, οὗ δυοῖν ἐχόντων βάρος καὶ τὸν ὄγκον ἴσον κάτω φέρεται θάτερον φύσει θᾶττον. There is some difficulty

heavy"; since "heavy" is to be oriented downward, what is "less heavy" will be less oriented downward and so will be moved more slowly than what is heavier. This usage is profoundly inadequate because, although it explains why heavy and less heavy things are moved downward more or less quickly, it fails to explain "light" things, i.e., air and fire, which are not by nature moved downwards but upwards.

This argument clearly presupposes conclusions established earlier in the *De Caelo* and *Physics*. Light and heavy are defined by their relation to up and down, which are constituted by place. "Up" is identified with the first boundary of the cosmos, and "down" with the middle of the cosmos; consequently, up and down are not merely conventional; rather, the cosmos is intrinsically determinate because it possesses a periphery and a middle.[5] Absolute light is not mere absence of heavy, but a tendency to be moved to the opposite place, i.e., the periphery of the heaven, which constitutes its proper determination; likewise absolute heavy is defined by its determination, the middle of the cosmos. Thus, even though light and heavy differ in their particular determinations, they are the same in that both are intrinsic orientations within things towards respective proper place, the periphery or the middle, within a cosmos that, because it is constituted by place as its boundary, must be determined in respect to up and down. Because, each element (except aether) must be either heavy or light, an account of the elements that treats them as heavy or light will provide a universal account of what differentiates them. Since all sublunar bodies are composed of the four elements, such an account will embrace all material body. Thus the account both considers of "proper terms" and applies universally to all things.

Accounts of things as relatively light and heavy consider only things that are heavy and so fail to take account of what is absolutely light, i.e., "up." In effect, such accounts fail to recognize the determinateness of the cosmos and the causal role played by that determinateness for all things within the cosmos: the elements (and hence all body) must be absolutely heavy or absolutely light. For this reason, explanations of "the heavy" in accounts that fail to consider "the light" must be wrong. Failure to recognize the causal role of the cosmos as determinate – as possessing "up" and "down" intrinsically – is the reason why none of Aristotle's predecessors

with the Greek of this sentence. A more, although not altogether, literal translation would read: "Of two things that are heavy and have equal bulk, the one which is light, or lighter, in relation to the other is the one which when they are carried downward, the other is by nature faster." The Didot edition reads: "Ad aliud autem leve ac levius, id esse dicimus quod aliquo natura deorsum fertur celerius, utrisque pondus habentibus aequalemque molem."

5 Lloyd connects the directional determinateness of the heavens in the *De Caelo* with the fact that Aristotle believes the heaven sphere is alive and that "right," "left," etc., are features of animals, "Right and Left in Greek Philosophy," 41.

dealt with these problems adequately.[6] Conversely, Aristotle's recognition of the causal role of nature as a principle of order and the cosmos as determinate explains the superiority of his own account. And because "inclination" is proposed as the name given to the activities and/or powers associated with being either heavy or light, it lies at the heart of natural elemental motion, of all body, and of the cosmos as determinate.

In addition to failing to define heavy or light, none of his predecessors, Aristotle claims, explains how the elements are differentiated from one another; hence, they also fail to explain the generation of the elements. As we shall see, from his own view, these failures form not a list of independent problems, but a whole of closely related parts – and hence closely related failures.

For Aristotle's view, the most telling criticism is his rejection of the "atomism" of Plato's *Timaeus*. Plato claims that all bodies are made up of identical parts and a single material; hence being heavier always depends on possessing a greater number of parts (*De Caelo* IV, 2, 308b5–12). So bodies differ in kind by virtue of possessing more or fewer parts.

Plato's account says nothing about what is light and heavy (308b12–13). For "fire is always light and moved upward while earth and all earthy things [are moved] downward and toward the middle" (308b13–15). If the relation between parts and being heavy were true, then the more fire, the heavier it would be and the more slowly it would be moved upward; but the opposite appears to be the case: the more fire, the lighter it is and the faster it is moved upward (308b15–19). The same account applies to air and water. If we posit a large amount of air and a small amount of water, then there would be more parts of air than water and the air should be heavier and go downward more readily than should the water, which would be lighter (308b21–26). But exactly the opposite occurs: a larger amount of air is carried upward more readily than even a small amount of water, and, indeed, any amount of air however large or small will always be carried upward from water.[7] These arguments recall Aristotle's earlier criticism: Plato fails to consider body as such, i.e., as heavy and light; consequently, he fails to recognize (and *a fortiori* to explain) the relation between body and the cosmos as determinate.

Aristotle turns in *De Caelo* IV, 2 to an "older" view of the problem of why fire goes up. This view provides clues as to what a successful solution (presumably Aristotle's own) to this problem must accomplish. First, it must account for natural upward motion, such as that of fire or air. The "fact" that a greater amount of fire or air rises more quickly than does a

---

6 I would suggest that the same remark applies to discussions that treat the heavy and light in Aristotle as "weight." For a clear example, cf. O'Brien, "Aristotle's Theory of Movement," 54: "Absolute weight, whether lightness or heaviness, belongs to a body."

7 Aristotle, *De Caelo* IV, 2, 308b26–28; what we call "evaporation" is included here; for a parallel passage, cf. *Physics* VIII, 4, 255b18–21.

smaller amount provides evidence (according to Aristotle) that upward motion is not somehow merely relative to downward motion; upward motion must be an independent determination given in nature and requiring its own definition. That is, some things, namely fire and air, are always moved upward if nothing hinders, whereas others, earth and water, are always moved downward. This determination, which Plato wholly fails to address, is what Aristotle means by "what is absolutely light and heavy."

The remainder of *De Caelo* IV, 2, focuses on views that explain differences in upward and downward motion by saying that void is imprisoned within bodies and thereby makes the body lighter (309a6–7). These views offer a strong advantage because, unlike Plato, they can theoretically explain composite bodies in which the heaviness does not correspond to the bulk – a small amount of bronze is much heavier than a large amount of wool (309a2–6). And in fact, those who argue that there is no void, for example, Anaxogoras and Empedocles, pass over the problem of the relation between heaviness and bulk in silence, giving no clues as to how they would resolve it (309a19–27).

Aristotle's own position concerning the differentiation among the elements now follows. He begins with a problem that most thinkers find troubling:

> on account of what the simple bodies are carried, some upward and some downward and some both upward and downwards; and after these things we must consider heavy and light and the things following from them, concerning their affections and on account of what cause each comes to be. (*De Caelo* IV, 3, 310a16–21)

The phrase "on account of what" is often translated "why"; but literally we need a "what" – in short a mover.[8]

As we have seen, Aristotle defines motion as the actualization of the potential by its proper actuality; thus, the identification of a proper mover for the moved explains its actualization. In *Physics* VII, 1, and VIII, 4, he argues that "everything moved must be moved by something"; the identification of a mover constitutes an explanation of motion in the moved because for Aristotle motion is by definition nothing other than a mover/moved relation.[9] So the question here, διὰ τί "on account of what," asks what serves as the actuality, or mover, when the simple bodies are moved to their appropriate places by nature. And here too the identification of a mover fully accounts for the motion of the elements. Consequently, it alone answers the question why.

---

8 Guthrie and Stocks both translate this phrase as "why," which I shall examine shortly. Moraux translates διὰ τί as "pour quelle raison." Elders comments on this line that "Aristotle wants to investigate the cause of the coming-into-being of the light and the heavy natural bodies," *Aristotle's Cosmology*, 342.

9 Aristotle, *Physics* VII, 1, 241b34; VIII, 4, 255b32–256a.

And the causal explanation of elemental motion begins directly. When each element is moved in respect to place, it does not differ from other kinds of change, e.g., increase and decrease or generation.[10] Most importantly, change does not occur by chance:

change not being into some chance thing by some chance. And likewise a mover is not some chance thing of some chance thing. But just as some thing altered and some thing increased are different, thus also what produces alteration and what produces increase. And in the same way we must regard also the mover according to place and what is able to be moved as not being some chance thing of some chance thing. If therefore what makes a thing heavy and what makes a thing light is a mover of what is down and what is up [respectively] and the potentially heavy and light are able to be moved, then being carried toward its place for each thing is being carried toward its form . . . [310a26–310b: καὶ οὐκ εἰς τὸ τυχὸν τῷ τυχόντι μεταβολὴν οὖσαν· ὁμοίως δὲ οὐδὲ κινητικὸν τὸ τυχὸν τοῦ τυχότος· ἀλλ' ὥσπερ τὸ ἀλλοιωτὸν καὶ τὸ αὐξητὸν ἕτερον, οὕτω καὶ τὸ ἀλλοιωτικὸν καὶ τὸ αὐξητικόν. τὸν αὐτὸν δὴ τρόπον ὑποληπτέον καὶ τὸ κατὰ τόπον κινητικὸν καὶ κινητὸν οὐ τὸ τυχὸν εἶναι τοῦ τυχόντος. – εἰ οὖν εἰς τὸ ἄνω καὶ τὸ κάτω κινητικὸν μὲν τὸ βαρυντικὸν καὶ τὸ κουφιστικόν, κινητὸν δὲ τὸ δυνάμει βαρὺ καὶ κοῦφον, τὸ δ' εἰς τὸν αὐτοῦ τόπον φέρεσθαι ἕκαστον τὸ εἰς τὸ αὐτοῦ εἶδός ἐστι φέρεσθαι . . .]

Aristotle goes on to explain the last point here, namely, how the place to which a thing is moved naturally is its form. Before proceeding, however, the claim that the mover and the moved are always in the same category of motion and are not related by chance requires consideration.

I turn to linguistic issues first because they are prior to substantive matters and because my translation differs in several important respects from the "standard" translations of Stocks and Guthrie. The issues at stake here have been addressed in the analysis of nature, but are crucial to the construal of this argument. First and most importantly, while both Guthrie and Stocks render φέρεται as active and φέρεσθαι as a noun, "motion," I translate both φέρεται and φέρεσθαι as passive. To suppress the passive force of φέρεσθαι here obscures the force of the mover/moved relation throughout the argument. Furthermore, as we have seen, Aristotle defines "motion" (the noun κίνησις does not appear in this passage) as a relation between movers and moved things. And this relation is the explicit topic of the argument here: "on account of what" are the simple bodies carried, some upward and some downward. Favoring a noun, which is more general, rather than the exact verb forms here suppresses this relation – itself quite explicit in the Greek – and consequently obscures the argument.

Differences in translation are even more serious for the concluding

10 Aristotle, *De Caelo* IV, 3, 310a20–23: "The case concerning each [element] being moved to its place must be regarded as similar to other forms of generation and change. For since there are three motions, that according to size, that according to form, and that according to place, we see in each of these the change coming to be from opposites into opposites and things intermediate and . . . "

sentence concerning place and form. Although the translations of Stocks and Guthrie differ slightly, their construal of the grammar is identical. Stocks reads: "Now, that which produces upward and downward movement is that which produces weight and lightness and that which is moved is that which is potentially heavy and light, and the movement of each body to its own place is motion towards its own form." And Guthrie reads: "We may say, then, that the cause of motion upwards and downwards is equivalent to that which makes heavy or light, and the object of such motion is the potentially heavy or light, and the motion towards its proper place is for each thing motion towards its proper form."

First, the sentence must be a conditional, presenting a clearly defined protasis and apodosis. Second, each of the two nouns following the μέν has a definite article [μὲν τὸ βαρυντικὸν καὶ τὸ κουφιστικόν], and the presence of the μέν plus these articles tells us that here we have the subject of the clause. Exactly the same point holds for the second clause – κινητὸν δὲ τὸ δυνάμει βαρὺ καὶ κοῦφον. The place of the δέ plus the article tells us that the subject must be "that which is potentially heavy, and light." Furthermore, κινητόν does not mean "moved" (which would require κινούμενον – Aristotle's regular word for what is moved), but "able to be moved." This phrase completes the protasis of the conditional sentence, and we see two contrasting parallel phrases: if on the one hand what makes a thing heavy and what makes a thing light is a mover and on the other what is potentially heavy and light is [something] able to be moved, then . . . The apodosis now follows.

The apodosis also consists of two parallel phrases, both possessing the same articular infinitive, τὸ φέρεσθαι, as their subject. To be moved to its place is to be moved to its form for each, i.e., the heavy and the light. An explanation of the identification of place and form constitutes the next step of the argument. But before turning to this step, we must consider the substance of this argument.

The relation of mover to moved returns us to the account of motion in *Physics* III. Most importantly, the actuality of the mover is the actuality of the moved, insofar as it is potential for that actuality (*Physics* III, 3, 202a15–20). So an oak is the actuality of an acorn, and the "acorn" is "potential oak." Furthermore, what is moved is always potential *for* that actuality. Because actuality is the fulfillment of the potential, what is potential is never "neutral" to its actuality; rather it is actively "aimed at" it.[11] Therefore, in the absence of hindrance, what is potential cannot fail to be moved by its proper actuality. And in this very strong sense, mover and moved are not related by chance: as actuality and potency *for* that actuality they are identical in actuality.

---

11 For two related senses of being "aimed at," cf. Aristotle, *Physics* I, 9, 192a20–23; *N. Ethics* I, 1, 1094a-3.

And the elements are no exception. For them too, what moves and what is able to be moved are not related by chance. First, the relation between potency and actuality presupposes two partners: (1) what is potential and as potential able to be moved by (2) what is actual and as actual able to move.[12] Given the relationship between potency and actuality along with the definition of motion as an actualization, a thing must be able to be moved before it can be actualized, i.e., actually moved. Consequently, Aristotle first explains what makes each element *able* to be moved.

The account is not difficult. An element is moved naturally because it is heavy or light; consequently, what makes it heavy or light makes it able to be moved. Because being heavy or light is the defining characteristic of body, what makes a thing heavy or light is its generator, i.e., what makes it be body. Hence, the generator is a mover toward the up and the down in an important sense: by making a thing heavy or light, the generator makes the element what it is potentially, e.g., potentially up or potentially down, and so makes it *able* to be moved up or down. Here is the first step in solving the problem of elemental motion: Aristotle defines the partnership between potency and actuality insofar as it involves a mover, i.e., the generator, in the sense of what makes a thing *able* to be moved, i.e., as generated an element is light or heavy and so potentially up or down.[13]

Given these two conditions – (1) that the generator is a mover in the sense of making a thing able to be moved and (2) that the potentially heavy or light is able to be moved – Aristotle explains being moved without further ado. For each thing, being moved toward its place is being moved toward its form:

therefore, where some one part is carried by nature, also the whole. And since place is the limit of what contains and it embraces both the up and the down and the outermost and the middle for all things that are moved, in some sort of way it becomes the form of what is contained, for what is carried toward its own place is carried toward what is like [to it]. [*De Caelo* IV, 3, 310b6–11: . . . ὥσθ᾽ ὅπου πέφυκεν ἔν τι φέρεσθαι μόριον, καὶ τὸ πᾶν. ἐπεὶ δ᾽ ὁ τόπος ἐστὶ τὸ τοῦ περιέχοντος πέρας, περιέχει δὲ πάντα τὰ κινούμενα ἄνω καὶ κάτω τό τε ἔσχατον καὶ τὸ μέσον, τοῦτο δὲ τρόπον τινὰ γίγνεται τὸ εἶδος τοῦ περιεχομένου, τὸ εἰς τὸν αὑτοῦ τόπον φέρεσθαι πρὸς τὸ ὅμοιόν ἐστι φέρεσθαι.]

"In some sort of way" place, i.e., up for fire, down for earth, respective place in the middle for air and water, is the form of the elements. At *Physics* IV, 4, 211b13–14, place is characterized as the limit of the containing body while form is the limit of the contained. Because place is the limit of what

---

12 Aristotle's most extended discussion of this relation occurs in *Metaphysics* IX, which will be discussed later.
13 Such is exactly the account at *Physics* VIII, 4, 255b8–17.

contains, it constitutes the form of the elements. In effect, this identification specifies the relation between place and the elements.[14]

The force of this argument becomes clear if we recall several points established earlier. (1) The elements are simple and undivided. For this reason, they cannot be self-movers: in a self-mover, the moved must be distinct from the mover. Being simple, fire is nothing other than to-be-oriented-upward, which makes it light. (2) Each element possesses one and only one place that is natural to it. And if something is in place, it must be an element or composed of the elements. (3) Being "by nature," each element contains an intrinsic source of being moved and being at rest, and the activity of each element is to be at rest in its proper place. Thus, to be in its proper place is the actuality at which the element is by nature aimed, the completion of its actualization and its activity. Fire, when it is in its proper place, may be called "actually up" or "actually light," whereas earth is "actually down" or "actually heavy" and water and air are aimed at their respective places in between.

At the same time, (4) place, serving as the limit of the containing body, renders the heavens intrinsically determinate as "up," "down," "left," "right," "front," and "back." Hence, the outermost heaven and the middle of the heavens are "actually up" and "actually down" respectively because they are constituted as such by place. Here is the identity between place and the elements: fire, resting in its natural place, is active, i.e., is "actually up," and the place where this activity occurs is "actually up." "Actually up" is both the form of fire and what it most fully is by nature. Because place renders the heavens determinate in respect to "actually up," "actually down," etc., it may be said in a sense to be the form of the elements.

This conjunction of form and actuality yields the mover of the elements. (5) Motion is the actualization of the potential *qua* potential by that which is actual; Aristotle identifies both moving causes and final causes as actual (and form fully developed). Hence, to identify form is to identify that toward which the elements are actively aimed – their actuality and mover.

The respective natural place of each element makes it to be actually what the generator makes it potentially. Fire is potentially up (earth down) and its respective place is actually up (actually down). Potency and actuality are related as moved and mover: they go together and work together not by chance but "by nature." Hence, this relation identifies

---

14 Waterlow seems to think that place is physical and constituted by the elements, *Nature, Change, and Agency in Aristotle's Physics*, 105; Gill claims that the elements stop because they are "compelled" to do so by the boundary supplied by place, *Aristotle on Substance*, 239. Both Waterlow and Gill fail to understand place as a principle resembling form and making the world determinate. Hence they also confuse the causal relations at work here and finally Aristotle's categories of natural and violent motion. I shall discuss these views more fully later.

proper place as the mover of each element by specifying the intimate relation between the elements and place.

Here is the solution to the problem "on account of what" the elements are carried, some upward, some downward, and some both upward and downward. As the first boundary of the all, place [ὁ τόπος] is "the where" [τό πού] of all things that are and are moved. As its first limit, place renders the cosmos determinate in respect to "up" and "down"; as rendered determinate, place within the cosmos, i.e., up (the outermost circuit of stars) and down (the center of the earth) serves as the form, the actuality, and the mover for each element – itself absolutely simple and undivided. Since all things within the cosmos are composed of the elements, all things are to this extent moved by proper natural place as their form: heavy things always go down and light always go up, etc.[15] Consequently, in this respect, *nothing* in the cosmos escapes the order of nature and the teleology of that order, i.e., its expression as an active orientation toward or rest in proper place.

Waterlow asks why up and down are where they are – why could up not "lie to the north, for instance"?[16] She replies:

it must, then, be the very character of the region itself considered simply as a part of space, independently of whatever physical object it holds. But how can different parts of physical space differ intrinsically from one another? This makes sense only on the assumption that the space of the world is shaped. Probably the simplest hypothesis is the one that Aristotle accepts, namely that it is spherical.[17]

But there is no such thing as "physical space" in Aristotle. Place resembles form and renders place within the cosmos formally determinate, i.e., determinate in respect to direction. There is neither a cosmos apart from, or independent of, this determination nor any further principle of determination (in respect to "where") required by the cosmos to account for why the center is down and the periphery up.

However, this is not the end of Waterlow's story. One page later, she reaches a remarkable conclusion:

Fire does not happen to move in the direction that we call 'up'. It moves thus *because* that direction is intrinsically different from any other, so making manifest the intrinsic difference of fire from the other elements. But the 'because' clause need not refer to an already constituted fact concerning a periphery regarded as some kind of distinct reality. It can mean, and I suggest does mean, that fire moves upward because *by so doing* it will form the physical periphery or spherical outer shell of the sublunary world.[18]

---

15 Cf. Aristotle, *Physics* IV, 1, 208a29–32; cf. also *Physics* II, 1, 192b18–20.
16 Waterlow, *Nature, Change, and Agency in Aristotle's Physics*, 103.
17 Waterlow, *Nature, Change, and Agency in Aristotle's Physics*, 104.
18 Waterlow, *Nature, Change, and Agency in Aristotle's Physics*, 105 (italics in original). I have suggested elsewhere that this view is at least as old as Philoponus, *Aristotle's Physics and Its Medieval Varieties*, 106–124. For a telling critique of Waterlow, cf. Furley's review in *Ancient Philosophy*, 108–110.

Again, Waterlow thinks of place as somehow physically constituted. Furthermore, by reversing the relation between place and the elements such that the elements cause, rather than are caused by, place, she not only replaces Aristotle's formally determinate cosmos with a material cosmos, she obscures the causal structure of the entire argument. When place as a cause disappears, the motion of the elements, which must be moved by something, remains unexplained. Indeed, his failure to understand the role of place as a limit constituting the world as determinate leads Philoponus to posit "inclination" as the mover of the elements.[19]

A further problem remains, however: "on account of what" are the elements generated as potentially heavy and light and so able to be moved upward or downward? This motion too cannot be random but must be of like to like. If we think of the elements not only in respect to place but also in respect to one another, the same relation holds. That is, the one that lies next to another – e.g., water lies next to air and air next to fire – are alike, and in fact the higher is to what is under it as form is to matter.[20] So water is potential air, air potential fire and so on. Because they are alike and the higher is form to the lower, water can become air and air fire. Therefore, to ask why fire is moved upward is like asking why the healable, being moved, changes *qua* healable and not, for example, *qua* white.[21]

The transformation of the elements into one another (in order) is explained more fully in *Physics* VIII, 4. Arguing that "everything moved must be moved by something," Aristotle distinguishes two senses of the word "potential" (255b17). Water is potentially light (the first sense of "potential") as it is oriented toward air; but it may also be potentially light (the second sense of "potential") after it has become air because it may be hindered from actually rising (255b19–21). When, for example, water evaporates within a closed container, the container prevents it from rising and so it remains potential in the second sense. As soon as the hindrance, e.g., a stopper, is removed, the air is moved upward toward its proper place. Both moments, (1) the transformation of an element into the next in order and (2) the consequent motion toward respective proper place (its

---

19 Cf. H. Lang, *Aristotle's Physics and Its Medieval Varieties*, 106–124. In Philoponus, *In Phys.* 195.29 ff., Davidson cites Philoponus's commentary on the *Physics* as the earliest work that "construes the nature of the elements as a motive cause," *Proofs for Eternity, Creation, and the Existence of God in Medieval Islamic and Jewish Philosophy*, 267. Zimmerman, "Philoponus' Impetus Theory in the Arabic Tradition," thinks Philoponus is doing what Aristotle "balked" at doing, 121. For a critique of Zimmerman, cf. Helen S. Lang, "Inclination, Impetus, and the Last Aristotelian," 221–260.

20 Aristotle, *De Caelo* IV, 3, 310b11; 14–15; cf. also *Metaphysics* IX, 7, 1049a20–22. Citing this text of the *De Caelo*, Machamer, "Aristotle on Natural Place and Natural Motion," 380, claims that "[w]ater is the form of the planet earth in that it provides the innermost containing body of the earth . . . and also in that it is cold like the element earth," implying that Aristotle is giving a sort of physical account here. But this view is impossible.

21 Aristotle, *De Caelo* IV, 3, 310b16–20. The same example appears at *Metaphysics* IX, 7, 1049a3.

actuality), are actualizations of a potential by that which is actual. The immediate actuality of water is air, and the actuality of air is its proper upward place. Hence proper place is the actuality of both potencies – the ultimate actuality of water and the immediate actuality of air.

Because respective proper place is the actuality of each element, the transformation of each element into the next in the series and its local motion toward its proper place are alike and appear together in the argument. In both moments, potency (identified with matter) is aimed at proper actuality and actuality (identified with form) is proper for that potency. The healable is moved *qua* healable and not *qua* white because it is moved insofar as it possesses specific potency, ability to be healthy, aimed at its proper actuality, health. And both the transformation and the locomotion of the elements are by nature and are no different from any other motion, i.e., any other potency/act relation: it presents a specific potency in relation to its proper actuality. When in *Metaphysics* IX Aristotle discusses the meanings of potential and actual, as well as the relation between them, he explicitly includes these problems (and some of the same examples) concerning the elements. But before turning to this argument, a further point about the elements here in *De Caelo* IV must be raised.

In one respect, it now appears, the elements do differ from other moved things. Other things seem to have a readily identifiable mover from the outside; so, for example, what is curable may be moved by a doctor or by medicine, and, furthermore, may be moved in "opposite" ways, i.e., what is curable may also be made sick (*De Caelo* IV, 3, 310b26–31). But although the heavy and the light resemble all moved things insofar as each must be moved by something and mover and moved are not related by chance, nevertheless, they seem uniquely to have within themselves a "source of change" (310b24–25). After developing the example of what is curable or able to be sick, Aristotle returns to the elements and their motions:

But what is light and what is heavy seem more than other things to have in themselves the source [of change] because their matter is closest to substance. And a sign [of this fact] is that locomotion belongs to separate things and comes to be last of the motions; therefore this motion is first according to substance. On the one hand, then, whenever air comes to be from water, and something light from something heavy, it goes toward the upward. At the same time, on the other hand, something light is, i.e., it no longer becomes but is there. And it is clear that being potentially, it goes toward being actuality there and toward such and how much of which the actuality is both of such a kind and to such an extent. And in respect of the same cause earth and fire are moved, when they already possess the source of motion and are to their places, if nothing hinders. For also nourishment and what is curable, whenever it is not clogged or hindered, are carried straightaway. And both the maker of the principle and what removes a hindrance or when a thing rebounds, moves, just as has been said in the earlier arguments in which we distinguished that none of these moves itself.

On account of what cause each of the things moved is moved and what is being moved to its place has been said. (310b31–311a14)

The account here turns on two closely related points. The first concerns the notions of potency and actuality, which we have seen in Aristotle's definition of motion and must now consider in relation to substance. The second concerns the phrase "source of change" [ἀρχὴ τῆς μεταβολῆς]. What does it mean? I have argued that in the definition of nature the noun phrase ἀρχὴ τῆς κινήσεως must be understood in light of the passive verb form κινεῖσθαι. But here, although Aristotle refers to the argument of *Physics* VIII, 4, that none of these things moves itself, no passive verb lies behind the noun, and *prima facie* the phrase "source of change" may seem less clear. The solution to this puzzle can be found in an examination of potency and actuality in relation to substance.

The Problem of Potency and Actuality for the Elements

In *Metaphysics* IX, Aristotle's most extended account of potency and actuality, he comments that these terms are important not only for motion but for substance as well.[22] As I shall argue, he makes the connection between motion and substance explicit, and this connection will allow us to understand the connection between the motion and substance of the elements in *De Caelo* IV, 3. Furthermore, the phrase "source of change" appears several times in the arguments of *Metaphysics* IX; there its meaning is unambiguous: it signifies potency in the sense of an ability to be changed. And this meaning too should be understood in *De Caelo* IV. Indeed, the account of potency and actuality in *Metaphysics* IX bears a striking relation to the argument of *De Caelo* IV in its problems, its examples, and its account of the elements. I turn to *Metaphysics* IX now and shall conclude that it develops an account of the actuality and substance of the elements that both agrees with the arguments of the *Physics* and *De Caelo* and provides clear grounds for understanding the present argument.

Metaphysics IX, 1, begins by announcing its topic: the investigation of what is in the primary sense requires an examination of potency and actuality (1045b26–35). The account will begin with potency in the strictest sense, which involves motion, even though this is not the most useful for the present investigation, and will then proceed to potency and actuality as they go beyond the problem of motion.[23] This procedure seems odd: why begin with what is not the most useful for the argument at hand?

---

22 Aristotle, *Metaphysics* IX, 1, 1046a1; cf. 8, 1050a4.
23 Aristotle, *Metaphysics* IX, 1, 1045b35–1046a2; on the first sense of potency, which agrees completely with this account, cf. *Metaphysics* V, 12, 1019a15–20.

Aristotle sets out the "strictest" sense of potency first because it is the most important (even if not the most useful for this investigation). He next takes up special problems concerning potency, e.g., those raised by the school of Megara, and secondary meanings of the term (*Metaphysics* IX, 2–5). He then turns to actuality, which is in every way prior to, better, and more valuable than potency (*Metaphysics* IX, 6–9). The book ends with some remarks on being and not-being (*Metaphysics* IX, 10).

Why does the discussion of potency, which is not prior in any sense, appear first? Potency is first in the sense of beginning a movement in that which is acted upon; that movement, which is more properly caused by actuality, is nothing other than the actualization of the moved. And Aristotle's *logos* here resembles just such an actualization: it sets out from what is most important for the development of the whole, defines the starting point, and solves problems about it; it then proceeds to the fulfillment of the account and the consequent view of the world, concluding with minor problems raised by it. The *logos* in this sense resembles his position itself and, as I shall now argue, yields his world and the order of nature that lies at its heart.

"Potency" has several meanings, all associated with being able, or possessing a capacity. They all relate to one primary meaning: a source of change in another or *qua* other (*Metaphysics* IX, 1, 1046a10–11). Aristotle spells out what he means by a "source of change" as one sort of potency and contrasts it with a second sense. First, potency is capacity for being acted upon, a source of change in the thing acted upon that makes it able to be acted upon by another or by itself *qua* other (1046a11–13). Indeed – and this point is identical to that of the *De Caelo* – without an ability to be moved, a thing would be immovable.

Potency as ability to be acted upon, *Metaphysics* IX, 1, continues, contrasts with its second meaning, ability to act. Potency in this sense is not in what is affected but in the agent (1046a26). Here the language of containing a "source of change" is replaced by the language of making. This kind of potency is present in that which is able to heat or in one who is able to build a house (1046a27–28). What is able to be acted upon and what is able to act are both called potencies because both imply ability or power, and this is the meaning of "potency."

These two kinds of potency are always distinct, however, and Aristotle concludes his introduction to potency by emphasizing the distinction between the potency of the patient, which contains a "source of change," and that of the agent, the builder or maker. This distinction produces an important conclusion: because potency presents two different abilities in things, "insofar as something is united by nature, it undergoes nothing itself by itself" [διὸ ᾗ συμπέφυκεν, οὐθὲν πάσχει αὐτὸ ὑφ' ἑαυτοῦ].[24] That

---

24 Aristotle, *Metaphysics* IX, 1, 1046a28; cf. *De Generatione et Corruptione* I, 9, 327a2. I argue earlier that language such as this is directed against Plato's definition of soul as that which moves itself by itself; cf. *Phaedrus* 245C ff.

is, insofar as a thing is one, it cannot be both agent and patient in the same respect at the same time.

This account of potency agrees not only with Aristotle's account of motion in *Physics* III, 1, but also with the argument of *Physics* VIII, 4, which considers the elements as "united by nature." In *Physics* VIII, 4, Aristotle, rejecting the view that the elements are self-moved, asks how "something continuous and naturally unified can move itself by itself" [ἔτι πῶς ἐνδέχεται συνεχές τι καὶ συμφυὲς αὐτὸ ἑαυτὸ κινεῖν;].[25] Such things, including the elements, are impassive; only insofar as a thing is divided is one part the mover and the other the moved (255a14–15). Hence, neither the elements nor anything else that is continuous can be self-moved, precisely because the mover must be distinct from the moved (255a16–17). In *Physics* VIII, the fullest description of each element emerges as Aristotle gives the example of a ball rebounding from a wall and concludes: "but it has a source of motion, not of moving or making, but of being affected" [ἀλλὰ κινήσεως ἀρχὴν ἔχει, οὐ τοῦ κινεῖν οὐδὲ τοῦ ποιεῖν, ἀλλὰ τοῦ πάσχειν] (255b30–31). Thus, the elements are united by nature and possess potency in the sense (and in the very language) specified in *Metaphysics* IX, 1: a source of motion not as a mover or maker, but as a thing being acted upon, i.e., moved.

On several important points, then, *Physics* VIII, 4, and *Metaphysics* IX, 1, agree. Being one and continuous, things "united by nature" cannot be self-movers because motion always implies a distinction between what is able to be moved and what is able to move. According to *Physics* VIII, 4, the elements are one and continuous in just this sense. Furthermore, as ability to be acted upon, potency must be in the thing affected, and the elements contain just such a source, i.e., a source of being changed.

Although we shall see further evidence in a moment, it is worth noting that ἀρχὴ τῆς κινήσεως and ἀρχὴ τῆς μεταβολῆς are used here to mean the same thing: a source of being moved or being changed. And this must be the meaning of "source of change" in *De Caelo* IV, 3, as well.[26] Quoting *De Caelo* IV, 1 and 3, Wildberg makes an important claim:

From these passages it becomes clear that the weight and lightness of a body and its respective motion have the same cause. Aristotle does not say that weight *causes* downward motion, but that the weight of a body and its downward motion possess the same cause. This, of course, is the nature of that body, inherent in matter, initiating motion toward the proper form and end: when air is generated from

---

25 Aristotle, *Physics* VIII 4, 255a12–13. In Potts and Taylor, "States, Activities, and Performances," 85–102, Taylor argues throughout that *Metaphysics* IX is consistent with the definition of motion in *Physics* III, 1.

26 Although I shall not take up the argument explicitly, I should note that Seeck finds these arguments in sharp disagreement; because there is no unmoved mover in the *De Caelo*, one is reduced, in effect, to an account of self-motion, "Licht-schwer und der Unbewegte Beweger (*DC* IV 3 and *Phys.* VIII, 4)," 214 ff. This view may also underlie the account given by Judson, "Heavenly Motion and the Unmoved Mover," 155–156.

water, it rises and a potentially light element becomes actually light and is light once it has reached its proper place.[27]

"The nature of that body, inherent in matter" initiates motion, not as a mover, but as a source of being moved. An answer to the question why the elements contain a "source of change" in this sense and why Aristotle also claims that their matter "lies closest to substance" will become clear later in the argument of *Metaphysics* IX.

Metaphysics IX continues the examination of potency by arguing that some potencies clearly belong to things having soul, and these are connected with reason, whereas others belong to soulless things, i.e., the elements (*Metaphysics* IX, 2, 1046a36–b4). Rational potencies can produce contrary effects, e.g., medicine can produce either disease or health, because the rational potency explains a fact essentially and its privation accidentally.[28] Again the issue (and the example) here returns us to *De Caelo* IV, 3. Where there is more than one possibility, more than one effect can be attributed to the potency: because it means possibility, potency as a cause is indeterminate.

For the elements, however, the case is different: one power produces one effect, e.g., the hot is capable only of heating (1046b5–7). Each element is unique in having only one potency and so being able to produce only one effect. For the elements too, potency means possibility and so is in principle indeterminate; but here the range of possibilities is restricted absolutely: there is only one. Again, the force of this point reappears in *Physics* VIII, 4.

In *Physics* VIII, 4, too the elements have but one potency: their locomotion is one-directional. In *Physics* VIII, Aristotle argues for the primacy of locomotion because it belongs to separate things and comes to be last of all the motions (*Physics* VIII, 7, 261a14–26). In *Physics* VIII, 4, he distinguishes the elements not from rational potencies, which can produce contrary effects, but from the related and broader category, animals, which possess potency to move in more than one direction (255a-12); each element is moved naturally in one direction only, which constitutes further evidence that the elements cannot move themselves (255a5–11). Although the details differ, *Physics* VIII, 4, and *Metaphysics* IX agree on the main point: the elements cannot be self-movers because each element has one potency only and thereby differs from beings (either rational beings or animals more generally) that possess potencies for more than one effect. As will become clear in a moment, this uniquely restricted potency makes the matter of the elements the closest to substance.

Metaphysics IX, 3, criticizes the position proposed by the school of

---

27 Wildberg, *John Philoponus' Criticism of Aristotle's Theory of Aether*, 43–44 (italics in original).
28 Aristotle, *Metaphysics* IX, 2, 1046b7–18. For a fuller account of rational potencies, cf. Freeland, "Aristotle on Possibilities and Capacities," 77–78.

Megara. By identifying potency and actuality, this school destroys both motion and becoming because a thing will be able to act only when it is acting and, likewise, what is not happening will be incapable of happening.[29] In short, prior to actual being, there is no possibility for an act or thing in any sense. So, for example, any art becomes impossible because there is no way to learn or acquire an art; indeed, this view fails to account for any kind of change or becoming (1046b35–1047a). What is not now has no possibility to be in the future.

This criticism raises the problem of the relation between ability (potency) and that for which it is an ability (actual being), for example, the ability to build or to perceive and actual building or perceiving; in *De Caelo* IV, 3, this relation is crucial to the account of what differentiates the elements and thereby contributes to their generation. Thus the two arguments must address the same problem. Here in *Metaphysics* IX, the solution appears immediately and is central to the larger problem of the elements, both their locomotion and their generation.

Because the elements must be a capacity "for something and at some time and in some way" [ἐπεὶ δὲ τὸ δυνατὸν τὶ δυνατὸν καὶ ποτὲ καὶ πώς], whenever the one capable of making and the one capable of being affected approach each other, the one makes and the other is affected [τὸ μὲν ποιεῖν τὸ δὲ πάσχειν] (*Metaphysics* IX, 5, 1048a-8). This relation need not always hold for rational potency because rational potency can lead to opposite results. However, if one of the results cannot obtain, then rational potency will resemble nonrational potency in the sense that the desired result must come about (1048a15–20). Indeed, when potency and actuality have been specified in this way, the usual caveat "if nothing external prevents" (or "hinders") (expressions appearing in *Physics* VIII, 4) is no longer necessary (1048a17–21).

The first point in effect reiterates the argument of *De Caelo* IV, 3: mover and moved cannot be related by chance. But here, the relation of potency to actuality is more completely expressed: when potency is fully determined to one result only – as it is with each element – and mover and moved "approach," potency cannot fail to be actualized: the element cannot fail to be moved.

This is the crucial point: potency in natural things is never either general or accidental; rather, it is always determined. Potency is always for something, at a given time and in some way. Because it is determined, potency is immediately related to that *for which* it is potential; in this sense it can never be neutral to, or independent of, the actuality that completes it. Even in the case of rational capacities, which possess potency for opposite results, when the potency is sufficiently determined, indeterminacy is eliminated and the outcome is fully determinate. In this case, rational

---

29 Aristotle, *Metaphysics* IX, 3, 1046b29–30; 1047a11–14.

capacities resemble the elements. Each element possesses only one natural potency. Consequently, within a potency/act relation each element presents what is always the case: potency that, as fully specified, must be actualized in one way only by its proper actuality. If nothing hinders, fire cannot fail to be moved upward and earth downward.

Two conclusions follow from Aristotle's account of potency as specified. (1) Potency is always relational, or oriented toward, that which completes it. To be acted upon is often called "passive potency," but it does not mean passive in the sense of being neutral to its mover; rather, it implies being acted upon because it is actively oriented toward a mover. Confusion in the literature on this point is so pervasive that an example may be useful. Irwin speaks of "passive potency" as if it were undirected: "A change in a subject requires a passive potentiality for being affected by external objects; . . . Nutrition might seem to be a passive potentiality, since it requires the subject to be affected in certain ways by the air, water, or soil in its environment."[30] But Gill, citing *Metaphysics* IX, 1, makes the orientation of potency very clear: "He characterizes the δύναμις of the moved as a source of passive change (ἀρχὴ μεταβολῆς παθητικῆς) by another thing or by the thing itself *qua* other . . . a δύναμις, whether active or passive, is always directed toward a definite end or actuality."[31]

Indeed, in *Physics* III, 3, Aristotle claims that potency receives its very definition from that which completes it (202a15–16). We see how active this relation is in his claims that the heaven "runs always" after its mover and that matter (or any mean) "is aimed at" its form or end.[32] Charlton notes that in *Physics* II, Aristotle associates nature with ὁρμή – active striving.[33] This active orientation is especially important for the elements because they are determined absolutely: each possesses potency for only one actuality. Hence, whenever possible, the potency cannot fail to act or be acted upon. This relation provides the grounds for a successful account of motion and, indeed, the very possibility of becoming; it forms the core of Aristotle's account of the elements and renders his position superior to that of the school of Megara.

(2) When Aristotle explains that on contact (or with sufficient specification) the agent must act and the patient be acted upon – in both cases ability must be exercised – he reveals his criterion for a successful account of motion. Motion always involves a relation between two things, the

---

30 Irwin, *Aristotle's First Principles*, 305. Cf. also Charlton, "Aristotle and the Uses of Actuality," 5. Given this confusion, as well as the subsequent history of the notion of passive potency, it may be better to speak of an ability to be affected.

31 Gill, "Aristotle on Self-Motion," 246. The quoted phrase is at 1046a12–13.

32 *De Caelo* I, 2, 270b23; *Physics* I, 9, 192a22–23; *N. Ethics* I, 1, 1094a-2. Again, works of art contrast with natural things because the former contain no ὁρμή to change as do natural things (*Physics* II, 1, 192b19).

33 Charlton, 92. Cf. *Physics* II, 1, 192b18–20; *Post. Anal.* II, 11, 95a1, *Metaphysics* V, 23, 1023a9, 18, and 23.

mover and the moved, and there is no such thing as motion apart from things (*Physics* III, 1, 200b33). Hence to explain motion is to provide an account of the relation between mover and moved. The identification of sufficiently specified potency as patient and an agent in proper relation to one another yields an agent that must act and a patient that must be acted upon. Under sufficiently specified conditions, an *ability* to act or be acted upon must act or be actualized and, as a result, motion in what is potential cannot fail to be produced by what is actual. In this sense, specification of the relation between mover and moved, i.e., act and potency, explains not only motion but also why motion *must* occur.[34]

This conception of motion is identical not only with that of *Physics* III but also with that of *Physics* VIII, of *Metaphysics* XII, and of the *De Caelo*. Aristotle defines motion as the actualization [by what is actual] of what is potential, insofar as it is potential; that is, motion is a relation between the potential and the actual.[35] In *Physics* VIII, 4, he concludes that "everything moved must be moved by something [other than itself]."[36] The argument requires an identification of a mover in contact with the moved for every kind of mover/moved relation. When nothing hinders, the mover cannot fail to produce motion or the moved to be moved – that is, neither can fail to exercise their abilities. This argument forms the first step in resolving an objection to the view (established in *Physics* VIII, 1) that motion in things must be eternal. The resolution ultimately identifies a first mover that must be unmoved and a first moved that must always be moved (*Physics* VIII 6, 260a14–15). Hence, Aristotle accounts for the eternity of motion in things by a series of demonstrations concerning the relation between movers and moved things. The *De Caelo*, as I have already argued, continues the work of the *Physics*. Finally, in the context of an examination of substance in *Metaphysics* XII, 7, Aristotle specifies the first unmoved mover, which produces motion as an object of love, as that which cannot fail to act, and the heaven as that which must always be acted upon, that which is moved eternally.

Aristotle now turns to actuality. Because actuality is the mover of what is potential, he must account for the actuality of the elements, if he is to explain elemental motion. Hence the account of actuality is crucial, and in the account of it in *Metaphysics* IX, the elements are of special interest. Indeed, this account, as I shall argue, allows us to understand *De Caelo* IV, 3. The account of actuality in *Metaphysics* IX begins not with a definition

---

34 Freeland emphasizes the necessity of this relation and argues that it accords with the account of generation in the biological works, "Aristotle on Bodies, Matter, and Potentiality," 399.

35 Aristotle, *Physics* III, 1, 201a10–11; VIII, 4, 256a2–3.

36 Aristotle, *Physics* VIII, 4, 256a2–3; the proposition is also argued in *Physics* VII, 1. There, as I have argued elsewhere, Plato and his account of self-motion are the main targets of this argument. Cf. H. Lang, *Aristotle's Physics and Its Medieval Varieties*, 35–62.

strictly speaking, but with a series of examples contrasting actuality and potency.[37] Actuality is identified with that which is building in contrast to what is only able to be built, with waking rather than sleeping (according to Aristotle sleeping is for the sake of waking but not the converse), with seeing rather than being *able* to see, and finally, with what has been shaped out of matter, e.g., a statue, rather than unshaped matter (*Metaphysics* IX, 6, 1048a35–b4). In short, potency is identified with ability or possibility, whereas actuality is associated with completion or achievement.

In the case of a statue, the distinction between being able to be shaped and being shaped seems clear. And "building" too clearly can be completed. But in what sense and when is seeing complete? Aristotle takes up this question immediately, and its answer is crucial for understanding first the relation between motion and actuality and finally the locomotion and generation of the elements.

Of actions having a limit, none is an end; rather, such actions are relative to the end (1048b18–19). Such things "are in motion" when they have not yet achieved "that for the sake of which there is motion" (1048b20–21). Because motion aims at, but has not yet reached, an end, all motion is by definition incomplete (1048b29). And completion always implies some difference between the motion and the final outcome. For example, one cannot at the same time walk and have walked (Aristotle thinks of walking as transportation, not exercise), build and have built, or come to be and have come to be (1048b30–33). In each case, the motion reaches an end or limit – a place, a building, a generation – and that end is different from the motion itself, precisely because it is something complete rather than incomplete and something that by virtue of being complete terminates the motion.

In an important sense, actuality contrasts with motion. Of actions that can be identified with actuality, the end is present, and there is no difference between the action and the end: "At the same time, we are living well and have lived well, and are happy and have been happy. . . . The same thing at the same time has seen and is seeing, or is thinking and has thought."[38] Although Aristotle does not explicitly define actuality, his point is clear: the end is in the action itself, and such action is complete in itself and so able to be continuous. Thus actuality, unlike motion, can be pursued for its own sake.[39]

---

37 Aristotle, *Metaphysics* IX, 6, 1048a25–34. Owens comments that this procedure is common in Aristotle, *The Doctrine of Being in the Aristotelian Metaphysics,* 404. Concerning the word "actuality", cf. Charlton, "Aristotle and the Uses of Actuality," 1.

38 Aristotle, *Metaphysics* IX, 6, 1048b33–34; 1048b25–26; on the importance of the present and perfect tense here, cf. Polansky, "*Energeia* in Aristotle's *Metaphysics* IX," 163–164. For a history of controversies concerning the tenses here, cf. Ackrill, "Aristotle's Distinction between *Energia* and *Kinesis*," 128–136.

39 On this point and its implications for ethics, cf. Polansky "*Energeia* in Aristotle's *Metaphysics* IX," 165–166, and Kosman, "Substance, Being, and *Energeia*," 127.

Motion and actuality, then, are in one sense the same and in another different. They are the same insofar as both involve potency and both require the completion of that potency by an end itself identified with actuality. But they also differ. For motion, the end differs from the potency and serves as its limit with the result that motion is always incomplete and terminates in something other than itself. For actuality, the end is contained in and the same as itself, with the result that activity can be continuous and in some cases is eternal.

If we return to *De Caelo* IV, 3, we may ask of the elements: are their motions incomplete, each terminating in something other than itself, or are their motions a kind of actuality, each containing its end and so able to be continuous? For Aristotle, the latter: their motions are a kind of actuality. The argument of *De Caelo* IV, 3, is worth repeating:

On the one hand, then, whenever air comes to be from water, and something light from something heavy, it goes toward the upward. At the same time, on the other hand, something light is, i.e., it no longer becomes but is there. And it is clear that being potentially, it goes toward being actually there and toward such and how much of which the actuality is both of such a kind and to such an extent. (311a1–5)

First, and we shall return to this point, there is no difference between becoming light and being moved upward. Rather, being moved upward is already contained in being light and so must be an actuality rather than a motion. Second, because being light is being moved upward, as soon as water becomes air, it is something light with the result that it no longer becomes but is there, i.e., upward; air will be upward to the fullest extent possible because to be potentially up and to be actually up are one and the same act. Indeed, this argument resembles that of *De Caelo* II, 14: the earth must be spherical because all earth goes down as far as possible (297b5–13). To be upward is the actuality of becoming light, and we shall see in a moment that the activity of each element is able to be continuous.

In *Metaphysics* IX, 7, the contrast between motion and actuality returns Aristotle to the relation between potency and actuality. Whereas earlier he considered "potential" as an ability and concluded that it must always be specified in relation to actuality, he now considers it as completed by actuality. As before, things involving rational thought contrast with the elements, i.e., things having the source of becoming [ἀρχὴ τῆς γενέσεως] in themselves (1049a5–18). The argument again concerns what is able to be moved (rather than able to move) and strengthens the earlier point about potency: the potential must bear an immediate relation to its actuality at some place and at some time.

The argument begins with a question and ends with an answer: is earth potentially a human? No. It must first become seed and even then – a further condition appears in a moment – it may not be a potential human

(1049a-2). Potential means "capable of" – someone capable of being healed is called "potentially healthy" (1049a2–4). In the case of objects of thought, whenever one wishes the result and nothing hinders, the result must come about. As we saw earlier, whenever a rational thought and its object are fully determined, only one outcome can result. And building works on the same model (1049a5–12). Without such capacity, change or becoming is impossible. Thus, "potential" accounts for change as a capacity immediately determined in relation to the result of the change, or becoming.[40]

The second case – things that do not involve rational thought – includes the elements and contrasts with those involving rational choice. When there is rational choice, the source of the becoming is outside the thing produced; but each element contains the source of becoming in itself (1049a12–13). When a thing has the source of becoming in itself, it is potentially "all those things (nothing hindering from the outside) it will be through itself" (1049a13–14). Now Aristotle returns to the initial question: is earth potentially a human? Earlier he said no, not until it becomes a seed and perhaps not even then. Now he completes his answer. "The seed must be deposited in something other than itself and change; but whenever it is already in such a state on account of its source, this is already potentially [a human]" (1049a15–16).

Here we reach an issue crucial to all three arguments, *Metaphysics* IX, *Physics* VIII, and *De Caelo* IV – to Aristotle's entire account of the elements. The elements contain the source of becoming in themselves. But what is the meaning of "source" and of the larger phrase "all those things . . . it will be through itself" – do they imply self-motion, i.e., a mover within each element? The short answer is no: the elements do not possess an intrinsic mover. The long answer takes us some distance in understanding the elements.

As we have seen, in *Metaphysics* IX, 1, potency means an ἀρχὴ τῆς μεταβολῆς when it is associated with an ability to be acted upon by another. And in this regard, *Metaphysics* IX is identical with the *Physics* and the *De Caelo*.[41] In *Metaphysics* IX, 1, the phrase ἀρχὴ τῆς μεταβολῆς is used exclusively in conjunction with potential as ability to be acted upon (1046a11–14). Hence, again in *Metaphysics* IX, 7, to speak of the potential as an ἀρχή, an ἀρχὴ τῆς κινήσεως, or μεταβολῆς, or γενέσεως is to speak of a source of being moved. As intrinsic and essential to a thing, this source of being moved is unique to things that are by nature, including the elements; indeed, as in *Physics* II, 1, possession of this source distinguishes natural from artistic things – artistic things require an extrinsic

---

40 Again, for the implications of this view for Aristotle's biology, cf. Freeland, "Aristotle on Bodies, Matter, and Potentiality," 401–404.
41 Kosman also explicitly connects the argument of *Metaphysics* IX to that of *Physics* III, 1 (and *Metaphysics* IX, 9), "Substance, Being, and *Energeia*," 128.

cause of being moved precisely because they lack the intrinsic principle possessed by natural things. This source of being moved, this potency, is actively aimed at its proper actuality with the result that if nothing hinders and what is moved is completely determined, it cannot fail to be moved by its proper actuality.

Likewise the phrase "all those things (nothing hindering from the outside) it will be through itself" refers to the ability to be moved and emphasizes the active orientation of potency toward actuality. This active orientation at once expresses the very meaning of nature and the sense of potency as determined toward its proper actuality. What renders the elements unique is that their potency is completely determined to only one outcome. So we may say, there is only one outcome that fire, in the absence of hindrance, will be through itself: it will be moved upward.

Here the point of the question "is earth potentially a human?" becomes clear. As we have already seen, the answer is no. Earth as such is not sufficiently determined to be called "potentially a human." Even when earth is determined into seed, it may not be sufficiently specified to become a human. Why not? Because it must be deposited into something other than itself and change. Only then is it potentially a human. Why? Because when potency, i.e., the source of being moved possessed essentially and intrinsically by natural things, is fully determined in relation to its actuality, no extrinsic cause is needed: the potency cannot fail to be actualized, the seed to become a human. Only at this point does potency become a source of becoming what it will be through itself. In short, only now is the requisite mover/moved relation completely determined; only now does it constitute an account of why motion, actualization of the potential as such by what is actual, must occur.

Aristotle completes his argument by emphasizing the immediacy of the relation between potency and actuality. When there is a series, what a thing is potentially must come immediately after it (1049a21–22). Earth is not potentially a human because seed and placement intervene between earth and human; but a properly placed seed is potentially a human because human is immediately "next" for the placed seed.

In *De Caelo* IV, 3, Aristotle says of the elements that more than anything else they possess their source of motion in themselves because their matter is closest to substance. At the opening of *Metaphysics* IX, he comments that "potency" and "actuality" extend beyond the realm of motion to substance. Hence he promises that his account of actuality will extend to substance, and the emphasis on the immediate relation between the potential and that of which it is potential brings him to the problem of substance and accidents. And I shall suggest in a moment that this problem in *Metaphysics* IX yields an understanding of the claim in *De Caelo* IV, 3, that the matter of the elements is closest to substance.

Generally, what is potential is "of a this"; the "this" is always substance, or what is determinate, whereas what is "of a this" is always accidental and indeterminate in the sense of being relative to the determinate (*Metaphysics* IX, 7, 1049a18–b3). Consequently, by definition what is potential is "of a substance," i.e., nothing other than an immediate relation to actuality, whereas actuality is more properly and directly identified with substance itself.

With this point, the account of what it is to be potential is complete (1049b-2). And it has been completed within the context of an account of actuality. Actuality contrasts with potency because it means fulfillment or completion, whereas potency means ability or possibility. Because potency means possibility, it is always relative to actuality. But it is not relative to any actuality, or to some actuality remote from itself, but only to what is "next in the series," i.e., that to which it bears an immediate relation. And the "mark" of this immediate relation lies in the fact that although what is potential is moved by its actuality, it is not thereby passive. To be potential is to be actively oriented toward proper actuality with the result that when the relation between potency and actuality is immediate, i.e., when nothing intervenes, potency cannot fail to be moved by its actuality. The features of this relation explain not only why motion occurs as it does – indeed, that in the absence of hindrance it must occur – but also the relation of potency to actuality as the relation of the indeterminate to the determinate, the accidental to substance.[42]

Metaphysics IX, 8, affirms this view and brings us to a remarkable claim about the elements: they imitate the heavens. And I shall suggest in a moment that *because* the elements imitate the heavens, their matter is closest to substance. Having established what it is to be potential, Aristotle returns to actuality and argues that it must be prior to potency.[43] And potency – the potency to which actuality is prior – is identified in things as broadly as possible: "every principle of motion or rest" (1049b7–8). Indeed, he adds, nature is in the same genus as what is potential because nature too is a source of motion (1049b8–10).

Here the connection to *Physics* II, 1, is explicit: nature is unique in possessing a source of motion, i.e., a source of being moved and being at rest in that to which it belongs primarily in virtue of itself and not accidentally.[44] And the elements, earth, air, fire, and water, are explicitly included among things that both are and are moved by nature (*Physics* II, 1, 192b10–11, 35–36). Consequently, the account of potency and actuality here in

---

42 I have argued elsewhere that hindrance plays an important role in Aristotle's physics because it accounts for why motion does not always occur. Such a separate cause is necessary because when fully specified, the relation between potency and actuality is so powerful that motion must, i.e., cannot fail to, occur. H. Lang, *Aristotle's Physics and Its Medieval Varieties*, 77–81, 100–101.
43 On the priority of act, cf. Owens, *The Doctrine of Being in the Aristotelian Metaphysics*, 406–409.
44 Aristotle, *Physics* II, 1, 192b21; for the phrase "source of motion and change," cf. also *Physics* III, 1, 200b12–13.

Metaphysics IX, 8, bears directly on things that are by nature, including the elements. And at this juncture of the argument, Aristotle extends the account of potency and actuality beyond motion and into the realm of substance.

Actuality is prior to potency in every sense. Most importantly, it is prior in substance because that which is prior (actuality) already has the form fully developed, whereas the other (the potential that develops or becomes) does not (*Metaphysics* IX, 8, 1050a4–7). Everything that becomes is moved toward a "source" – source not in the sense of "source of motion," but in the sense of an end or that for the sake of which (1050a8–9). Source in the sense of "end," is actuality and is prior. Becoming is for the sake of the end whereas the actuality is an end; potency is acquired for the sake of this, and this order cannot be reversed: we possess sight for the sake of seeing, and theoretical science so that we may theorize (1050a9–12). Indeed, matter is potentially in order that it may proceed to form; and when it is actually, it *is* form (1050a15–16).

This relation of potency to actuality also accounts for motion (1050a16–20). And the reason why is clear: the action is an end, and the actuality is an action; indeed, the word "actuality" derives from the word "action" and points to the fulfillment of a thing (1050a21–23). In the identification of actuality with action, Aristotle identifies both its priority and its primary meaning.

Actuality, as Aristotle now makes clear, is most closely associated with activity, and activity means not capacity, but the full exercise of capacity; thus, the ultimate purpose of sight is to see and nothing else (1050a23–25). However, in some cases, e.g., the art of building, not an activity but a product, e.g., a building, results. Even here, the product is more an end than is mere potentiality, the ability to build, and the actuality is more in the product (1050a26–29). Aristotle connects this point to one made earlier: movement is in the thing being moved, the statue or the building, and when there is no "product" apart from the agent, e.g., sight, the actuality is in the agent. Consequently, "the substance, namely the form, is actuality" (1050b2–3).

We shall see shortly that there is an even more important sense in which actuality is prior to potency. But the identification of actuality with substance and ultimately form completes the first set of distinctions and so the first moment in the account of the priority of actuality. And this moment is important for the account of the elements.

The priority of actuality is established not by a deductive argument, but by a series of identifications. Actuality is an end, i.e., the end for the sake of which potency is acquired (and from which it gains its definition). As an end, actuality is primarily identified with activity, and actuality as activity is complete in itself precisely because it contains its own end. Actuality is also identified with form. Against the background of *Physics* II and

Metaphysics XII, this dual identity – actuality is both form and end – comes as no surprise. In *Physics* II, Aristotle identifies form as the end when it acts as a principle of motion, and actuality is regularly identified as a cause of motion, i.e., the mover (*Physics*, II, 7, 198b2). God in *Metaphysics* XII is pure form, pure actuality, and acts as the end of the heavens.[45]

So *Metaphysics* IX reaffirms the explicit conjunction of the end as a principle of motion and form as the primary sense of substance.[46] As end (or form), actuality is properly identified with activity: activity is always the end when the actuality is in the agent, and secondarily with the product, the result of the activity of the agent.

Actuality is prior to potency in an even higher sense, however: eternal things are prior in substance to perishable things, for nothing eternal is potentially (*Metaphysics* IX, 8, 1050b7–8). Potential means "able to be," which includes the ability (capacity or possibility) to be the opposite of the present case. Hence, the ability to be includes the possibility of not being, and any substance that contains potential (in respect to substance) thus contains the capacity not to be. In short, such things are perishable. Conversely, anything that is imperishable by definition cannot be potential (contain the possibility of not being) in respect to substance. All such things must be actually.[47] (Thus, god is pure actuality [*Metaphysics* XII, 7 1072b25–29].)

Aristotle concludes that what is eternal, what is necessary, and what is moved eternally cannot be potential in respect to substance; what is moved eternally is potential in respect to "whence" and "whither," and such matter makes it capable of being moved. But because they possess no potential in respect to their substance, "the sun and the stars and the whole heaven act always and there is no fear that they may sometimes stand still" (*Metaphysics* IX, 8, 1050b22–23). Hence, he concludes, unlike things having potency in respect to substance, the heavens do not tire of their activity because their movement is not connected with the possibility for opposites (1050b23–25).

Activity is attributed to god as pure unmoved actuality and the first mover of the heavens in *Metaphysics* XII;[48] in *Metaphysics* IX activity is also attributed to the heavens, i.e., the first effect of god. But there is no contradiction because potency means capacity, or ability, and always implies a direct relation to actuality, while actuality is in every way prior. Most importantly, actuality is prior as constituting eternal substances, e.g., the sun, the stars, and the whole heaven. Therefore, in respect to substance

---

45 Aristotle, *Metaphysics* XII, 6, 1072a5–10; 7, 1072a25–27; 1072b5–7, 28.
46 The primacy of form as a thing's nature is affirmed in *Physics* II, 1, and as substance in *Metaphysics* VII, 3, 1029a5–7, 27–32; 8, 1033b17–19.
47 On the implications of this account for form and sensible substance, cf. Halper, "Aristotle's Solution to the Problem of Sensible Substance," 671.
48 Aristotle, *Metaphysics* XII, 7, 1072b22–29; 1073a5–11.

they are like god: pure and eternal actuality without possibility for not being. Because actuality is identical with activity, these substances may be called "always active." Herein lies the reason why the heaven will never stand still and moves without labor: its motion is continuous and without possibility for not-being.

This is not to say that these substances do not have potential, here associated with matter. They do. But their potential is limited solely to their capacity to be moved. The heavens are moved around in a circle and are thus moved in respect to whence and whither.[49] Only by virtue of this potency are the stars and sun, the whole heaven, moved (by god, as we know from *Metaphysics* XII). The actuality of the stars is their substance, their form, and this actuality is itself devoid of potency of not-being. By virtue of this actuality alone, the heaven "always acts," ἀεὶ ἐνεργεῖ. For this reason, the same word applies to both the heavens and god: it applies to both insofar as each contains actuality without any possibility for not-being.

Now Aristotle embraces the elements within the realm of the "ever active":

But indeed the beings involving change, such as earth and fire, imitate the imperishables. For these things too are ever active; they have their motion, in virtue of themselves and in themselves. [*Metaphysics* IX, 8, 1050b28–30: μιμεῖται δὲ τὰ ἄφθαρτα καὶ τά ἐν μεταβολῇ ὄντα, οἷον γῆ καὶ πῦρ. καὶ γὰρ ταῦτα ἀεὶ ἐνεργεῖ· καθ' αὑτὰ γὰρ καὶ ἐν αὑτοῖς ἔχει τὴν κίνησιν.]

I shall suggest in a moment that here we have the reason why Aristotle says of the elements in *De Caelo* IV, 3, that their matter is closest to substance. But first we must consider the meaning of these lines in *Metaphysics* IX, 8.

Most obviously, as we have seen, the elements by nature possess potency for only one result, i.e., each is oriented toward its proper natural place. Fire is potentially upward, earth downward etc.; thus in the absence of hindrance, fire cannot fail to be moved upward and earth downward. Because the potency of the elements is completely determined, the elements "imitate" the heavens by being moved in one way only (each its own way) and by being moved necessarily (assuming nothing hinders). In effect, full determination of their potency makes the motion of the elements resemble the activity of the heaven – they "always act" and always act in the same way.[50] Indeed, the argument implies that like the heavens, the motion of each element would not be laborious because none possesses potency for its opposite.

---

49 Cf. also *Physics* VIII, 6, 260a14–19; 9, 265a24–27; *Metaphysics* XII, 6, 1072a15–18; 7, 1072a21–23.

50 Owens puts the point rather differently: "In this way corruptible things *imitate* the incorruptible (*Metaphysics* Θ 8, 1050b28–39) by being always in act as far as the limitations of movement allow," *The Doctrine of Being in the Aristotelian Metaphysics*, 404. But the notion of imitation here must rest on the active orientation of potency for actuality, not on the limitations of movement.

248 THE ELEMENTS

Yet even if one accepts this account of elemental motion, serious questions remain. The elements change or move (Aristotle uses both words) in two different ways. (1) Locomotion: each element is moved naturally to its respective proper place, e.g., up for fire, down for earth.[51] But when an element is in its natural place, it rests there – for "to rest" is immobility for a thing that is subject to motion.[52] What then is the relation between motion, rest, and being "always active"?

There is also the second type of motion, (2) transformation into one another: water becomes air, air fire, and so on. In the *De Generatione et Corruptione* (II, 10, 337a1–7), Aristotle says that the transformation of the elements into one another imitates circular motion because it too completes a cycle. Citing this argument, Ross says of such imitation in *Metaphysics* IX, 8: "It is doubtful whether this refers to the natural movement of fire upwards, and of earth downwards, or to the constant tendency of the elements to change into one another."[53] More recently, Kahn rejects Ross's claim and restricts imitation to the motion of each element to its natural place because this motion "is an expression of their essential natures."[54] But on this account, how are we to understand either the transformation of the elements into one another or the claim of "imitation"? Perhaps Kahn is echoing a gloss by Owen that the elements "μιμεῖται in the properly Platonic sense that they are and are not ἄφθαρτα (immortal by constantly changing into one another – GC 337a1–7), not as an explanation of their behavior, which stems from their own nature."[55]

This interpretation raises more problems than it solves. Aristotle criticizes Plato's account of physical things imitating the forms because "imitation" [ἡ μίμησις] is, he says, a "name" with no clear meaning (*Metaphysics* I, 6, 987b11, 13). In short, if the elements "imitate" the heaven, what can "imitate" mean here, and why should Aristotle be exempt from the charge that he lays at Plato's door – using a "name" with no clear meaning? Furthermore, given that there are two kinds of motion, locomotion and transformation, why does Aristotle claim that the elements have only one potency? They would appear to have at least two – potency for locomotion and potency for transformation.

I shall start, so to speak, at the end, i.e., the transformation of the elements into one another. The elements form a cycle in which each element is potentially the next member of the series – water is potential air, air potential fire, etc. The serial relation within the cycle cannot be violated,

51 Aristotle, *De Caelo* I, 8, 276a22–30; IV, 3, 311a5–7; *Physics* V 6, 230b10–15.
52 Aristotle, *Physics* III, 2, 202a3–5; *De Caelo* III, 2, 300a28–29; 300b5–6.
53 Ross, *Aristotle's Metaphysics: A Revised Text with Introduction and Commentary* II, 265–266. Citing Ross, Gill restricts the meaning of imitation to locomotion, but does not explain why, *Aristotle on Substance*, 235.
54 Kahn, "The Place of the Prime Mover in Aristotle's Teleology," 189.
55 Burnyeat et al., *Notes on Books Eta and Theta of Aristotle's Metaphysics*, 145.

e.g., water is not potential fire. Thus this potency is "one-directional" and is potential in the strict sense established by the answer to the question "is a seed a potential human?" To understand this relation and hence the transformation of the elements, we must consider their natures more closely.

The very name "fire" entails being light, and "light" entails being oriented upward; the very name "earth" entails being heavy, and "heavy" entails being oriented downward. Hence, "to be light" and "to be oriented upward" are neither consequences of being fire nor in any way different from being fire; rather, they are what it is to be fire, i.e., different names for the same entity. And so on for the other elements. Because by definition each element is oriented toward its proper place, potency to become that element is potency to be moved to a certain place, up for fire, down for earth, and to their respective places in the middle for water and air. So, when air is potentially fire, it is also potentially toward the respective place of fire, i.e., up. Consequently, the orientation of each element toward the next in the series is nothing other than an active orientation toward the natural place of that element; transformation is not a different or separate potency from potency in respect to proper place. For this reason, the elements must be thought of as possessing one potency only (and in this regard contrasting, for example, with rational potency). And this view is expressed in both the *Physics* and *De Caelo*.

Physics VIII, 4, explicitly discusses the transformation and locomotion of the elements in terms of potency and actuality. Potency, Aristotle says, is spoken of in many ways, and, consequently, there is confusion about elemental motion (255a30–33). And he compares the elements to someone learning a science (255a34-b5). In both cases, actualization is achieved in two steps. One who is ignorant but capable of learning must acquire knowledge before being able to exercise it. Even after knowledge is acquired, something, e.g., drunkenness or sleep, may prevent its active exercise. But if the knowledge has been acquired and nothing prevents, active exercise follows directly. So if someone calls for a doctor, a "potential doctor" not in the first sense (ignorant but able to learn) but in the second sense, i.e., knowing and so able to practice, will, one hopes, step forward immediately and actually practice medicine.

The elements, Aristotle says, are like this (255b5–31). Water is potentially light in the sense of being able to become air; and when it has become air, it exercises the activity of being light and rises, if nothing, e.g., being in a closed container, prevents it from doing so. In such a case, the water would remain at the "intermediate stage" of actualization, air that is being held away from its proper place. But like mind, the elements are immediately actualized and made active once they are at the intermediate stage of actualization. This immediate activity (in the absence of hindrance) cannot fail to occur because each element possesses potency for one and only one result, i.e., being in its proper place. So Aristotle says in *Physics* VIII, 4: "Thus not only when a thing is water is it potentially light, but when it has become air it may be still potentially light [because of a hindrance] . . . but if someone removes the hindrance, it acts and becomes always higher.[56]

Consequently, the elements imitate the heavens in two ways: (1) their actualization terminates not in a product, but in an activity, and (2) the "cannot fail to occur" character of their actualization imitates the activity of the heavens that have no potency in respect to substance. Given the contrast between motion and actuality, elemental motion must be identified with actuality, i.e., containing its end in itself and able to be continuously. And the account of *Physics* VIII, 4, agrees with that of *De Caelo* IV, 3. When air, which is light, comes to be from water, which is heavy, the air is immediately moved upward (assuming nothing hinders) as far as possible (*De Caelo* IV, 3, 311a1–5).

But potency is by definition relative to actuality. How can we identify the actuality of the elements? First and unequivocally, potency is potency for what is next in the series. So, for example, if water is potentially air, air is the actuality of water; but air is potentially upward and so when water becomes air, it must be potentially upward and (if nothing hinders) is carried there immediately. Being up is the actuality of air and so of water, which is potentially air. When air is up, it naturally rests in its proper place. In short, to rest in its proper place is the ultimate actuality and hence the activity of the elements. Indeed, Aristotle could not be more explicit:

Whenever, therefore, air comes to be from water, i.e., light from heavy, it goes toward the upward [place]. And it is light immediately, namely it no longer becomes but there it is. [*De Caelo* IV, 3, 311a–3: ὅταν μὲν οὖν γίγνηται ἐξ ὕδατος ἀὴρ καὶ ἐκ βαρέος κούφον, ἔρχεται εἰς τὸ ἄνω. ἅμα δ' ἐστὶ κοῦφον, καὶ οὐκέτι γίνεται, ἀλλ' ἐκεῖ ἔστιν.]

Odd though it may sound to us, for the elements "to rest" is not cessation of activity, but its opposite: resting in its proper place (and so resisting being moved out of it) is for each element a flourishing, an activity: the activity that completes its motion.[57]

---

56 Aristotle, *Physics* VIII, 4, 255b20–21. Polansky argues that all *energeiai* are in souls, "*Energeia* in Aristotle's *Metaphysics* IX," 165–169. But he does not consider the elements as such, neither the claim that they possess only one potency, nor the assertion that they imitate the heavens and always act, nor this explicit comparison of mind and the elements.

57 Machamer, "Aristotle on Natural Place and Natural Motion," 379, considers two texts (*Physics* IV, 5, 212b30–35, and *Metaphysics* 5, 4, 1014b16–26) and concludes that "[t]he body once in its natural place has *natural rest* in the same sense that a part rests naturally in a whole" (italics in original). Such an account fails to explain rest as an activity. Cohen, *Aristotle on Nature and Incomplete Substance*, 42, suggests that if an element reaches its natural place, "it can no longer exercise a disposition to move" toward that place; a rock, e.g., at the center would possess its disposition only *en dunamei* and hence "at the

Here we have a full sense of "imitation." The motion of each element is not the same as that of the heavens because the heavens have no potency in respect to substance and consequently are ever active in a sense that cannot fail. The elements possess potency for only one result. But with the elements, unlike the heavens, a hindrance can intervene between potency and actuality and so prevent actualization, e.g., columns hold up a heavy roof that would otherwise fall downward (*Physics* VIII, 4, 255b25). Insofar as the elements possess a possibility of not being actualized, they are unlike the heavens. If nothing hinders, however, then, like the heavens, each element cannot fail to be moved to its proper place. Because each possesses potency for only one result and because its potency is "aimed at," i.e., immediately oriented toward, its respective actuality, the elements cannot fail to be actualized and so to be active. Thus the elements imitate the heaven because both are "ever active."

We may return to the question of self-motion. When Aristotle says that the elements are "ever active" because they have their motion by virtue of themselves and in themselves, does he mean that they are self-moving? No. This phrase refers to his definition of things that are by nature: some source and cause of being moved and being at rest in itself primarily in virtue of itself and not accidentally. And the elements, imitating the heavens, fully meet this definition. Each possesses a principle of being moved and being at rest, i.e., its ability to be actualized and when actualized to be active, and each possesses its principle immediately and by virtue of its definition.

Aristotle's account of potency and actuality in *Metaphysics* IX is consistent with his account of nature in *Physics* II, his account of motion as an actualization in *Physics* III, and his argument in *Physics* VIII, 4, that each of the elements, like all natural things, is moved by something [other than itself]. He rejects Plato's account of self-moving motion because it breaks the law of noncontradiction; his own account both in the *Physics* and the *Metaphysics* separates what is moved (potency) and what moves (actuality). Consequently, Aristotle gives the notion of "imitation" his own meaning: when there is only one potency (and nothing intervenes), to be moved is another way of being ever active.

When each element is in its respective proper place, it ceases to become and simply is: actualization is complete and activity ensues. Hence for each element to be in its proper place is to rest, i.e., to exercise its proper activity such as being up for fire or down for earth. The cause of actualization must be actuality, and the actuality of each element is the natural place toward which each is actively oriented, i.e., for which it possesses inclination.

center of the cosmos, rocks are rocks in name only." This view fails to understand either the notion of rest or the potency/act relation that defines motion. And Cohen goes on in the next line: "We can avoid this difficulty . . . by denying that Aristotle holds that fire, by its nature, has a natural motion."

Here a question traditionally associated with *Physics* VIII, especially *Physics* VIII, 4, should be considered for the *De Caelo* as well. If natural place is the actuality and mover of the elements, then is it a moving cause or a final cause? Duhem identifies it as a final cause – a tradition that may go back to Averroes – and he is joined more recently by both Wolff and Sorabji.[58] More commonly, the mover of the elements has been identified as a moving cause.[59] But the arguments of *Physics* VIII, 4, *Metaphysics* IX, and the *De Caelo* operate exclusively in terms of a potency/act relation, at least in part because this relation defines motion. And both moving causes and final causes produce motion by being actual relative to what is potential. Thus to ask whether a mover, or actuality, produces motion as a moving cause or a final cause requires a specification of actuality beyond the problem, the topic, at stake in these arguments concerning the motion of the elements. This argument requires an identification of actuality, and Aristotle provides it.

The argument of *De Caelo* IV, 3, concludes with a reference to *Physics* VIII, 4, and with the assertion that what made the element from the beginning and what removes a hindrance are also causes. Two points here. (1) *Physics* VIII, 4, unambiguously concludes that "everything moved must be moved by something," and the reference to *Physics* VIII, 4, confirms that proposition here in *De Caelo* IV, 3. The mover of what is potential must be proper actuality, here identified as the respective proper place of each element. (2) What moves from the beginning is the generator, and what removes a hindrance is an extrinsic, or accidental, cause. The first cause we have already seen: the generator that grants potency and by specifying it makes an element able to be moved by its proper place, itself the ultimate actuality for each element. At the same time, the element next in order serves as its immediate actuality. For example, when water is generated and becomes air, the motion is of like to like because air is the immediate actuality of the water, although place is the ultimate actuality. What removes a hindrance is an accidental cause of motion (*Physics* VIII, 4, 255b25–27). By removing a hindrance, this cause establishes contact between mover (actuality) and moved (potency); on contact the moved cannot fail to be actualized by its proper actuality.[60]

Two serious questions remain. The identification of proper place as the mover of each element in *Physics* VIII, 4, is at least as old as the Byzantine commentators – we find it in Philoponus, Simplicius, and

---

58 Duhem, *Le Système du monde*, 209. A Hebrew text of Averroes is given but not identified by Wolfson, *Crescas' Critique of Aristotle*, 141, 337–338, n. 22. Cf. also Wolff, *Fallgesetz und Massebegriff*, 48, 71–79, and Sorabji, *Matter, Space, and Motion*, 222.

59 Le Blond, *Logique et méthode chez Aristote*, 383–392; also Buckley, *Motion and Motion's God*, 58–59. Weisheipl criticizes Duhem and the identification of natural place as a final cause, in "The Principle *Omne Quod Movetur Ab Alio Movetur* in Medieval Physics," 31–32.

60 Cf. H. Lang, *Aristotle's Physics and Its Medieval Varieties*, 79.

Themistius.[61] But we do not find it explicitly in *Physics* VIII, 4. If the elements are moved by place, why does Aristotle fail to say so when he argues that they too are moved by another? The answer to this question lies in identifying the presuppositions at work in this argument.

The primary thesis of *Physics* VIII is announced in its opening line: is motion in things eternal, or did it begin and will it one day stop, leaving nothing in motion (250b11)? Aristotle argues that it must be eternal (*Physics* VIII, 1), but he raises and briefly replies to the three most serious objections to this thesis (*Physics* VIII, 2). These objections, he says, require more extended consideration, which occupies the remainder of the *logos* (252b29–253a21).

The first two objections require an account of the structure of the cosmos: why is it that some things never move, some things always move, and some things both move and rest (*Physics* VIII, 3, 253a28–29)? The argument concludes with a specific answer to this question: some things are moved by an eternal unmoved mover and so are always in motion, other things are moved by that which is itself in motion and so they vary, being sometimes moved and sometimes at rest (*Physics* VIII, 6, 260a15–19). And the argument that the elements too must be moved by something appears in this context. (The last objection from *Physics* VIII, 2, is addressed in *Physics* VIII, 7–10, but does not concern us here.)

The resolution of the objection that occupies *Physics* VIII, 3–6, works by displaying the internal structure of the cosmos and demonstrating that within that cosmos the elements operate like all other natural things: they are moved by something. This is the crucial point: in demonstrating the structure of the cosmos in respect to motion in things, Aristotle already presupposes that the cosmos is determinate, i.e., orderly in respect to "where." So, for example, when he begins the argument that "everything moved is moved by something," he asserts his categories of motion, e.g., violent and natural, as already given. As we have seen, violent and natural motions are defined in relation to natural place. Therefore, he assumes his account of place and, even more importantly, the causal role of place within an orderly cosmos.

Herein lies the reason why Aristotle does not identify place in *Physics* VIII, 4. It has been assumed as a cause and the cosmos is assumed to be orderly. That is, because place acts as a limit, the cosmos exhibits up and down, which are the respective natural places of the elements and define natural and violent motion in the elements. Does this assumption make Aristotle's argument circular? No, it does not. The argument establishes that everything, including the elements, is moved by something, and, at the conclusion of the argument, Aristotle's assumption of place identifies

61 Philoponus, *In Phys.* 830.14–16; Simplicius, *In Phys.* 1216.2–7; 1213.3–6, 11; Themistius, *In Phys. Paraph.*, 218.11, 24–29; 219.5–23.

what that mover is so immediately and completely that it goes without saying.

This point raises a second serious question. Everything moved must be moved by something, and the mover must be together with the moved in the sense that nothing intervenes between them.[62] But if one grants that the elements are moved by their respective natural places, in what sense are mover and moved together? If a stone, for example, is held "high up" and then released in what sense is it "together" with the center of the cosmos (and the earth)?

As a cause, place resembles form because it renders the cosmos determinate in respect to up, down, etc. And "up" and "down" as absolute may be identified with the periphery and the center. But "up" and "down" are not only absolute but also relative. So, for example, one place is "up" (or "down") relative to another even if that place is not the periphery (or center) itself.

Place and the elements cannot be "together" as two discreet physical things are together because place is not another physical thing. Rather, place is a limit, resembling form, and proper place within the cosmos is a formal determination with up and down, the periphery and the center, as the most extreme moments of this determination. In short, the cosmos is determined throughout in respect of "where." Therefore, all things composed of the elements, both natural and made, are always in place and being in place always have an immediate relation to natural place: there is no need to establish a "connection" between an element and its proper place because each element and its place are always together (in the absence of hindrance). Indeed, each element is always moved toward its natural place unless the motion is prevented by a hindrance, such as happens with pillars holding up a roof. The force of the point here is best expressed by the ubiquitous relation between place and the elements: place renders the entire cosmos determinate in respect to "where," and each element is by nature an active inclination toward its respective proper place.

I have already suggested that this relation exhibits the order of nature and is itself teleological in the fullest sense possible. A final point may now be made explicit: because place is a single unique determinative principle and all things composed of the elements are oriented toward their proper places, the order of the cosmos as a whole and the order of its parts are identical. The relation of each element to its proper place is one of potency to actuality. As we have seen, there is but one definition for both potency and actuality, that of actuality, because what is potential is nothing other than a relation to the actual. The determination granted by place to the cosmos is what the elements are "by nature" determined to be.

---

62 Aristotle, *Physics* VII, 2, 243a32–34; cf. V, 3, 227a23 ff.

In *Metaphysics* IX, Aristotle says of the elements that they imitate the heavens because they are ever active and have motion in themselves and by virtue of themselves.[63] Here, I would suggest, we find the clue to his claim in *De Caelo* IV, 3, that "what is light and what is heavy seem more than other things to have in themselves the source [of change] because their matter is closest to substance" (310b30–32). Each element is a single determined potency – itself identified with matter – for nothing other than locomotion toward its respective natural place and activity, i.e., to rest in that place. Each element possesses the highest kind of motion, locomotion; and each possesses it in and of itself because it is nothing other than a completely determinate active orientation toward its natural place; hence (in the absence of hindrance) active orientation cannot fail to be actualized and terminates in an activity that is the highest kind of actuality. Here lies the reason why the elements are "things that are by nature" and why their potency lies closest to substance: each is absolutely simple, being nothing other than active orientation toward its natural place, i.e., each contains an intrinsic principle of being moved and being at rest that in the absence of hindrance cannot fail to be actualized by proper actuality, i.e., the natural place toward which each is actively oriented.

In her recent treatment of *De Caelo* IV, 3, Gill also argues that this argument is consistent with *Physics* VIII, 4; but she reaches quite different conclusions:

Since motion toward its place is movement of like to like, a dislocated element already has the character that it will have when it reaches its own location; it is not changed by its motion. . . . Natural motion develops an element into what it already is.

. . . The elements move automatically upward or downward according to their natures, but they stop moving only because they are compelled to stop. Fire is not programmed to stop at the periphery; if there were no boundary contained by the fifth element, fire would continue its upward progression. Similarly, the downward progress of earth is limited when it reaches the center because it can proceed downward no further. And although no cosmic boundaries mark the intermediate regions, the motion of the elements is confined at the top by the dominant element that fills the adjacent region. . . .

. . . The elements do seem to have an active principle, but in fact, since they are totally simple, their passive principle adequately explains their natural motion. The elements need no external mover to set them in motion but move "straightaway" if unimpeded. All that requires explanation is why the elements stop moving, and to explain this Aristotle says that their motion is confined by the adjacent element. In a certain way, then the confining element is the form for the element confined, since it limits and thus regulates the motion. But what is truly interesting . . . is the claim that the matter of the elements is "closest to substance."

---

63 Aristotle, *Metaphysics* IX, 8, 1050b28–30; cf. also *Metaphysics* IX, 7, where Aristotle discusses the transformation of the elements into one another. This association of transformation and locomotion exactly parallels the argument of *De Caelo* IV, 3.

In fact, the elements simply are their material natures, and it is presumably because those material natures alone can sustain natural motion that Aristotle suggests that elemental matter is closest to substance. Unlike all other generated bodies, the elements need no active cause to direct their activity or to preserve them. All this their material nature can do on its own.

Of all material bodies that seem to be substances and fail, the elements come closest to succeeding. . . . The elements are heaps because they are formless matter. And although, as pure matter, the elements nearly succeed as genuine substances, substancehood is fairly awarded to the autonomous organic unities that the elements serve.[64]

My disagreements with this account are both fundamental and far-reaching.

Although we agree that the motion of an element toward its proper place is a motion of like to like that actualizes what an element is, we disagree about virtually every other feature of the argument. (1) In *Physics* VIII, 4, Aristotle is at pains (whether he is successful or not is a separate issue) to show that the elements must be moved by something other than themselves. Because they require a mover, they do not move "straightaway" in the sense of automatically.

(2) Likewise, they do not stop moving because they are compelled to do so. A hindrance stops an element from moving because it intervenes between potency and actuality so that they are no longer together. To think of a limit as "stopping" an element in the sense of compelling it, confuses two points. It confuses (a) limits, which are formal, with hindrances, which are material, and (b) natural motion, which occurs whenever respective potency and act are together, with violent motion, which represents compulsion from the outside and is away from natural place. Each element stops moving (in the absence of hindrance) only because it has achieved its proper goal – the goal toward which it is actively aimed and rests: in that place it does not become but is (*De Caelo* IV, 3, 311a2).

(3) The notion of a limit is not a physical boundary, but a formal determination; likewise, there is a sense in which as pure potency oriented toward its respective proper place, each element might be thought of as material, but matter in this sense is not "stuff." There is no sense in which the elements can be thought of either as "heaps" or as "formless matter."[65]

(4) Each element is both totally simple and nothing other than an active orientation toward its respective proper place and caused by that proper place as potency is caused by actuality. "To be caused" does not, for Aristotle, mean "to be passive." Rather, for natural things (unlike what

---

64 Gill, *Aristotle on Substance*, 139–140.

65 Obviously, a full examination of this problem lies beyond the bounds of this study. Suffice it to say that for Aristotle in nature (unlike art) there is no such thing as a heap or bits of stuff, but only informed matter. (Lear, e.g., uses the phrase "various bits of matter," *Aristotle: The Desire to Understand*, 17.) For a further treatment of this problem, cf. H. Lang, *Aristotle's Physics and Its Medieval Varieties*, 23–34.

is by art), it means "to be actively oriented toward a cause, proper actuality."

Lastly (5) and perhaps most importantly, there is absolutely no sense or indication in *De Caelo* IV, 3, that the elements are somehow "failing" to be substance. Aristotle could hardly be more positive about the elements and their movements. His account here (as he sees it) explains the motion of the elements, each toward its proper place, and shows why when each reaches its place, it no longer becomes but is. This is the case because each element possesses an active ability to be caused, i.e., moved, by its proper place.

And the proper name for this principle, or active orientation, is ῥοπή, inclination. When each element is out of its natural place, inclination is its potency, i.e., intrinsic ability to be moved by proper place as its actuality. This orientation immediately expresses itself as determinate: as heaviness for earth and lightness for fire. And this intrinsic ability is nothing other than an active orientation toward actuality, i.e., ability to be moved by respective proper place. When actualization is complete, each element rests in its natural place because its inclination has been fully actualized and now expresses itself as actuality of the highest order, i.e., activity to be up (for fire) or down (for earth). And when each element rests in its natural place, it resists being moved out of that place and can be so moved only by force applied from the outside. In this sense and on the basis of these distinctions alone the full meaning of inclination emerges as a name for the activities of the elements. Aristotle began his account at *De Caelo* IV, 1, by saying that no name has been given for the activities of the elements, except perhaps inclination; but other thinkers made use of these activities without properly defining them (307b32–308a4). His distinctions present a full defining account of both the elements and their natural motion and place as their form.

Several further distinctions concerning the elements now follow. As we have seen these points before, their summary may be brief. First, Aristotle defines the differences and what results from them for the elements. Earth is absolutely heavy, i.e., always goes downward, whereas fire is absolutely light, i.e., always goes upward (311a16–29); air and water possess both contraries and so go to places between the middle and the periphery. These differences account for differences among things such as wood and lead (311a30–311b13).

The remainder of *De Caelo* IV, 4, is devoted to proving that there is something absolutely light and absolutely heavy. Bodies are defined as by nature moved always either up or down, unless hindered (311b15–17). This "being moved" brings us to the cosmos as determined and so possessing a center and a periphery (311b19–312a11). *De Caelo* IV, 4, concludes by reiterating that what surrounds resembles form whereas what is surrounded resembles matter (312a12). This point is true in all the categories including place. Place is the first containing limit that renders the cosmos determinate whereas the elements are contained within the cosmos. In short, the elements as moved and the cosmos as possessing a center and periphery (constituted by place) fit together and work together.

What is above is closer to form, and what is below is closer to matter.[66] This point raises the question of the potencies (and matters) of the different elements. The discussion of the matter of the four elements (De Caelo IV, 5) need not be considered at length. Indeed, I wish to make only one point about it. Like the potency of the elements that we have already seen, matter for each of the elements is identified with its ability to be moved upward, downward, or to the middle, respectively, and this ability in its turn is connected to being light or heavy, which in turn is connected to being in or out of respective proper place. In this sense, the explanation turns conceptually on the distinctions that have operated throughout the account of the elements in the De Caelo.

The last and most specialized argument of De Caelo IV concerns whether shape is a cause of motion. And this argument agrees completely with the earlier argument concerning earth. Shape does not itself cause motion but makes the motion faster or slower. And here the book concludes. The account of the elements, their natures and activities, is complete.

This account could not be more coherent, as evidenced by Aristotle's position. He defines nature in Physics II, 1, and takes up motion in Physics III, 1, because we cannot know nature if we do not know motion. Defining motion as a potency/act relation, he considers common and universal terms, including place, without which motion seems to be impossible – and these considerations occupy Physics III through VI.

In Physics III, 1, Aristotle suggests that an account of proper terms will come later. I have suggested that a direct account of the elements does not appear in the Physics because they do not possess a common or universal nature; rather, each is specified by its own nature – fire to go upward, earth downward, etc. – and so the elements taken together comprise the proper terms suggested in Physics III, 1. And in the De Caelo, nature for each of the elements is constituted by inclination; by being heavy or light, inclination orients each element toward its respective natural place, its actuality, where it is ever active and so imitates the heavens. Place, as universal and common, constitutes the cosmos as determinate, and the determinate cosmos – the cosmos constituted as actually up, actually down, etc. – is the actuality of each element, each being nothing other than its particular inclination, its particular active orientation to be moved toward or rest in its proper place.

The elements are treated under the rubric of an investigation of body as such that includes magnitude. Magnitude signals that body is rendered

---

66 Aristotle, De Caelo IV, 4, 312a15–16; cf. Physics VIII, 10, 267b7–8.

determinate by form. That is, magnitude is body's way of expressing shape and determination, and in this sense magnitude always presupposes form. Indeed, according to Aristotle, not only is there no matter, or body, without form, but matter "runs after" form (*Physics* I, 9, 192a23–24). And each element in relation to its natural place is active in just this sense. Throughout the *De Caelo*, each element is regularly defined and referred to as heavy or light and being heavy or light expresses itself as inclination, i.e., active orientation toward the natural place proper to the element – either the periphery or the center.

As place is a sort of container, so the elements must all be contained in place. And as there can be only one cosmos – a cosmos constituted by place, although not itself in place – so all the elements must be contained within this cosmos. Furthermore, because the cosmos is constituted by place as its limit, and all the elements must be contained within it, outside the cosmos there is nothing, neither matter, nor place, nor even the possibility of matter or place. (In *Metaphysics* XII, god, the unmoved mover, is separate from the cosmos; but this is an entirely different problem. As pure actuality, a thinking on thinking, god is neither like place, i.e., a limit *of something*, nor, like the elements, oriented toward something. Rather god is quite different from both place and the elements. Hence, in this respect, there is no necessary contradiction or shift from the arguments of the *De Caelo* to that of *Metaphysics* XII.)

In respect to "where," or the question why all things that are must be "somewhere," the arguments of the *Physics* and *De Caelo,* working together, constitute a significant part of the investigation of physics as a science precisely because they complement each other: the *Physics* yields an account of the determinative principle of the all without specific consideration of what is contained in the all, and the *De Caelo* yields an account of things contained within the all, including the way in which these things necessarily presuppose proper place as a determinative principle, i.e., "where" things are and are moved. This returns us to the methodological starting point of this study: because Aristotle defines these topics differently, he does not specifically or systematically combine them; but they are related logically by the sense in which the topic of the *De Caelo,* body as such, by its definition presupposes a cause such as that of *Physics* IV, i.e., place that resembles form.

Physics as a science would, on Aristotle's definition in *Physics* II, be incomplete without *both* accounts. Place resembles form because, acting as the limit of the first containing body, it renders the cosmos directional and hence one in form and number. The sublunar elements are moved, each to its natural place, because they have but one potency: each is absolutely simple, being nothing other than inclination to be moved toward or rest in its respective proper place. On the one hand, there is but one cosmos; on the other hand, all natural things as well as artifacts within that cosmos are composed of the elements; therefore, the investigations of place and of the elements constitute an account of the cosmos in respect to "where" things are and are moved.

The cosmos, because it is bounded by place as its limit, is one in form and number, and all things are one in respect to place. So the place of earth is down, and earth by virtue of its inclination is actively oriented downward. Indeed, for this reason, the earth must be immobile, at the center, and spherical. Inclination in the case of earth is an intrinsic ability to be moved downward, defined by place in the sense that place defines "down," i.e., the center of the cosmos as bounded by place.

And earth is not different from the other elements. Each element must have an inclination in respect to heavy and light as evidenced by the natural motion of each element. (1) The natural motion of each element is one because it is toward the proper place in which the element rests by nature, e.g., down for earth, up for fire. Unnatural motions are manifold because they may be in any direction away from the element's proper place. As each element possesses one proper place, so each element possesses one proper motion. For this reason, natural motion is always definite and orderly.

(2) The elements are moved to the same place according to kind because it is the very nature of each kind to be moved to its respective proper place. Natural motion is not imposed extrinsically nor is it in any way arbitrary relative to each element. Just the opposite is true. The elements are among things that are by nature and so possess an intrinsic source of being moved and being at rest. Hence motion toward proper place, of which there can only be one, is nothing other than the expression of the very nature of each element according to its kind.

(3) All body must be either heavy or light. Earth and fire are absolutely heavy and light (respectively) whereas water and air are so relatively. To be light means to possess inclination upward or toward the periphery, whereas to be heavy means to possess inclination downward or toward the center. All body, by being heavy or light, presupposes a determinate cosmos, i.e., a cosmos defined as up or down by the first boundary, i.e., place. And being heavy or light means expressing an active orientation toward or away from the center.

(4) On the one hand, what possesses no inclination will be unmoved, i.e., mathematical. On the other hand, body cannot be derived from mathematical, i.e., nonphysical, parts (*contra* Plato and Pythagoras). Body immediately entails magnitude, and magnitude is not mathematical but material. And matter is that which is moved. Hence to be body is to be moved and to possess inclination. Thus, it again appears that inclination is the source of being moved in each element according to its kind, and all things, both natural and artifacts, are composed of the elements.

(5) What differentiates the elements is (a) what generates each element

and so makes it able to be moved and (b) its actuality, which is also form that acts as a mover toward which ability to be moved is aimed. As with all the arguments of the *De Caelo,* the crucial point lies in the identification of each element with being heavy or light and the definition of being heavy or light as being by nature moved toward the center (away from the periphery) or toward the periphery (away from the center). But the natural motions of the elements, i.e., each being moved to its respective place, cannot occur by chance. Indeed, when Aristotle defines motion as the actualization of the potential as such, he specifies that potency and act must be one in definition. An acorn and an oak have the same definition, i.e., oak, and an acorn is nothing other than a potential oak. And these distinctions provide the key to explaining both what differentiates each element by generating it and what differentiates each as making it actually moved.

(a) What generates an element makes it potentially heavy, i.e., earth, or light, i.e., fire. But to be heavy or light means nothing other than to be oriented toward the middle or toward the periphery. Thus, what generates the element enables it to be moved by nature either toward or away from the center. The same point may be expressed by saying that whatever generates the element makes it potentially up or down. Because each element is absolutely simple, whatever makes each potentially up or down differentiates each by enabling it to be moved naturally toward its proper place.

This argument plays a special role in establishing Aristotle's view of nature as orderly: it eliminates any possibility of chance or randomness in the relation of a moved thing to its mover. A thing cannot be moved unless it already possesses potency in that respect. Only something heatable can be heated, only something curable can be cured, etc.; likewise, what is cured is so by virtue of being curable and not by virtue of being heatable. And the generator grants to the moved potency by virtue of which it is oriented toward actuality. Thus, the generator is not a mover in the same sense as is the actuality that actualizes the given potency. Rather, the generator is the cause that, by granting a thing its potency, enables it to be moved [by proper actuality]. Hence, the generator accounts for motion by accounting for that which is moved insofar as it possesses a given potency, i.e., an active orientation for proper actuality.

(b) Actuality in the fullest sense differentiates each element. In the absence of hindrance, what is potential cannot fail to be moved by its proper actuality. Such is the force of Aristotle's definition of motion as an actualization. Actuality differentiates what is potential in respect to it by giving that potency its definition and thereby making it what it is in the fullest sense. And for each element, to be moved toward its respective proper place is to be moved toward form and, hence, actuality; when an element arrives at its proper place, it ceases to become and simply is: to reach its proper place is to reach its form. And respective proper place serves as the goal and form of each element.

When a thing reaches its actuality, actualization is complete and activity ensues. And the activity of each element is to rest in its respective proper place. Such activity presents the fullest expression of being for each element precisely because this activity fully expresses what each element is, without possibility of further development, and, consequently, as fully determinate. Inclination, we may conclude, is the name for the principle within each element by virtue of which it is moved toward its respective proper place, which is actualized as it approaches that place, and which expresses itself in the activity of rest when each element is in its respective proper place. For this reason and in this sense, inclination is the very nature itself of each element, the matter of the elements comes closest to substance, and we may say that the elements "imitate the imperishables; for these things too are ever active" (*Metaphysics* IX, 8, 1050b29–30).

With the account of inclination (and a brief criticism of Democritus's inferior view), *De Caelo* IV concludes. The problem that remains unsolved, i.e., that of the generation of the elements from one another, is taken up in the *De Generatione et Corruptione*. But this problem lies beyond the bounds of a study of the order of nature in respect to place and the elements.

A number of questions still must be addressed. They concern the relation of the arguments of the *Physics* and those of the *De Caelo*. Beyond particulars of the text are questions concerning Aristotle's position as it emerges across these different texts. And beyond establishing Aristotle's position as such is the question of evaluating that position both in itself and in its broader historical context. To these I turn now.

Part III

Nature As a Cause of Order

8

The Order of Nature in Aristotle's Physics

With the account of the nature and activities of the elements, the analysis of place and the elements is complete. Taken together "the where," i.e., place and not void, and inclination, i.e., the active orientation of each element toward its respective proper place, exhibit the order of nature in Aristotle's physics. Nature is always a cause of order, and that order is constituted within the world by two principles. Place is the first limit of the containing body and renders the cosmos orderly in respect to direction; thus the cosmos exhibits "up," "down," "left," "right," "front," and "back" immediately and intrinsically in itself. Second, inclination constitutes the very nature of each element as an intrinsic source of being moved toward its proper place, e.g., up for fire and down for earth; consequently, elemental motion is never random or irregular because, in the absence of hindrance, each element cannot fail to be moved toward (and to rest in) its proper place.

These principles solve a number of problems and so express the order of nature in three ways. (1) Place renders the cosmos determinate in respect to "where," while each element is ordered to its proper place. Place and the inclination of each element work together to produce the order of nature in respect to the intrinsically directional motion of the elements. (2) The order of nature as expressed by the relation of the elements to place is teleological because it is immediate, intrinsic, and characterized by an active orientation of the moved toward a mover that is at once its form and actuality. (3) The order of nature as a whole, constituted by place as a single, unique, formal principle for all things, is identical with the order of nature in all its parts, constituted by the elements as determined each toward its proper place so that in the absence of hindrance each cannot fail to be moved toward and rest in that place.

Place

In *Physics* IV, Aristotle turns first to the problem of place, because, he says, everyone thinks that things that are must be "somewhere" [πού]. πού, "where" or "somewhere," is one of the categories of being, and the first and highest kind of motion is motion according to place. The real question is "where are things that are and are moved?" "Place" and "void" are both proposed as answers, and place is the true answer and so something without which motion is impossible. The reason why is clear: by rendering the cosmos determinate in respect to up, down, left, right, front, and back, place as actuality causes the motion of the elements from which all natural things and artifacts are composed; the void, being a principle of indeterminacy, wholly fails to cause motion in things and so must be rejected.

Aristotle's account of place may be summarized here. He begins with evidence that place *is:* "where" there once was one thing, we now find another; the elements exhibit natural locomotion; and the six directions, i.e., up, down, etc., are given in nature and are not just relative to us. This evidence raises the problems that any successful account of place must solve. And Aristotle clearly thinks that his account succeeds.

First, place has three dimensions. Indeed, it is in the same genus as body, but cannot be body. This apparent contradiction is solved by the definition of place as the first limit of the containing body. As a limit of this body, place must be both three-dimensional and distinct from the containing body itself.

The second problem is as difficult as the first: of what in things is place a cause? It cannot be one of the four causes, i.e., form, matter, moving cause, or final cause. Indeed, as these are the causes of things that are in place and place itself cannot be another thing *in* place, we should not expect place to serve as a cause in any of these four ways. Rather, place is a cause as a principle of determination in respect to direction. Because place limits the cosmos, the cosmos contains an immediate up, the first or outermost heaven, and an immediate down, the center of the cosmos, which coincides with the center of the earth, as well as an immediate middle, the place between the outermost heaven and the center. Being constituted by place as a limit, "up," "down," and "middle" serve as the respective natural places of each element. Because place renders the world determinate, place is something without which motion is impossible.

In the *De Caelo,* the causal role of natural place emerges unambiguously. Each element is oriented toward its respective proper place: up for fire, down for earth, and the respective proper places in the middle for air and water. Furthermore, to achieve its proper place is for each element to achieve its form and actuality: once in its proper place it ceases to become and is actually. As a limit, place causes all motion by rendering the cosmos determinate and so producing the proper place for each element within the cosmos. Proper place within the cosmos causes the motion of each element from which all things are composed, just as any actuality causes the actualization of the potency naturally oriented toward it.

There are, however, further problems about place itself. Zeno had raised the problem of whether place must have a place, but Aristotle gives this problem short shrift. Place, being a limit and constitutive principle, explains why all things that are and are moved, i.e., all natural things and artifacts, are in place, but it itself need not be in place. Hence, there is no infinite regress in which "place" also requires a place.

Zeno's objection – "Is place *in* place?" – raises the question of the various meanings of "in." Most importantly, a thing can be in itself neither accidentally nor essentially; but a whole can be in itself in the sense that its parts are in the whole. For example, given a whole such as an amphora of wine, both parts, the amphora and the wine, may in a sense be said to be "in" the whole. And in this sense, place can be in the whole comprised by place and the first containing body. But place is not "in" the whole as another thing contained and so itself requiring a place; rather, it is "in" the whole as its container – its limit.

Aristotle's own constructive account begins with the four characteristics rightly thought to belong to place because any account worthy of the name must be able to explain these characteristics. (1) Place is what first surrounds without being part of the thing contained. As the first unmoved limit of the containing body, Aristotle's definition obviously accounts for this characteristic. Being the first limit, place surrounds the first body, but is not itself another part of that body. (2) Place is neither less than nor greater than the thing contained. The limit is a formal not a material constitutive part, and so is contained in the whole accidentally. Indeed, looked at as a quantity, place and the first containing body are continuous in the sense that nothing intervenes between them. Therefore, Aristotle's account meets this requirement as well. (3) Place is separable from what is in place. The limit is separable in definition from the limited; in this sense they are contiguous rather than continuous. Indeed, these three characteristics are straightforward and Aristotle's definition of place as the first unmoved limit clearly addresses them.

The last characteristic of place is the most difficult. (4) Every place has up and down and each element is carried to its proper place, in which it rests, and "this makes either up or down." To explain this characteristic requires the account of the elements and inclination. Although I have largely treated the arguments of *Physics* IV and *De Caelo* IV as separate, the coherent position produced by my analysis shows that we may speculatively relate the problems of place and inclination.

Place, as we have seen, renders the cosmos determinate in respect to direction. Hence every place within the cosmos must be up, down, etc., with the outermost sphere being absolute up and the center absolute down. Each element possesses its own unique inclination by virtue of which the element is oriented and moved toward its proper place. This inclination is the very nature of the element. Because each element and its respective proper place relate as potency to actuality, in the absence of hindrance each element must be moved toward its proper place where it rests, i.e., is actually and actively. And "this makes either up or down" both in the sense of the constitution of the cosmos and in the sense of the dynamic orientation (and hence natural motion) toward the proper place of each element of which all things within the cosmos are made.

The problem of place would never arise were it not for the fact of locomotion. A thing may be moved either essentially or accidentally, and accidental motion returns us to the problem of how one thing can be "in" another, which in turn raises the issues of "touching," "contiguity," and "continuity." Considered as container and contained, place and what is in place differ in definition and so touch but are always divided; hence, they are contiguous and not continuous.

Having established the distinctions required by an explanation of place, Aristotle turns to the four candidates for place, rejecting three and arguing for the fourth and last. (1) Place cannot be form, or shape, although they share important features: both surround and are limits. But form is the limit of the contained, whereas place is the limit of the containing body. (2) The interval between extremities cannot be place because it cannot be separate and would involve an infinite regress of places. The real problem with the interval (as with the void as well) is that it is indeterminate and so fails to serve as a principle of determinacy. (3) Matter is the least interesting candidate for place. It can neither be apart from nor surround the object in place whereas place must be both.

Place is (4) a limit, indeed, it is the unmoved limit of the first containing body. Because it is a limit – and thus distinct from what is limited – place cannot share the characteristics of the limited. Place must be unmoved and first. And as first, i.e., the unique single boundary of the whole, place constitutes the cosmos as determinate in respect to direction with the result that the cosmos has an outermost part, the heaven, and a center, which coincides with the center of the earth, and a middle between the center and the heaven. Being a constitutive principle in this sense, place is more important than what is in place just as a cause is always more important than its effects.

Defining place as a limit raises the problem of the exact meaning of "limit." As the definition of place, "limit" has historically often been taken to mean "surface." However, place cannot be a surface. Aristotle never calls it a surface, and, although all surfaces are limits, not all limits are surfaces. Indeed, surface lacks the characteristics crucial to place. A surface cannot be three-dimensional, it cannot be unmoved, and it must itself have a limit. For these reasons, surface, unlike place, must be in place. Furthermore, surface functions as a limit by dividing the body of which it is a surface from what is outside the body, for example, water or air. Finally, surface is an object of mathematics, not physics.

Although surface and place are both limits, not all limits are alike, and place is all that surface is not: three-dimensional, unmoved, itself without a limit, and a constitutive principle. As a limit, place is unique. It limits not by dividing – there is nothing outside of place from which to divide the cosmos as contained – but by constituting the cosmos as directional. Hence, it is "first" in a sense that no other limit can share. As first, place cannot itself have a limit and so is not *per se* in place. As the first constitutive principle, place is the limit of the first containing body. As the limit of this body, place is in the same genus as is body and so is both three-dimensional and an object not of mathematics but of physics. Furthermore, unlike surface, place as a limit must be unmoved because it constitutes all place within the cosmos; because the cosmos is unique, place as a limit defines the "where" within which all motion must occur.

In *Physics* IV, 1, when Aristotle raises the question of "the where [τό πού] of all things that are and are moved," he does not mention his categories of being. But "where" appears regularly in the list of categories. Thus the account of place as "the where," along with the rejection of the void, which fails in every way as a cause of motion, is an account of this category. "Where" as a category of being applies to all things that are by nature, and it does so in a way that is unique. Indeed, "where" constitutes a category of being because it is ubiquitous and unique. And in this account of place, we see its force as a category of being, "the where" of all things that are and are moved.

In respect to "where" as a category of being and place as an account of that category, a systematic problem arises concerning the relation of place and things that are in place that has not been addressed earlier. Matthen and Hankinson claim that "In Aristotle's system, parts that maintain the nature of the whole are typically treated as deriving their nature from the whole, and as posterior to it in being (however that is to be understood)."[1] And they go on to argue that the elements are posterior to the sphere "that constitutes the totality."[2]

I have argued throughout that place is a cause in the sense that it renders the cosmos determinate and that each element is caused by its respective natural place insofar as, in the absence of hindrance, each cannot fail to be moved toward that place. In this sense, place and the elements exercise a cause/effect relation. This is not to say, however, that the elements are in any way ontologically derived either from place or from the first sphere. In fact, Aristotle goes on to argue that the elements are generated from one another – a generation that forms an eternal continuous cycle that imitates circular locomotion (*De Generatione et Corruptione*, II, 10, 337a1–6). Furthermore, as he often insists,

1 Matthen and Hankinson, "Aristotle's Universe," 426.
2 Matthen and Hankinson, "Aristotle's Universe," 430.

"being falls immediately into the categories," and there is no category from which the others are derived.[3] The elements do not derive their being from another, but, like all things that are by nature, are properly identified as substances, the first category of being. They imitate the heavens, and, in this respect, their matter comes closest to substance in the highest sense.

As a principle of determination for the cosmos, place does not cause the being of the elements; rather, it causes the cosmos to be determinate in respect to the category of "where" and thereby at once causes *both* the motions of the elements (each to its respective place) *and* the order and regularity of that motion. Each element possesses one natural place and hence one and only one natural motion (and rest), i.e., motion toward (or rest in) that place, which in the absence of hindrance must always be actualized. When actualization is complete, the element "rests" in its natural place: it no longer becomes but is.

In *Physics* IV, the account of place raises a number of issues for the heaven as that which is first contained by place. Within his account of place, Aristotle asks how the heaven is and is not in place (*Physics* IV, 5, 212b12–23). Earlier in the text he discusses the various meanings of "in," using the example of an amphora containing wine (*Physics* IV, 3, 210a27–210b22). These distinctions apply to and (on Aristotle's view) solve the problem of how the heaven is in place. Place does not exist apart from the heaven; rather, together they form a whole (the heaven) comprised of container (place) and contained (first body). In short, the account of place and the heaven is an explanation of how the whole and its parts are and are not "in" place: the whole as such is not in place, but the parts are in place (*Physics* IV, 5, 212a32–b3).

This discussion is strictly limited to the problem of the whole of which place is the limit, i.e., the heaven comprised by place and first body. It is *not* an account of the first body as such. Indeed, because Aristotle separates his analysis of common and universal things from that of proper things, the first containing body is not examined in *Physics* IV. The definition of place as a limit requires a reference to the limited, first body, and how the two form a whole, the heaven, that must itself as a whole either be or not be in place. But concerning problems of the limited taken in and of itself, Aristotle says "let us take these up later."[4] And, as I have argued, the particulars concerning the first body and inclination appear in the *De Caelo*.

---

3 The best-known argument on this point is undoubtedly *Metaphysics* IV, 2, 1003a32–1004a9, cf. esp. 1004a5. For other examples, cf. *Metaphysics* V, 7, 1017a23–30; VII, 4, 1030b11; XIV, 2, 1089a7.

4 Aristotle, *Physics* IV, 5, 213a4–5; properly speaking, the particulars of this promise are identified in *De Generatione et Corruptione* I, 3.

The Elements

Aristotle opens the *De Caelo* by confirming that we are in the science of physics, and he takes up the parts of the cosmos. He argues that there must be a fifth element in addition to the four sublunar bodies. The fifth element is, so to speak, the first, or primary, body and, as I have argued, the containing body of which place is the first limit. Thus in the *De Caelo*, we arrive at something proper that complements place as common and universal. That is, place is the limit of the first containing body and so taken together place and the first body form the first heaven that contains all things that are by nature and by art. The formal part of the heaven is examined as something common and universal in the *Physics* whereas the first body is examined as something proper in the *De Caelo*.

In this sense, the *De Caelo* presents the converse of the *Physics*. Just as we find no account of the elements in *Physics* IV, so too we find no direct examination of either god or place in the *De Caelo*. And for the same reason: in the *De Caelo* we have an examination of proper things whereas god belongs to an examination of substance (*Metaphysics* XII) and place to an examination of common and universal things (*Physics* III-VI).

The fifth element, aether, is prior to the other elements and so is more divine. Hence it is different and separate from them. It is moved only by circular locomotion, the first and highest motion. And it is determined in this respect by the first limit, place. Experiencing only the first and highest motion, circular locomotion, the fifth element must be without a contrary. In fact, it is unique. Its motion is both complete and eternal; as complete, it is prior in nature to the incomplete. Therefore, this element is akin to the divine.

The power of place is remarkable, and the body of which it is a limit is the most divine of all bodies. In short, the characteristics of place as a limit are, insofar as is possible, shared by the first contained body: unique, eternal, and prior in nature. Hence place and the first body go together and work together as container and contained, that which renders the cosmos determinate and the first determinate part. Within the first heaven, place and the first body, the entire cosmos is rendered determinate in respect to "where," and the directions "up," "down," etc., are given immediately and intrinsically.

I have argued that place is unique. It is unique not as the first or highest member of a group, or part of a whole. It is unique as a single constitutive principle of the cosmos. By identifying place as a unique constitutive principle of the whole, we reach the meaning of the phrase "common and universal." Place is not common and universal as an abstraction, whether mental or real, taken apart from particulars; rather, it is common and universal in the sense of being a single unique principle of determination for the cosmos as a whole and hence for all things that are by nature and, being by nature, are within the cosmos.

And in just this sense, the first body is not common and universal; rather, it is the first of the elements and the highest part of the cosmos. It is that which is first rendered determinate by place. Although it is an element, the fifth element is, unlike the sublunar elements, unique, eternal, and without a contrary. Consequently, in the fifth element, the first body, we reach the first of "proper things." And proper things are just this: unique and specific each of its kind, but members of a group, which may be considered as parts of that group. So, for example, consideration of the four sublunar elements as a group proves the necessity of a fifth element to complete the group, or class, of elements. On the one hand, each element possesses a different, unique, and specific nature; on the other, an account may be given of these natures as such and there is no need to treat each element separately. Because the fifth element is first and is unique among the elements, it is treated first in the *De Caelo*. Nonetheless, it is an element and so a member of this group, and thus its examination appears in the *De Caelo* – among proper things – and not in the *Physics* – among common and universal things.

After examining the fifth element, aether, Aristotle takes up the four elements – earth, air, fire, and water – and problems concerning them. The first argument that there cannot be more than one world turns on the differentiation of the world into "up," "down," etc., and the orientation of each element towards its natural place. The claim that there is more than one world violates this relation and so must be rejected. The second argument that there cannot be more than one world rests on the assumption that each element is formally the same everywhere. Hence each element must have one and only one natural motion. And in each case, this natural motion is defined solely by reference to natural place, which must also be one in form and number. So, for example, fire is always moved upward by nature and up can only be the outermost circumference of the heavens, earth is always moved down, i.e., toward the middle. In short, place within the cosmos is defined by place, the first unmoved limit of the containing body, and the elements are irrevocably locked together in a relation that becomes clear only in *De Caelo* IV. Given the universality of place in relation to all things and the uniqueness of the cosmos, this relation, although particular to each element, embraces all things that are and are moved by nature.

In the arguments that follow, inclination appears unambiguously as the very nature of the elements. Nature is by definition a source or cause of being moved and being at rest in that to which it belongs primarily by virtue of itself and not accidentally. The elements are absolutely simple. Therefore the inclination of each element must constitute it completely: inclination is an ability to be moved by nature toward (or rest in) its respective proper place. And the elements are nothing more or other than this. In the constitution of the elements – and they comprise *all* things that are and are moved by nature – we find the conjunction of place, nature intrinsic to things that are, and a simple principle of being moved.

When an element is in its natural place, becoming is complete and it is actually. Rest, for each element, is nothing other than the activity of being in its proper place. For the proper place of each element is, Aristotle says, like its form, and reaching proper place is to rest, i.e., simply to be as the highest activity of each element. For this reason an element can be removed from its proper place only by force applied from the outside. Because each element is simple and its motion must terminate in the activity of being in its proper place, the elements may be said both to imitate the imperishables and to possess matter that is closest to substance. That is, the elements are ever active and contain in themselves a source of being moved.

This view of the elements and their relation to place brings us to the final and broadest moment of this analysis. First, throughout the *Physics* and *De Caelo* (and I would suggest throughout the whole of Aristotle's corpus) the arguments are causal. Place is a term without which motion seems to be impossible whereas the void must be rejected because it fails in every way to serve as a cause of motion. When in the *De Caelo* we turn to the elements, we turn from the cause of motion to that which is caused, from the agent to the patient. And again the argument is causal. Because the elements are simple, they have but one possibility, one orientation, i.e., toward their actuality. And actuality is the cause of actualization in the potential. Hence to identify a potency/act relation in which only one outcome may obtain is to specify the necessity of that outcome in the absence of hindrance. And it specifies the relation as teleological. The cosmos is rendered determinate by place and is comprised of elements, each of which is fully and exclusively determined toward its proper place. And as a result of this construction, the fifth element is more divine than the others whereas the others, by their constant orientation and activity, imitate the divine.

Place and the Elements: Aristotle's World

God and nature do nothing in vain. The order of nature may properly be thought of in three ways. (1) It is exhibited fully in the structure of the world, the determinate world with its determined parts. Thus we find a single unique whole rendered determinate by a formal constitutive principle, place; all things that are and are moved by nature are contained within this whole and are in respect to "where" fully determined by place – each by its proper place within the whole. Among natural things – and artifacts as well insofar as they are made of natural things – nothing escapes the mark of the determination of place and hence nothing lies outside the order that is the very nature of the cosmos.

(2) Nature is an intrinsic principle of being moved and being at rest in that to which it belongs essentially and not accidentally. The order of nature is also an order intrinsic to each thing that is by nature, i.e., the intrinsic and immediate orientation of every potency to its proper actuality. Indeed, that the source of motion must be intrinsic to the moved is part of its very definition as natural. The cosmos presents nature writ large, so to speak. And the order of the cosmos is not "fed in" from the outside. It is intrinsic and immediate: part of the very constitution of the cosmos and each thing within the cosmos insofar as it is and is moved. And place is the principle, the common and universal term, that constitutes the cosmos as determinate and hence orderly in respect to "where." And place does so immediately and eternally – hence it is something common and universal.

But each element is by nature as well. In the elements we find nature writ small, the nature not of the whole but of each part in all its specificity within that whole. And each element possesses a formal nature: each element is intrinsically ordered to its respective natural place as to its form or actuality. And nature in this sense too, i.e., in the sense of being determined, is an immediate and intrinsic source of order.

(3) Nature in things is defined as form (plus a reference to matter). And form is a thing as actual. Therefore, the definition of a thing as by nature and the definition of a thing as moved are identical: form or actuality. For the elements, that form or actuality is proper natural place – the particular place constituted by place as the first unmoved limit of the heavens rendering the world determinate. Hence there is no difference between the order granted to the cosmos by place as a determining principle and nature as an intrinsic source of being moved and being at rest: they are one and the same.

Indeed, there is no difference between the order of nature and the teleology of nature. Aristotle's teleology is often identified with his account of "final causes" as if, apart from them, the rest of his physics (or philosophy more generally) were not teleological. Such an account is another version of the view that Aristotle's philosophy is a proto-mechanistic view with teleology, in the guise of final causes, added on. But such a view is more than misleading: it is false.

In respect to nature, Aristotle's teleology is expressed by the immediate, intrinsic, dynamic, never-failing orientation of what is moved (and identified with potency and matter) toward its actuality (which is identified as form, activity, and the highest expression of being). Actuality is at once the mover and the definition of what is moved. In short, the order of nature is nothing other than the orientation of each element toward the place that is its form and actuality – natural places within the cosmos that are defined by place as the limit of the first containing body – and this relation *is* Aristotle's teleology of nature.

Aristotle's Immediate Historical Context

The force and originality of Aristotle's view of nature as an immediate and intrinsic cause of order may be seen by briefly contrasting his view with that of his predecessors – first, of course, Plato. Form according to Plato is separate from the world of becoming, which we may think of (not altogether accurately) as the counterpart to Aristotle's natural world. Consequently, form must be placed into matter by an extrinsic agent, soul or the demiurge. Furthermore, because they do not go together naturally, form must be imposed on matter while matter must be persuaded to accept form. There is no natural orientation of matter to form, and, indeed, for Plato the very hallmark of the realm of becoming is its instability.

Although a full discussion lies beyond the bounds of this study, we may note that Aristotle's account of nature exactly reverses that of Plato. Aristotle and Plato do agree about art – because in works of art no intrinsic relation obtains between matter and form, and an extrinsic cause is required to combine them. But for Plato, the world of becoming is first and foremost a work of art, one produced by a master craftsman. For Aristotle, however, nature contrasts with art precisely because matter is intrinsically, dynamically, and, by its definition, aimed at form; consequently, no artist is required to impose form on matter. Indeed, the aptitude of matter for form lies at the heart of Aristotle's teleology.

Here too is why Plato's becoming and Aristotle's nature are *not* counterparts: Plato must explain the origin of order, or partial order, in becoming, whereas Aristotle must explain why nature sometimes fails to be orderly. For this reason, Aristotle's principles of disruption in nature, e.g., hindrance or privation, have no direct analogue in Plato. Likewise Plato's extrinsic sources of order or motion in becoming, soul or the demiurge, have no direct analogue in Aristotle because for him nature is intrinsically orderly. For Plato, the demiurge imposes form on matter and soul, "loosing its wings," descends into body from without, whereas for Aristotle the heaven runs after the divine, and body desires soul as its first entelechy. Aristotle complains that Plato fails to explain motion, and, given Plato's accounts of soul and the demiurge, we might speculate that the complaint rests on Plato's "failure" to provide an *intrinsic* source of motion and order. But then Plato does not see the realm of becoming as intrinsically orderly. The world as intrinsically orderly constitutes Aristotle's vision of nature, the cosmos, and all natural things because they are "by nature."

Aristotle's account of nature as always and everywhere a cause of order also contrasts with the account of proponents of the void, both the materialists and the Pythagoreans. The question that opens *Physics* IV, 1, asks

"what is 'the where' of things that are and are moved?" There are two possible answers to this question: place or void. Until the void – the candidate
for "the where" proposed by Aristotle's opponents – is eliminated, place
has not been established as the *exclusive* answer to this question. Aristotle
argues first that void fails to stand as a coherent concept and then that it
fails in every way to serve as a cause of motion. With this rejection, place
is in "first place": "the where" for all things that are and are moved.

Aristotle first defines void as a kind of place, i.e., empty place. Hence,
the logical status of the void is a bit odd. It is at once a competitor of place
posed as the category "where" and a special case of place, i.e., place with
nothing in it. Aristotle's rejection of the void as a cause of motion presupposes his account of place because as a concept void presupposes
place. The point can be expressed more generally: void is indeterminate
(and hence a source of disorder), and what is indeterminate always presupposes what is determinate (and hence a source of order). Consequently, both in fact and in principle, it is impossible to examine the void
without presupposing a principle(s) of determination, which in this case
is place (and ultimately motion and nature).

Aristotle takes up accounts of those who propose the void as a cause
of motion to show that in every case it fails or is unnecessary. Most importantly, the void is an incoherent concept. It cannot function as a concept without producing a contradiction. Indeed, those who propose it do
so only because they confuse void and place.

The status of void as a concept is crucial for Aristotle's arguments that
reject it. Because the void fails to serve as a coherent concept, arguments
that take it as a meaningful term originate exclusively in a starting point
provided by Aristotle's opponents. Hence these arguments for the necessity of a void have no constructive relation to Aristotle's account of nature, the world, or things in motion.

The logic of his strategy appears clearly if we recognize the sense in
which nature is, for Aristotle, a source of order intrinsic to the cosmos as
a whole and each thing that is by nature. As a concept, the void is absolutely at odds with this notion of order. Hence it cannot operate co-
herently within an account founded on this notion of order because to
do so would involve either giving up the view that order is primary or tol-
erating a direct contradiction at the core of the analysis. Aristotle is most
certainly unwilling to do the former, and the latter would vitiate his entire project. Hence, although he must address the problem of the void,
he can do so only indirectly.

The materialists provide several arguments in support of their view
that a void is necessary for motion. But these are easy to refute. More serious arguments are found in the Pythagorean account, which utilizes the
void as a principle of separation and limit.

Among their arguments, as we have already seen, the most serious concerns motion, void, and "that through which." It begins by asserting that that through which (e.g., through water or earth) causes a thing to be moved faster (or slower, although Aristotle does not say so) and concludes that a body would be moved through a plenum and through a void in the same time. A body may also be moved faster because of a greater inclination of heaviness or lightness, but in a void, all bodies, regardless of their inclination, would be moved equally fast. And Aristotle concludes that the void cannot serve as a cause of motion. Indeed, assuming a void produces the opposite result of that anticipated by its proponents: it renders motion impossible. And a series of brief arguments against the void completes Aristotle's rejection of it.

The question, *mutatis mutandis,* must be raised concerning the relation of the refutations of the void to Aristotle's account of inclination. These arguments make no direct contribution to Aristotle's account. However, I would suggest now that they do make an indirect contribution.

The arguments of the *De Caelo* explicitly refer to the cosmos as intrinsically orderly. Furthermore, when Aristotle argues that there can be only one heaven, he assumes this order as a relation between each element and its respective place. Finally, in the account of the motion of the elements, the determinate relation between each element and its place completes the account. In short, Aristotle's account of elemental motion is possible only within a world rendered determinate in respect to direction by a cause such as place; the assumption of a void, which is by definition indeterminate, as something universal and common without which motion would be impossible is not so much "wrong" or "false" as meaningless within the conceptual framework of Aristotle's account of nature and motion in things. And this failure of accounts of the void – the failure to be meaningful – reveals the ubiquitous order of nature both within Aristotle's world of nature and within his physics as the science of this world.

In fact, the complete sense in which the void fails to come to the starting gate of a meaningful argument can be seen in another very important sense. The plenum may be taken as the opposite of the void. Void means empty, plenum full. And full and empty would seem to be mutually exclusive and exhaustive alternatives. Thus it has become virtually a commonplace to think of Aristotle as committed to a plenum because he is in effect forced to this commitment by his rejection of the void.

But there is something odd here: Aristotle's constructive account of elemental motion – indeed, his account of *all* natural motion in things – rests on the intrinsic orientation of the moved, which is potential to its actuality. Indeed, the order of nature is teleological because it is based on an intrinsic relation of moved to mover. Not only is there no discussion of a plenum but the external relations among things that would be constituted by a plenum are all but absent in his constructive account. And the reason for this is not hard to see: the void and the plenum are mutually exclusive and exhaustive alternatives within a materialist account of the cosmos. As such, both represent, albeit in opposite terms, extrinsic relations among things and between things and "the where." By virtue of being extrinsic, these relations always imply an element of indeterminacy. Consequently, both the void *and* the plenum fail to provide the causal account required by Aristotle's notion of nature, the cosmos, and things that are by nature. He himself makes the point explicit, claiming without explanation that "void," "full," and "place" are not the same, but are three different "essences" (*Physics* IV, 6, 213a18–19).

This issue appears most sharply in the refutation of the void based on the claim that the "that through which" causes the body to be moved faster. Traditionally, this argument is taken to refer to varying speeds through different media because of the different resistance offered by the media. I have argued that neither "speed" nor "medium" nor "resistance" are concepts appropriate to this argument precisely because they all indicate extrinsic relations among things. And if one assumes that these concepts are operative here, the argument becomes obscure.

If however the argument is read as resting on the intrinsic relations that constitute Aristotle's basic conception of nature and motion in things, then the argument becomes coherent. The images of a "that through which" and of a medium, representing some sort of plenum, may look the same, but they do not operate in the same way as causes. One is extrinsic while the other is intrinsic. And in his constructive account, Aristotle has no use for the causal conceptions associated with extrinsic relations, be they those of a void or those of a plenum.

For Aristotle, the mutually exclusive and exhaustive alternatives are that which is orderly and a principle of determination (place) and that which is disorderly and a principle of indeterminacy (void – and in principle plenum too). He rejects the void as an incoherent concept precisely because on his view nature is always and everywhere an intrinsic source of order. And the alternative – the alternative that he chooses and that thereby operates throughout all his arguments concerning the parts of the whole – is a principle of formal determination from which nothing within the cosmos escapes: all place within the cosmos is either up, down, left, right, front, or back while each element, out of which all things both natural and artifacts are made, is oriented toward its proper place. Indeed, this order presents the teleology of nature: all natural things (and artifacts insofar as they are made of natural things) are oriented toward their proper place, and hence activity, by an intrinsic relation that never fails (but can be hindered from the outside).

In short, place is in sole possession of the category "where." Place is the principle of determinacy in respect to this category; place, not void, is a term without which motion seems to be impossible because place alone renders the cosmos determinate in respect to "where" things are and are moved. Place alone can play the causal role required by Aristotle's account of the elements and their respective natural motions. Even ancient myth with its vision of a river encircling the cosmos bears witness to the success of place and Aristotle's account of it.

Aristotle's account of place meets the conditions for the best possible account. It defines its object, addresses alternate views, explains why what is rightly held to be true is in fact true, and, finally, provides a pathology of wrong opinions concerning the where of all things that are and are moved. The same may be said of the account of the elements. These two accounts, we may conclude, when taken together, express the order and teleology of nature in the determinate world of Aristotle's physics.

The Systematic Character of Aristotle's Physics

A variety of problems remain, the most important of which are methodological. As I have argued, Aristotle takes up a specific problem and pursues his analysis of it with little (sometimes no) regard for related problems raised in the course of his analysis. In this respect, his procedure resembles that of Plato, even if their problems differ. (Indeed, this "problem-specific" procedure of philosophy seems to characterize classical Greek philosophy more generally.) As Plato asks what is justice, what is temperance, what is courage, so Aristotle asks what is nature, what is motion, what is "the where." Beyond place and the elements, are further topics, e.g., what is time, what is the soul.[5] As a consequence, Aristotle's analysis is defined by and conducted within the domain established by the problem that he specifies initially. Understood in this way, his arguments are not only successful, but elegant.

However, Aristotle provides few clues concerning the relationships *among* the answers to his specific problems. He never addresses these relationships directly because they themselves never form a specific problem and so are never a topic of investigation. On the one hand, a systematic sense of philosophy outside these specific problems is neither Aristotle's interest nor his project – and in this respect he is part of the tradition of Greek philosophy that also includes Plato and the Presocratics. (Systematic philosophy across problems may well be an invention of Hellenistic philosophy associated with the project of synthesizing Plato and Aristotle.[6]) But on the other hand, contemporary readers may also find the call of a later tradition, a conception of philosophy that does consider systematic relations among topics, irresistible. The absence of explicit considerations of this sort in Aristotle's arguments undoubtedly suggests for modern readers what they would characterize as a limitation of his project.

5 Aristotle, *Physics* IV, 10, 217b29; *De Anima* I, 1, 402a5–9.
6 Pfeiffer, *History of Classical Scholarship*, 152.

Hence I shall return to the conclusions established earlier and with them address systematic issues, recognizing that they are not Aristotle's own. I shall first take up the relation between the accounts, that of something universal and common and that of proper things, and finally considerations concerning Aristotle's physics as a science, in particular, the task of evaluating both his arguments and his larger philosophic project as such.

I shall suggest that Aristotle's accounts are in fact amenable to such speculation, even though it is not an explicit topic in his writing. Indeed, my analysis explains why this is the case: Aristotle's solutions to the problems that he defines and solves (to his own satisfaction) – those examined here and others as well – invariably presuppose the view that nature is always and everywhere a cause of order. Hence the solutions to the various problems that define Aristotle's arguments presuppose this view, and this consistency yields a ground for speculation concerning systematic relations among Aristotle's various arguments.

In short, the coherence of Aristotle's larger philosophic project, which renders it in some sense systematic, comes as a consequence of subordinating his arguments to problems that ultimately rest on a single presupposition: nature is everywhere a cause of order. *Physics* II and *Physics* III are separate *logoi* because they are defined by different problems. *Physics* II concerns things that are by nature, including their definition, the science that studies them, and the causes at work by nature (along with refutations of alternate views). *Physics* III ostensibly concerns first motion (*Physics* III, 1–3) and then the infinite (*Physics* III, 4–7).

The relation between the topics of these *logoi* is made explicit in the opening lines of *Physics* III, 1: if nature is not to remain hidden, then motion – together with the common and universal things without which it seems to be impossible – must be considered. The relation between the infinite and nature is also made explicit as Aristotle opens his discussion of the infinite: "since the science of nature concerns magnitude and motion and time, each of which must be either infinite or limited . . . the one who considers the things concerning nature will theorize concerning the infinite" (*Physics* III, 4, 202b30–35). But neither nature nor the relation between either nature and motion or nature and the infinite appears again in *Physics* III. There is certainly no explicit summation or return to nature when the accounts are complete.

Setting aside the problem of the infinite, which lies outside the bounds of this study, the relation between motion and nature seems clear: nature, defined as a principle or cause of being moved and being at rest, is presupposed throughout the account of motion. In this sense, the account of motion is subordinated to the definition of motion. Aristotle explicitly argues that there is no such thing as motion apart from things, and these things can only be things that are by nature or are composed of natural things. Hence nature is implicitly part of the account of motion all along – there is no need to return to nature with the completion of the account of motion because we never left it.

In short, the domains of nature and motion are coextensive. Within the science of physics, the study of motion is a study of motion intrinsic to things, and anything that possesses such a source of motion is by definition a natural thing. For this reason, even though *Physics* II and *Physics* III, 1–3, ostensibly consider different topics – nature and motion – the conclusions of these arguments not only are consistent but form a coherent position.

Motion, as Aristotle defines it, is the actualization of the potential (insofar as it is such) by that which is actual. Hence, there is no motion apart from things – indeed, motion is best thought of as a development of the moved. And this development of the moved, which must be by nature (there are no natural things that are not so moved) is produced by its proper actuality for two reasons. (1) Potency and actuality are identical in definition, namely, the definition of the actuality. Thus, although Aristotle elsewhere specifies causes of motion as moving cause and final cause, both these causes produce motion as something actual relative to something potential. And in this sense they confirm Aristotle's account of motion because its definition rests exclusively on the notions of potency and actuality.

(2) Far from being passive or neutral to actuality, potency is actively oriented toward its proper actuality. Thus, whenever possible, i.e., when nothing hinders or blocks the relation between them, potency is "aimed at" its proper actuality and actualization will occur.[7] Furthermore, insofar as these relations characterize motion in things, they characterize it as intrinsic and so are characteristic of all things that are by nature. In this sense, the topics of *Physics* II and III, although separate in definition, are in fact coextensive in Aristotle's determinate world.

At the beginning of the *De Caelo,* Aristotle indicates that he remains within the science of physics but will now take up problems concerning the parts of the whole. The arguments primarily concern the elements – first among these, the fifth element, aether, that constitutes the heavens and the heavens themselves as a part of the whole. Aristotle emphasizes the link between physics as a science – it is the science of things that are by nature – and body as its object. When he opens *Physics* II, 1, he includes the elements among things that are by nature, and in the *De Caelo* we reach an account of them. Again, given the explicit identification of the elements as natural things, we have every right to expect the account of them in the *De Caelo* to be fully consistent with the account of things that are by nature in *Physics* II and that of motion in *Physics* III. And I have

---

7 Aristotle, *Physics* I, 9, 192a23–24; *N. Ethics.* I, 1, 1094a1–2.

argued that they are consistent because the account of the elements is subordinated to the accounts of nature, motion, and place.

Indeed, the case is even stronger. All things that are by nature – not only the elements but also plants, animals, their parts, and all artifacts as well – are composed of the elements. If someone tumbles out a window, the body goes down however uplifting the individual's thoughts. Artifacts are no different: when something heavy is dropped, it always goes downward. Insofar as something is made from the elements, it exhibits the natural motion of the elements and must in this respect be included among things that are and are moved by nature. Thus the domain of the account of the elements is coextensive with the accounts of nature and motion.

In *De Caelo* III, 2, being heavy or light is the primary characteristic of body, the proper object of physics, and inclination appears unequivocally as the principle of being moved associated with being heavy or light (301a22–26). An element rests in its proper place, and this rest too is nothing other than an expression of its inclination. Indeed, Aristotle later adds, if something has no inclination, it will be unmoved and mathematical (*De Caelo* III, 6, 305a25–26). Such a thing is neither natural nor a proper object of physics. Thus, by definition, inclination both constitutes the nature of each element and is exactly what we would predict on the basis of *Physics* II, 1, i.e., a principle of being moved and being at rest.

Beyond these important particulars, the coherence among Aristotle's arguments can be understood more broadly. Aristotle's arguments in *Physics* VIII and *De Caelo* III take up problems raised by natural motion as Aristotle defines it. Within these arguments we find his so-called theory of projectile motion. But "projectile motion," e.g., the motion of a stone after it leaves a thrower's hand, involves not natural motion, but violent motion – motion away from a thing's natural place, which must be produced extrinsically. This problem is marginalized – indeed, it never forms a topic in Aristotle's physics – precisely because it lies outside of nature conceived as an intrinsic source of being moved. Consequently, on this view the systematic coherence between *Physics* II and *Physics* III as accounts of nature and things in motion also extends to *Physics* VIII and *De Caelo* III, although each of these arguments is also defined by its own specific problem.

Finally, in *De Caelo* IV Aristotle provides a full account of inclination as the nature and activity of each element: each element is constituted by and expresses (in the absence of hindrance) an intrinsic orientation toward its respective proper place in which it rests. I have argued that this relation must be understood in light of Aristotle's definition of motion in things found in *Physics* III. Indeed, the relation of each element toward its respective natural place is exactly what is required on the basis of this definition: an active orientation of potency toward the actuality that constitutes its definition. For this reason, when each element reaches its proper place, it no longer becomes but *is* there and cannot by nature be moved from this place. In *Metaphysics* IX, Aristotle comments that the relation of potency to actuality is important beyond becoming – it is important for being too. And this argument too is fully consistent with that of the *De Caelo*. Each element possesses but one potency that, in the absence of hindrance, cannot fail to be actualized until activity ensues; for this reason, the elements may be said to imitate the heavens because they are ever active, containing in themselves the source of their motion.

Considered as formal arguments, the topics of the *De Caelo*, the *Physics*, and *Metaphysics* IX, differ. Nonetheless, Aristotle's accounts of nature, motion, and the inclination of the elements, as well as potency and actuality, are in full agreement. And I suggest why in the conclusion of my analysis of the arguments of the *De Caelo:* the account of the *De Caelo* presupposes the definitions of nature, motion, and place arrived at in the *Physics*. In its turn, motion is defined as a potency/act relation, and these terms form the topic of *Metaphysics* IX – a topic important not only for becoming but for being, i.e., not only for motion but for nature. Thus these accounts go together not because Aristotle has set out to build a system but because they share a single starting point, a commonality of purpose and so of related concepts. And within these topics the order of nature appears and is expressed by place as a constitutive principle and the elements as oriented each toward its proper place.

The substantive relation between place, the heavens, and the elements raises another important methodological question for Aristotle's physics: what relation obtains between the accounts of common and universal things and the accounts of special things? I argued earlier that the analysis of proper things, such as inclination, is subordinated to that of common and universal things, and in this sense the latter are logically prior. Here I should like to consider the accounts as such.

The most common view of this relation is that the account of what is logically prior, i.e., the account of place as something common and universal, somehow comes to be applied to the world and so produces an account of special things, i.e., what is particular and so logically posterior.[8] Indeed, this view, i.e., that the arguments of the *De Caelo* are somehow derived from, or special applications of, the principles of the *Physics,* may explain why the *De Caelo* has received considerably less attention than has the *Physics*.

I am proposing a quite different view, however. Place (I suggest that this point applies also to the other things named in *Physics* III, 1) is something common and universal not only in a logical sense but also in an ontological sense – it is unique and a determinative principle. Hence, its

8 For two clear and recent examples, cf. Matthen and Hankinson, "Aristotle's Universe," 430, and Lettinck, *Aristotle's* Physics *and Its Reception in the Arabic World,* 185.

priority to proper things cannot be entirely understood as a logical priority. Although the definition of place is presupposed by the account of the elements, this account is not somehow an application of place to the world.

On the other side, the arguments concerning the parts of the whole, including inclination, do not represent "applied accounts." The elements are not in any sense ontologically derived from place – they are substances in their own right. Consequently, the account of them is as an investigation in its own right: it comprises a part of the science of physics, and it possesses an announced topic and subject. All of these arguments, those concerning place *and* those concerning the elements, are fully and properly found within the same science, physics, the investigation of things that are and are moved by nature.

The priority of place as something common and universal is a priority of being. Place is prior because it is a unique constitutive principle and so a cause rather than an effect, a source of determination rather than what is determined. It is the category "where." And the examination of place is an examination of this principle as such and according to its definition. However, by concluding his account of place with the problem of how the heaven is and is not in place, Aristotle indicates that place is being examined as it is by nature – as it is together with the heaven – and not as an abstract principle. That is, although the account of place establishes its definition and is in this sense formal, it is nonetheless an account of place conceived as constituting the whole comprised of place as a limit, first body as limited, and the heaven, composed of place and first body, as what is first in place. Thus, Aristotle's account of place is not an account of place operating in abstraction from things. Rather, as an account, it possesses a specific category as its topic, "where," that turns out to be place, and it examines that topic as it is, i.e., a constitutive limit of the heaven. As a unique constitutive principle, place limits the cosmos, rendering it directional and so causing the motion of all things within the cosmos. Hence, place is something common and universal not because of the nature of the account of it given in *Physics* IV, but because of what place is.

Likewise, the arguments of the *De Caelo* take up specific topics and examine these topics not in abstraction from things, but as they involve the parts of the whole. Here we find problems that involve inclination because the things being examined, primarily the elements, are proper things. Inclination is a "proper thing" because there is not one, but as many as there are elements. Each element possesses a single natural motion that is by definition directed toward its respective proper place. Thus, each inclination is specific to its element and at the same time a member of a group. The different kinds of inclination can be considered as a group because they are identical insofar as each constitutes the nature of its element: each is a proper source of being moved and being at rest, and each is caused by its respective actuality, i.e., proper place. Hence, the account of inclination (and of the different inclinations of the different elements) is not one that applies abstract explanatory principles to particular subjects, but one that has as its topic a subject that is specialized in respect to each element.

As inclination emerges in the different arguments of the *De Caelo*, it turns out to be the very nature of the elements and a source of being moved that is actively oriented toward its proper place. As such, inclination is the complement of place. It is determined and an effect, i.e., that which is moved. And it is determined toward and moved by its respective proper place within the cosmos, which is defined as directional and so constituted by place as the first unmoved limit of the heaven. Hence, the various arguments in which inclination appears constitute an account of an effect that taken in itself forms a topic of investigation.

This account of the relation between the arguments of the *Physics* and those of the *De Caelo* provides a pathology for a striking feature, and puzzle, found in the *De Caelo:* although place does appear at the conclusion of the argument in *De Caelo* IV, Aristotle "fails" to provide a full-scale account of the cause of the natural motion of the elements. As we see in *Metaphysics* IX, potency is a cause in the sense of being a source or ability to be moved. And the arguments of the *De Caelo* have as their topic the parts of the whole in this sense, i.e., as being moved. Hence the arguments of the *De Caelo* constitute an examination of moved things as causes in this sense, i.e., as potency. Place is not a cause in this sense and so it does not belong within the domain of this argument any more than the elements belong within the arguments of *Physics* IV.

But at the conclusion of the account of inclination, natural place reappears. It reappears because the elements and their proper places are related as potency and actuality. Proper place is the actuality of each element because place is the determinative cause in respect to where all things that are by nature are and are moved. We understand place in this way on the basis of the account of *Physics* IV. Thus although the topics of the *Physics* and the *De Caelo* are different and so by definition require separate investigations, the accounts are also complementary and render the larger position coherent: place is a determinative principle, and it renders the world determinate. For this reason, place reappears as the nature of the elements within the account of inclination in *De Caelo* IV.

Two conclusions follow. (1) The accounts of *Physics* IV (of common and universal things) and the *De Caelo* (of proper things) do not differ as accounts. Each considers a specified problem with a specific being as its topic: place and the parts of the whole. Furthermore, each considers its topic as it actually is, rather than in abstraction from the world, and each considers its topic as proper to physics, the science of things that are by nature.

(2) The difference between the accounts, including the priority of the account of *Physics* IV, rests exclusively on the different natures of the objects being examined. *Physics* IV, the account of place, considers a cause, a unique determinative principle of all things that are by nature, whereas the *De Caelo* considers the parts of the whole, i.e., that which is determined, each to its proper place. Hence, the relation between the accounts cannot rest on a difference between them as accounts, as would be the case were one an examination of a principle and the other an application of a principle. Rather, it rests on differences between the objects being examined and the different status of each: the one is a cause, something common and universal, and the others are effects, proper things.

The Problem of Evaluating Aristotle's Physics

Of course, this is not the end of the story. That Aristotle's arguments meet the criteria that he has established and that they can be shown to work together makes them, as I have argued, coherent and, beyond coherent, elegant. But coherence and elegance do not make an account true; in this case, they have not even made it persuasive for many philosophers. And this conjunction of coherence, truth, and persuasion raises the problem of evaluating *both* Aristotle's position – including the starting point, the problems, and the arguments on which it rests – *and* the situation of his position within the larger framework of the history of philosophy.

For each text considered in this study, I have insisted first that the problem that Aristotle defines (and so presumably solves) be identified and that the arguments he presents as solutions be considered on their own terms. In criticizing Hussey's analysis – itself an outstanding example, indeed a model, of the acontextual method – of Aristotle's arguments concerning the void, I suggested, *contra* Hussey, that if we are to understand Aristotle's position, we must take his arguments on their own terms rather than first presupposing a quite different model of physics, e.g., Newtonian, and then evaluating Aristotle's arguments in terms of this other model. In short, only by grasping on their own terms the concepts at work can we understand the arguments themselves.

And I would extend this principle further. Not only must we analyze Aristotle's arguments in terms of his concepts, but even the definition of the problems at stake within the science of physics have similarly to be understood in Aristotle's terms. So, for example, when analyzing Aristotle's so-called account of projectile motion, I argue that the problems of motion may vary from one physics to another; therefore, to understand Aristotle's physics, we must be able to identify the problems distinctly defined by it. Indeed, I conclude that Aristotle never directly addresses the problem of projectile motion because his definition of physics and things that are by nature largely marginalizes this problem. Hence neither Aristotle's arguments nor the success or failure of his physics in providing a solution to a given problem provides a standard for evaluating his physics as a science or for comparing it to other conceptions of physics.

But this view poses a serious problem of its own. On the one hand, if we take Aristotle strictly on his own terms, then physics as the science of things that are and are moved by nature need not include any problem that Aristotle himself does not recognize. It need not include, for example, an account of projectile motion, insofar as a projectile is moved extrinsically and violently. Consequently, Aristotle can be neither praised nor blamed for the absence of a full account of projectile motion because, given his definition of physics, projectile motion is an unnatural and so unimportant phenomenon.

On the other hand, if we require of any physics that it give an account of such an "obvious" phenomenon, then Aristotle's "account" of projectile motion (or the lack of it) appears to be a failure and so renders his physics a failure in this respect. But the apparent failure rests not on Aristotle's physics, but on the *requirement* for an account of projectile motion: it both privileges (without apparent justification) a particular phenomenon and demands that the arguments of *Physics* VIII and *De Caelo* III answer a question that is entirely foreign to Aristotle's physics.

Both alternatives are unsatisfactory. On the first, there is no way to evaluate an argument (or physics) except on its own internal grounds and so no way to compare or evaluate arguments (or different physics) in relation to one another. Indeed, opposite accounts may be held to be equally valid and hence equally satisfactory. And in this sense, larger issues of evaluation turn out to be at best subjective, at worst simply meaningless.

On the second alternative, different accounts may be compared and evaluated, but the grounds for doing so are arbitrary. Such arbitrariness always implies not a comparison of two accounts – such comparison requires criteria independent of both – but an evaluation of one in terms of the other. And again projectile motion is a case in point. It is a primary feature of Newtonian physics; thus, to insist on its primacy is also to insist on this physics. Any subsequent evaluation of Aristotle based on this primacy does nothing more (or less) than hold his view accountable to the conceptions of Newtonian physics. And again, it is hard to see how such evaluation can be meaningful or what its claims can amount to. Hence both these alternatives, albeit for different reasons, fail to provide satisfactory grounds for the evaluation of the arguments (and the physics) that have been examined here.

There is, however, a third, more complex, ground for an evaluation of Aristotle's position. It involves several steps, beginning with the clarification of coherence as a criterion of evaluation. For Aristotle's physics – and

Aristotle is but one member of a very long tradition – coherence is a necessary condition without which there would be no position worth evaluating. And many commentators, Waterlow, for example, fail to find Aristotle's position coherent and so ultimately conclude that he and his
physics are in deep trouble. As I have argued throughout my analysis of
the *Physics* and *De Caelo,* coherence within and across the various arguments comprising the science of physics – and metaphysics insofar as becoming also entails being – drives Aristotle's view and indeed produces
his remarkable vision of the cosmos. In fact, it seems to me that the continuing interest of philosophers and historians of science in his arguments, long after their every feature has been rejected as false, is in large
measure accounted for by the remarkable coherence of these arguments
and, consequently, of the position presented by them. The power of Aristotle's arguments lies not in their specific content (e.g., place, a determinate cosmos, the elements, inclination), but in the patterns and elegance exhibited by their formal structure. And to detect this power, the
arguments must be grasped on their own conceptual grounds because
this power is nothing other than their coherence – a coherence impossible to understand without the kind of analysis provided here.

In this sense, coherence is the ultimate internal criterion of Aristotle's
position. Consequently, it is necessary to recognize this coherence to *establish* the position – without it there is no position to evaluate. But for this
very reason, i.e., because coherence serves to establish a position and so to
evaluate whether or not a position has been established, it cannot serve as
a criterion for evaluating that position once established. And the reason
why is obvious: a criterion of evaluation is that by which we judge a position as a coherent whole, and no criterion can at the same time and in the
same respect form both that which is judged and that by which it is judged.
Because coherence is clearly the former, it cannot also be the latter.

A position includes not only the solutions but also the problems taken
as central to the investigation. As the first step of this analysis, I have insisted on an identification of the topic or problem at stake because it defines the domain of Aristotle's arguments and ultimately his position. But
as the problem of projectile motion demonstrates, there is nothing particularly given, or absolute, about any phenomenon or problem. The
problem of projectile motion looks central to physics after Copernicus
because of a host of prior decisions – just as it looks marginal to Aristotle
because of his decisions concerning nature and motion. In short, problems as well as solutions are embraced within coherence as the ultimate
internal criterion of a physics. And the same conclusion follows again: it
is a necessary condition for any physics that it be able to solve the problems defined within it to be successful as a science; and we may find it successful, partially successful, or unsuccessful. Nevertheless, if we conclude
that a given physics is coherent and successful in solving the problems

posed within it, such success can never serve as the basis of an independent evaluation of that physics either in itself or in relation to other conceptions of physics. And we may note for a problem such as projectile motion that this point holds not only for Aristotle but also for Galileo and Newton.

However, in one important respect the status of solutions and the status of problems differ from one another: solutions cannot in and of themselves be compared to one another whereas problems can. Aristotle's account of motion as an actualization cannot be directly compared to an account of a body moving through a medium because the accounts share no common terms or conceptions. They fail to share anything in common because solutions are determined by the problems that they are intended to solve, and these accounts are intended to solve quite different problems.

But the problems within different conceptions of physics can be compared. In fact, the ability to notice that a problem central for one physics is marginal for another already constitutes a form of comparison between two conceptions of physics. And such comparison rests on a recognition that as a systematic investigation, no particular physics solves every problem. Whereas Aristotle, as we have seen, marginalizes projectile motion in favor of problems concerning things that are by nature, Galileo and Newton explain projectile motion (among other things) – but they do this by marginalizing virtually all problems concerning intrinsic natures. Hence choosing a physics is first an expression of preference in respect to the problems to be solved and only subsequently (and consequently) a preference for solutions or positions developed to solve them.

Presumably one should be able to explain why a given problem is important or interesting – and more so than others. But on closer inspection, this challenge turns out to be exceedingly difficult. The question comes to this: is it possible to evaluate the problems posed by a given conception(s) of physics so as to determine which are more important and which less – and to do so in a way that it is not arbitrary? Again, such an evaluation would require criteria independent of any particular conception of physics, and it is difficult to know what such criteria would be. On the one hand, to find one set of problems more interesting or more important than another may be no more than the expression of arbitrary prejudice or preference in the guise of comparison. For example, problems may become interesting when they are defined as central and uninteresting when they are marginalized.

But on the other hand, some features in the broader pattern of Aristotle's definition of problems and his position do seem to be open to independent assessment. His position requires that the earth be stationary at the center, whereas for all Copernicans it moves. Aristotle marginalizes the problem of projectile motion, whereas for the Copernicans it must and can be solved. And surely we believe that in respect to these issues, the Copernicans are "right" and Aristotle "wrong."

More systematic problems are harder but also instructive. I have not considered the implications of Aristotle's distinction between things that are by nature and those that are by art, but in his various accounts, problems of artifacts and technology are entirely subordinated to those concerning things that are by nature. For example, Aristotle concludes in *Physics* II, 8, that in some cases art completes what nature cannot bring to completion whereas in other cases art imitates nature (199a15–21). The achievements of modern technology (for better *and* for worse) seem to belie such a relation between artifacts and natural things. *Prima facie* it seems that the evidence granted by technology is somehow – now, if not originally – independent of a commitment to a particular conception of physics.

As work in the history of science (like that of Lloyd on ancient science and Biagioli on Galileo) demonstrates, science is not a closed logical system, and the problems defined by a science derive not only from a first starting point, they also express and/or reflect a multitude of cultural interests and values. Hence a given starting point may produce a variety (although a limited variety) of different problems depending on what independent values are also at work. An account of how one evaluates (or compares) a conception(s) of physics should be able to account for this apparently independent value. I hope to make a suggestion on this point shortly.

But first, a further point about the problems defined within a physics must be clarified. The definition of a problem for physics – or the designation of a phenomenon as important – is itself in part the outcome of a prior commitment: the commitment to a first principle. And, as I have argued throughout, for Aristotle's physics this prior commitment is the claim that nature is everywhere a source of order and its order is immediately theological.

The problems of physics thus occupy a "middle ground." They determine the domain and concepts that will operate in solutions, but they are themselves determined in part by a prior philosophic commitment. Insofar as they represent values or cultural choices, we can detect differences among them and form comparisons that may serve as the basis for an evaluation. But insofar as problems are themselves determined by a prior philosophical commitment, they are no more directly comparable than are solutions. And for the same reason: insofar as problems are determined by different prior commitments they cannot share common terms or concepts and hence cannot be compared.

And here we reach the prior commitment, i.e., the assumption that defines the problems and, ultimately, the solutions of physics. This commitment defines the primary subject matter of physics, namely, things that are by nature, which in its turn raises the problem of motion, etc. Two points about this commitment in Aristotle. (Although the problem lies beyond the bounds of this study, I believe that these points are true of any conception of physics.)

(1) The commitment to a first principle, when held consistently – as I have argued it is in Aristotle's physics – serves both as a source for the problems that a physics defines as central (or marginal) and the ground for a successful solution to those problems. As such, this first principle (along with a consistent commitment to it) defines the problems, solutions, and hence the coherence of the account. (2) This commitment functions as a first principle in the sense that it is not another element proven, or established, by arguments within the physics. Rather, it is determinative of the content – both problems and solutions – of physics as an enterprise. Consequently, Aristotle's claim that nature is everywhere a cause of order is never itself directly proven by his physics. Like the claim that there are things that are and are in motion – things without which physics is impossible – (*Physics* I, 2, 185a13–20), the claim that nature is always and everywhere a source of order is a first starting point: it is a presupposition without which Aristotle's physics would be impossible, and it must be taken as given. It is thus *both* unproven (in fact, unprovable) *and* outside the content of physics as a science. Physics, according to Aristotle, is the investigation of things that are by nature, things that involve body as such – things that are distinct from artifacts on the one hand and mathematicals and god on the other. And the assumption that nature is everywhere a cause of order establishes and is presupposed by this subject matter. Hence it is never itself proven, but is always at work within his physics.

If Aristotle's conception of nature lies outside his physics as the science of things that are by nature, then what exactly is its status? First, I would suggest that for Aristotle the logical status of his notion that nature is a cause of order is identical to the assumption of things in motion: it is impossible to have physics without such an assumption, and it is impossible to assume the opposite as a foundation for a competing notion of physics. To assume that there is no motion terminates physics. Likewise, as we saw in the rejection of the void and the account of inclination, to assume or to work within a view that assumes that disorder is prior to order in nature is to assume something impossible and something that must be rejected immediately. In this sense, to say that "nature is everywhere a cause of order" is to state an assumption without which physics (as Aristotle conceives it) cannot operate.

Second, although the proposition "nature is everywhere a cause of order" is not directly proven in Aristotle's physics, this view of nature is established by at least two of the investigations included in Aristotle's *Metaphysics*. Thus, we find in *Metaphysics* IX an examination of potency and actuality that establishes the relation of becoming to being and in so doing presents an account of nature that concludes with the claim that the elements imitate the heavens. In *Metaphysics* XII, 10, we also find an account of nature and of the cosmos as a whole: not all things are randomly together, but all, the fish, the fowl, and the plants, are ordered toward some one (1075a15–19). These arguments bear witness to the status of Aristotle's claim that nature is everywhere a cause of order: it is a part of his broadest philosophic commitment, it is established by his metaphysics, and it is a conception without which his physics would be impossible.

I have suggested that a large measure of our continuing interest in Aristotle's physics derives from its coherence. A further measure of our interest derives from the success of his physics in solving the problems posed by his view of nature as a cause of order. And now I would conclude that the largest measure of our continuing interest in Aristotle's physics derives from the substance of his first commitment, the commitment that is not a part of physics, but founds it as a science. Like so many claims in Aristotle, this view of nature seems to speak directly to our experience of nature and even to nature itself.

Indeed, this view of nature renders Aristotle's physics unique. The radical identification of nature with order is largely rejected both by Aristotle's contemporaries and in the history of ideas. Nature is almost universally conceived of as containing an important element of randomness or chance. Whereas for Aristotle matter is actively oriented toward form, according to Plato (and all the Neoplatonists), matter resists formation; whereas for Aristotle the stars run after god, for Plato god imposes form on matter. In fact, all philosophers who use, in some sense or another, the notion of a creating God introduce into physics the view that nature could be other than it is because God created it out of his own free will and thus could have created it differently. And of course, a principle of randomness is central to modern physics (and philosophy).

And here we reach the heart of the problem not only for Aristotle's physics but also for any physics insofar as it rests on a first commitment. Because that principle cannot be proven in the sense of being derived from prior principles, it remains "blind." And as such the assumption that nature is a source of order and the assumption that it contains an element of chaos or randomness are equal. If such principles are "established," they are so only *ex post facto,* in an indirect way by the successful outcome of the arguments determined by them as suppositions.

The difference between accounts appear here. Finally, the problems posed by Newtonian physics produce more and better *ex post facto* evidence – in the form of arguments, experiments, appeal to experience, and so on – to validate them (and their starting point) than does Aristotle's physics, and Einstein's problems produce more than either Aristotle or Newton. Finally, it is only here – i.e., in *ex post facto* evidence – that a given conception of physics can generate confirmation of its independent value.

Three points about this value. (1) Because first starting points are "blind," there is no – not even the slightest – guarantee at the outset that one will arrive at an account of the world itself. Aristotle, as I have argued, is fully consistent in his commitment to his first assumption about nature. And this consistency yields and renders coherent the problems that he defines and the solutions that he develops so single mindedly and that together constitute his physics. And he was wrong about everything. The earth is not immovable at the center, the cosmos is not inherently directional, the elements do not always (in the absence of hindrance) go to their natural place. Some commentators think of Aristotle's physics as right about some things but wrong about others; thus, one can sort through it to separate the wheat from the chaff. But as I have argued, such a process fails to see the coherence of his account because it necessarily presupposes an arbitrary standard of right and wrong. Indeed, such a process finally destroys Aristotle's account. Aristotle's account is completely wrong because, finally, his initial assumption and all that follows from it cannot, in its account of the world, generate confirmation. And, perhaps even more importantly, he is wrong in thinking, as I have argued he does, that the opposite assumption – that nature is not a cause of order – cannot found physics as a science. Arguments for the void are not void. That his arguments retain their power and that his conception of nature seems to speak to our experience testify to their coherence and intuitive appeal. But not to their truth.

(2) If starting points are "blind" in the sense I suggest here, is their assumption arbitrary? Relative to the conception of physics, its problems and their solutions generated by this assumption, it must be arbitrary in the sense of unjustified. Science is a high-risk enterprise. However, in another sense, the choice of a starting point is anything but arbitrary. The first starting point of Aristotle's physics, like that of any conception of physics, expresses the values of one's most deeply held views and, often, the values of the cultures of the individuals. For example, modern technology, for better and for worse, seems to provide independent evidence for Newtonian and Einsteinian physics. But when we assume that the productivity of technology bears witness in this sense, we do so in part because as a part of the values expressed by modern culture, success means that something works. Plato's views of the realm of becoming clearly reflect his views of Athenian politics and the death of Socrates. And Aristotle's view of nature too expresses his own most deeply held values and his vision of what is best in Athenian politics and culture: order must be prior to disorder. And there is nothing arbitrary about the expression of values held in this way as the foundation of physics.

(3) But this point raises the inevitable – and most difficult – question: does not this talk of value return us to a sort of subjectivism when the truth of a statement such as "Nevertheless, it [the earth] moves" cries out for an account of objective truth? The short answer is no. The long answer – which of necessity remains but a suggestion here – concludes this study.

The talk about value returns us to subjectivism only if we presuppose a distinction between absolute objectivity, i.e., truth outside any reference to an individual or a culture, and subjectivism, i.e., anything that falls short of absolute objectivity. But there is no such thing as absolute objectivity because it is impossible to stand outside all culture. For example, it is impossible to express an idea without using language, and all language is cultural. But the notion of absolute objectivity is false in another, more important, sense: by establishing a false model of truth, it inhibits the ability to realize objective truth within culture. Copernican physics, for example, does not require that we stand independently of the earth to understand that "it moves"; "it moves" is true "absolutely" for the one who is moving along with it. Likewise, the motion of the observer is both unavoidable and plays a central role in Einstein's physics.

And the expression of value, especially the value expressed by the first assumption of a physics is not unlike the truth of science. In Aristotle's hands, it represents a complete commitment that can spell itself out or validate itself only through the formation of problems and solutions – in short, by a complete position. And the view arrived at by that position validates or fails to validate itself as true through an *ex post facto* expression of cultural value.

On the one hand, the evaluation of the expression of cultural value cannot take place without first establishing the position at stake in all its coherence (or lack thereof). If we fail to perform this initial work with care and precision, our results can only be arbitrary and will tell us nothing of the choices represented by cultural values, whether our own or those of others, or the truths revealed by them. And such has been the task of this study. On the other hand, evaluating a position on the basis of its broadest philosophical commitments is a quite different project than establishing what this position is on the basis of its internal coherence. And the broader project, i.e., evaluation, remains untouched by this study.

I close then with a suggestion. The final measure of our fascination with Aristotle's arguments lies in the way in which they express his own values and those of his cultural commitments – metaphysical, political, ethical, rhetorical. Only in evaluating these will we come to understand not only Aristotle's physics but also the history of ideas, the history of culture, the choices represented by that history, and, thus, ourselves and our own choices.

Bibliography of Works Cited

Texts and Translations of Aristotle

Aristote: Du Ciel. Texte établi et traduit par Paul Moraux. Paris: Budé, 1965.

Aristote: Physique. Texte établi et traduit par Henri Carteron. Vol. I. 2d ed. Paris: Budé, 1956.

Aristote: Sur la nature (Physique II). Introduction, traduction, et commentaire par L. Couloubaritsis. Paris: Vrin, 1991.

Aristotelis Categoriae et Liber Interpretatione. Ed. L. Minio-Paluello. Scriptorum Classicorum Bibliotheca Oxoniensis. Oxford: Clarendon Press, 1949.

Aristotelis De Anima. Ed. W. D. Ross. Scriptorum Classicorum Bibliotheca Oxoniensis. Oxford: Clarendon Press, 1956.

Aristotelis De Caelo. Ed. D. J. Allan. Scriptorum Classicorum Bibliotheca Oxoniensis. Oxford: Clarendon Press, 1936.

Aristotelis Metaphysics. Ed. W. Jaeger. Scriptorum Classicorum Bibliotheca Oxoniensis. Oxford: Clarendon Press, 1957.

Aristotelis Opera Omnia. Ed. A. F. Didot. Vol. 2, in Greek and Latin. Paris: 1874.

Aristotelis Physica. Ed. Carolus Prantl. Lipsiae: Teuvner, 1879.

Aristotelis Physica. Ed. W. D. Ross. Scriptorum Classicorum Bibliotheca Oxoniensis. Oxford: Clarendon Press, 1950.

"Aristotle's *Categories,* Chapters I–V: Translation and Notes," by J. L. Ackrill. In *Aristotle: A Collection of Critical Essays,* ed. J. M. E. Moravcsek, 90–124. Garden City, N.Y.: Anchor Books Doubleday, 1967.

Aristotle's Categories and De Interpretatione. Translated with notes by J. L. Ackrill. Oxford: Clarendon Press, 1963.

Aristotle's De generatione et corruptione. Translated with notes by C. J. F. Williams. Oxford: Clarendon Press, 1982.

Aristotle's De motu Animalium: Text with Translation, Commentary, and Interpretive Essays. Edited, translated, and with commentary by M. Nussbaum. Princeton: Princeton University Press, 1978.

Aristotle's Metaphysics. Revised text, with introduction and commentary by W. D. Ross. Vol. II. Oxford: Clarendon Press, 1924.

Aristotle's Metaphysics. Translated with commentaries and glossary by H. G. Apostle. Bloomington: Indiana University Press, 1966.

Aristotle's Physics. Revised text, with introduction and commentary by W. D. Ross. Oxford: Clarendon Press, 1936.

Aristotle's Physics. Translated with commentaries and glossary by H. G. Apostle. Bloomington: Indiana University Press, 1969.

Aristotle's Physics: Books I, II. Translated with introduction and notes by William Charlton. Oxford: Clarendon Press, 1970.

Aristotle's Physics: Books III and IV. Translated with notes by Edward Hussey. Oxford: Clarendon Press, 1983.

De Caelo. Trans. J. L. Stocks. Oxford: Clarendon Press, 1922.

The Complete Works of Aristotle: The Revised Oxford Translation. Ed. J. Barnes. 2 vols. Princeton: Princeton University Press, Bollingen Series LXXI-2, 1984.

On the Heavens. Greek-English. Translated with introduction by W. K. C. Guthrie. The Loeb Classical Library. Cambridge: Harvard University Press, 1939.

The Philosophy of Aristotle. New translations by J. L. Creed and A. E. Wardman. New York: New American Library, 1963.

Physica. Trans. R. P. Hardie and R. K. Gaye. Oxford: Clarendon Press, 1930.

The Physics. Trans. Philip H. Wicksteed and Frances M. Cornford. Vol. I. The Loeb Classical Library. Cambridge: Harvard University Press, 1934.

Ancient and Medieval Authors

Albertus Magnus. *Physicorum Libri.* Vol. 3 of *Alberti Magni Opera omnia,* ed. Borgnet. Paris: Vives, 1890–99.

Burley, Walter. *Burleus super octo libros Phisicorum.* Venice, 1501. Reprinted in facsimile as Water Burley, *In Physicam Aristotelis expositio et quaestiones.* Hildesheim and New York: Georg Olms Verlag, 1972.

Diels, H., and W. Kranz, *Die Fragmente der Vorsokratiker.* 10th ed. Berlin: Weidmann, 1961.

Empedocles: The Extant Fragments. Edited with an introduction, commentary, and concordance by M. R. Wright. New Haven and London: Yale University Press, 1981.

Euclid, *The Thirteen Books of the Elements.* Translated from the text of Heiberg, with introduction and commentary by Sir Thomas L. Heath. Vol. I. New York: Dover Publications, 1956.

Herodotus, *Historiae.* Ed. Carolus Hude. 3d ed. Scriptorum Classicorum Bibliotheca Oxoniensis. Oxford: Clarendon Press, 1940–41.

Homer, *Homeri Opera.* Ed. David B. Monro and Thomas W. Allen. 3d ed. 2 vols. Scriptorum Classicorum Bibliotheca Oxoniensis. Oxford: Clarendon Press, 1908–20.

Kirk, G. S., J. E. Raven, and M. Schofield, *The Presocratic Philosophers: A Critical History with a Selection of Texts.* 2d ed. Cambridge: Cambridge University Press, 1983.

Ioannis Philoponi. *In Aristotelis Physicorum libros tres priores commentaria.* Ed. H. Vitelli. Vol. 16 in the Prussian Academy edition, *Commentaria in Aristotelem Graeca.* Berlin: G. Reimer, 1887.

In Aristotelis Physicorum libros quinque posteriores commentaria. Ed. H. Vitelli. Vol. 17 in the Prussian Academy edition, *Commentaria in Aristotelem Graeca.* Berlin: G. Reimer, 1888.

Place, Void, and Eternity. Philoponus: Corollaries on Place and Void. Trans. David Furley, with *Simplicius: Against Philoponus on the Eternity of the World.* Trans. Christian Wildberg. Ithaca: Cornell University Press, 1991.

On Aristotle's Physics 2. Trans. A. R. Lacey. Ithaca: Cornell University Press, 1993.

Pindar. *Pindari Carmina cum fragmentis.* Ed. C. M. Bowra. 2d ed. Scriptorum Classicorum Bibliotheca Oxoniensis. Oxford: Clarendon Press, 1947.

Plato. *Opera Omnia.* Ed. Ioannes Burnet. Scriptorum Classicorum Bibliotheca Oxoniensis. Oxford: Clarendon Press, 1900–1907.

Proclus. *In primum Euclidis Elementorum librum commentarii.* Ed. G. Friedlein. Bibliotheca Scriptorum Graecorum et Romanorum. Teubneriana Hildesheim: G. Olms, 1967. This work has been translated as *Proclus: A Commentary on the First Book of Euclid's Elements.* Translated with introduction and notes by Glenn R. Morrow. With a new foreword by Ian Mueller. Princeton: Princeton University Press, 1992.

Simplicius. *In Aristotelis Physicorum libros quattuor priores commentaria.* Ed. H. Diels. Vol. 9 in the Prussian Academy edition, *Commentaria in Aristotelem Graeca.* Berlin: G. Reimer, 1882.

On Aristotle's Physics 4.1–5, 10–14. Trans. J. O. Urmson. Ithaca: Cornell University Press, 1992.

Simplicius: On Aristotle's Physics 7. Trans. Charles Hagen. Ithaca: Cornell University Press, 1994.

Themistius, *In Aristotelis Physica Paraphrasis.* Ed. Henricus Schenkl. Vol. 5, pars 2, in the Prussian Academy edition, *Commentaria in Aristotelem Graeca.* Berlin: G. Reimer, 1900.

Thomas, *In Octo Libros Physicorum Aristotelis Expositio.* Ed. P. M. Maggiolo. Turin-Rome: Marietti, 1954. This text has been translated by R. J. Blackwell, R. J. Spath, and W. E. Thirlkel as *Commentary on Aristotle's Physics.* London: Routledge and Kegan Paul, 1963.

Modern Authors

Ackrill, J. L. "Aristotle's Distinction between *Energia* and *Kinesis.*" In *New Essays On Plato and Aristotle,* ed. R. Bambrough, 121–141. New York: The Humanities Press; London: Routledge and Kegan Paul, 1965.

Algra, Kiempe. "'Place' in Context: On Theophrastus Fr. 21 and 22 Wimmer." In *Theophrastus: His Psychological, Doxographical, and Scientific Writings,* ed. William W. Fortenbaugh and Dimitri Gutas, 141–165. Rutgers University Studies in Classical Humanities, Vol. 5. New Brunswick and London: Transaction Publishers, 1992.

Allan, D. J. "Causality Ancient and Modern." *Proceedings of the Aristotelian Society,* supp. vol. 39 (1965): 10–18.

Ashley, Benedict M. "Aristotle's Sluggish Earth, Part I: The Problematics of the *De Caelo.*" *New Scholasticism* 32 (1958): 1–31.

Barnes, Jonathan. "Life and Work." In *The Cambridge Companion to Aristotle,* ed. J. Barnes, 1–26. Cambridge: Cambridge University Press, 1995.

Berti, E. "Les Méthodes d'argumentation et de démonstration dans la *Physique* (apories, phénomènes, principes)." In *La Physique d'Aristote et les conditions d'une science de la nature,* ed. F. De Gandt et P. Souffrin, 53–72. Paris: Vrin, 1991.

Biagioli, M. *Galileo Courtier: The Practice of Science in the Culture of Absolutism.* Chicago: University of Chicago Press, 1993.

Bickness, P. J. "Atomic *Isotacheia* in Epicurus." *Apeiron* 17 (1983): 57–61.

Blair, George. "Unfortunately It Is a Bit More Complex: Reflections on Ἐνέργεια." *Ancient Philosophy* 15 (1995): 565–580.

Bonitz, H. *Index Aristotelicus* 2d ed. Graz: Akademische Druck-U. Verlagsanstalt, 1955.

Bos, A. P. *On the Elements: Aristotle's Early Cosmology.* Assen: Van Gorcum, 1973.

"*Manteia* in Aristotle, *de Caelo* II 1." *Apeiron* 21 (1988): 29–54.

Cosmic and Meta-Cosmic Theology in Aristotle's Lost Dialogues. Leideni, New York: Brill, 1989.

Broadie, S. "Nature and Craft in Aristotelian Teleology." In *Biologie, logique et métaphysique chez Aristote: Actes du séminaire C. N. R. S. - N. S. F. Oleron 28 juin - 3 juillet 1987,* ed. et publiés par Daniel Devereux et Pierre Pellegrin, 389–403. Paris: Editions du C.N.R.S., 1990.

Brunschwig, J. "Qu'est-ce que la *Physique* d'Aristote?" In *La Physique d'Aristote et les conditions d'une science de la nature,* ed. F. De Gandt and P. Souffrin, 11–40. Paris: Vrin, 1991.

Buckley, M. J. *Motion and Motion's God.* Princeton: Princeton University Press, 1971.

Burnyeat, Myles F. "The Sceptic in His Place and Time." In *Philosophy and History: Essays on the Historiography of Philosophy,* ed. Richard Rorty, J. B. Schneewind, and Quentin Skinner, 225–254. Cambridge: Cambridge University Press, 1984.

Burnyeat, Myles, et al. *Notes on Books Eta and Theta of Aristotle's Metaphysics.* Oxford: Clarendon Press, 1984.

Carteron, H. "Does Aristotle Have a Mechanics?" In *Articles on Aristotle.* Vol. 1: *Science,* ed. J. Barnes, M. Schofield, and R. Sorabji, 161–174. London: Duckworth, 1975.

Case, Thomas. "The Development of Aristotle." *Mind* 34 (1925): 80–86.

"Aristotle." *Encyclopedia Britannica* 1910. 2:501–522. Reprinted (with omissions) in *Aristotle's Philosophical Development: Problems and Prospects,* ed. W. Wians, 1–40. London: Rowman and Littlefield, 1996.

Charlton, William. "Aristotelian Powers." *Phronesis* 32 (1987): 277–289.

"Aristotle and the Uses of Actuality." In *Proceedings of the Boston Area Colloquium in Ancient Philosophy,* ed. John J. Cleary and Daniel C. Shartin. Vol. 5, 1–22. New York: University Press of America, 1991.

Cherniss, Harold. *Aristotle's Criticism of Presocratic Philosophy.* Baltimore: The Johns Hopkins Press, 1935.

Aristotle's Criticism of Plato and The Academy. Vol. 1. Baltimore: The Johns Hopkins Press, 1944.

Chroust, A.-H. "The Miraculous Disappearance and Recovery of the Corpus Aristotelicum." *Classica et Mediaevalia* 23 (1962): 50–67.

"The First Thirty Years of Modern Aristotelian Scholarship (1912–1942)." In *Aristotle's Philosophical Development: Problems and Prospects,* ed. W. Wians, 41–65. London: Rowman and Littlefield, 1996.

Cleary, John J. "Mathematics and Cosmology in Aristotle's Development." In *Aristotle's Philosophical Development: Problems and Prospects,* ed. W. Wians, 193–228. London: Rowman and Littlefield, 1996.

Cohen, Sheldon. "Aristotle on Heat, Cold, and Teleological Explanation." *Ancient Philosophy* 9 (1989): 255–270.

"Aristotle on Elemental Motion." *Phronesis* 39 (1994): 150–159.

Aristotle on Nature and Incomplete Substance. Cambridge: Cambridge University Press, 1996.

Cooper, John. "Aristotle on Natural Teleology." In *Language and Logos: Studies in Ancient Greek Philosophy Presented to G. E. L. Owen,* ed. M. Schofield and M. Nussbaum, 197–222. Cambridge: Cambridge University Press, 1982.

Corish, Dennis. "Aristotle on Temporal Order: 'Now,' 'Before,' and 'After.'" *ISIS* 69 (1978): 68–74.

Corte, M de. "La Causalité du premier moteur dans la philosophie aristotélicienne." *Revue d'histoire de la philosophie* 5 (1931): 105–147. Reprinted in Marcell de Corte, *Etudes d'histoire de la philosophie ancienne: Aristote et Plotin,* 107–175. Paris: Desclée de Brouwer, 1935.

Dancy, R. M. "On Some of Aristotle's First Thoughts about Substance." *Philosophical Review* 84 (1975): 338–373.

Davidson, Herbert. *Proofs for Eternity, Creation, and the Existence of God in Medieval Islamic and Jewish Philosophy.* Oxford: Oxford University Press, 1987.

Denyer, Nicholas. "Can Physics Be Exact?" In *La Physique d'Aristote et les conditions d'une science de la nature,* ed. F. De Gandt and P. Souffrin, 73–83. Paris: Vrin, 1991.

Drabkin, I. E. "Notes on the Laws of Motion in Aristotle." *American Journal of Philology* 59 (1938): 60–84.

Drummond, John J. "A Note on *Physics* 211b14–25." *New Scholasticism* 55 (1981) 219–228.

Duhem, P. *Le Système du monde: Histoire des doctrines cosmologiques de Platon à Copernic.* Vol. 1: *La Cosmologie hellénique* 2d ed. Paris: Hermann, 1954.

Düring, I. "Notes on the History of the Transmission of Aristotle's Writing." *Göteborgshögskolas arsskrift* 56 (1950): 37–70. Reprinted in *Aristotle and His Influence: Two Studies.* Vol. 24 of *Greek and Roman Philosophy: A Fifty-two Volume Reprint Set,* ed. Leonardo Taran. New York and London: Garland, 1987.

Easterling, J. H. "Homocentric Spheres in *De Caelo*." *Phronesis* 6 (1961): 138–153.

Edel, Abraham. "'Action' and 'Passion': Some Philosophical Reflections on *Physics* III, 3." In *Naturphilosophie bei Aristoteles und Theophrast; Verhandlungen des 4. Symposium Aristotelicum veranstaltet in Göteborg, August 1966,* ed. I. Düring, 59–64. Heidelberg: Verlag, 1969.

Elders, L. *Aristotle's Cosmology: A Commentary on the De Caelo.* Assen: Van Gorcum and Co., 1960.

Feyerabend, Paul. "Some Observations on Aristotle's Theory of Mathematics and of the Continuum." In *Farewell to Reason,* 219–246. London: Verso, 1987.

Frank, E. "The Fundamental Opposition of Plato and Aristotle." *American Journal of Philology* 61 (1940): 34–53; 166–185.

Frede, M. "The Original Notion of Cause." In *Essays in Ancient Philosophy,* 125–150. Minneapolis: University of Minneapolis Press, 1987.

"The Definition of Sensible Substance in *Met. Z*." In *Biologie, logique et métaphysique chez Aristote: Actes du séminaire C. N. R. S. - N. S. F. Oléron 28 juin - 3 juillet 1987,* ed. et publiés par Daniel Devereux et Pierre Pellegrin, 113–129. Paris: Editions du C.N.R.S., 1990.

Freeland, Cynthia A. "Aristotle on Possibilities and Capacities." *Ancient Philosophy* 6 (1986): 69–89.

———. "Aristotle on Bodies, Matter, and Potentiality." In *Philosophical Issues in Aristotle's Biology*, ed. A. Gotthelf and James G. Lennox, 392–407. Cambridge: Cambridge University Press, 1987.

Freudenthal, Gad. *Aristotle's Theory of Material Substance: Heat and Pneuma, Form and Soul.* Oxford: Clarendon Press, 1995.

Friedman, Michael. *Kant and the Exact Sciences.* Cambridge: Harvard University Press, 1992.

Friedman, Robert. "Matter and Necessity in *Physics* B 9 200a15–30." *Ancient Philosophy* 3 (1983): 8–11.

Furley, David J. "Aristotle and the Atomists on Motion in a Void." In *Motion and Time, Space and Matter*, ed. Peter K. Machamer and Robert G. Turnbull, 83–100. Columbus: Ohio State University Press, 1976.

———. Review of *Nature, Change, and Agency*, by S. Waterlow. *Ancient Philosophy* 4 (1984): 108–110.

———. "The Cosmological Crisis in Classical Antiquity." In *Proceedings of the Boston Area Colloquium in Ancient Philosophy*, ed. John J. Cleary. Vol. 2, 1–19. New York: University Press of America, 1987.

———. *The Greek Cosmologists.* Vol. 1: *The Formation of the Atomic Theory and Its Earliest Critics.* Cambridge: Cambridge University Press, 1987.

———. *Cosmic Problems: Essays on Greek and Roman Philosophy of Nature.* Cambridge: Cambridge University Press, 1989.

———. "Self-Movers." In *Self-Motion: From Aristotle to Newton*, ed. Mary Louise Gill and James G. Lennox, 3–14. Princeton: Princeton University Press, 1994. Reprinted from *Aristotle on Mind and the Senses: Proceedings of the Seventh Symposium Aristotelicum*, ed. G. E. R. Lloyd and G. E. L. Owen, 165–179. Cambridge: Cambridge University Press, 1978.

Garber, Daniel. *Descartes' Metaphysical Physics.* Chicago: University of Chicago Press, 1992.

Gill, Mary Louise. "Aristotle's Theory of Causal Action in *Physics* III, 3." *Phronesis* 25 (1980): 129–147.

———. *Aristotle on Substance: The Paradox of Unity.* Princeton: Princeton University Press, 1989.

———. "Aristotle on Self-Motion." In *Aristotle's Physics: A Collection of Essays*, ed. Lindsay Judson, 246–254. Oxford: Clarendon Press, 1991. Reprinted in *Self-Motion: From Aristotle to Newton*, ed. Mary Louise Gill and James G. Lennox, 15–34. Princeton: Princeton University Press, 1994.

Gilson, Etienne. *History of Christian Philosophy in the Middle Ages.* New York: Random House, 1955.

Goldin, Owen. "Problems with Graham's Two-Systems Hypothesis." In *Oxford Studies in Ancient Philosophy*, ed. Julia Annas. Vol. VII, 203–213. Oxford: Oxford University Press, 1989.

Goldstein, Bernard R., and Alan C. Bowen. "A New View of Early Greek Astronomy." *ISIS* 74 (1983): 330–340.

Gomez-Pin, Victor. *Ordre et substance: L'enjeu de la quête aristotélicienne.* Paris: Centre National de la Recherche Scientifique: Editions anthropos, 1976.

Graham, Daniel W. *Aristotle's Two Systems.* Oxford: Clarendon Press, 1987.

"Two Systems in Aristotle." In *Oxford Studies in Ancient Philosophy*, ed. Julia Annas. Vol. VII, 215–231. Oxford: Oxford University Press, 1989.

"The Development of Aristotle's Concept of Actuality: Comments on a Reconstruction by Stephen Menn." *Ancient Philosophy* 15 (1995): 551–564.

"The Metaphysics of Motion: Natural Motion in *Physics* II and *Physics* VIII." In *Aristotle's Philosophical Development: Problems and Prospects*, ed. W. Wians, 171–192. London: Rowman and Littlefield, 1996.

Grant, Edward. "The Principle of the Impenetrability of Bodies in the History of Concepts of Separate Space from the Middle Ages to the Seventeenth Century." *ISIS* 69 (1978): 551–571.

Much Ado About Nothing: Theories of Space and Vacuum from the Middle Ages to the Scientific Revolution. Cambridge: Cambridge University Press, 1981.

In Defence of the Earth's Centrality and Immobility: Scholastic Reaction to Copernicanism in the Seventeenth Century. Vol. 74, part 4, Transactions of the American Philosophical Society Held at Philadelphia for Promoting Useful Knowledge. Philadelphia: The American Philosophical Society, 1984.

ed. *A Source Book in Medieval Science*. Cambridge: Harvard University Press, 1974.

Grene, Marjorie. *Portrait of Aristotle*. Chicago: University of Chicago Press, 1963.

"About the Division of the Sciences." In *Aristotle on Nature and Living Things: Philosophical and Historical Studies Presented to David M. Balme on His Seventieth Birthday*, ed. A. Gotthelf, 9–13. Pittsburgh: Mathesis Publications, 1985.

Groot, Jean De. *Aristotle and Philoponus on Light*. Garland Publishing, 1991.

"Philoponus on Separating the Three-Dimensional in Optics." In *Nature and Scientific Method: Studies in Philosophy and the History of Philosophy*, ed. Daniel O. Dahlstrom. Vol. 22, 157–174. Washington D.C.: The Catholic University of America Press, 1991.

Guthrie, W. K. C. "The Development of Aristotle's Theology." *Classical Quarterly* 27 (1933): 162–172; 28 (1934): 90–98.

Introduction to *On the Heaven*, by Aristotle. Cambridge: Harvard University Press, 1939.

"Notes on Some Passages in the Second Book of Aristotle's Physics." *Classical Quarterly* 40 (1946): 70–76.

History of Greek Philosophy. Vol. I. Cambridge: Cambridge University Press, 1971.

Hahn, Roger. "Laplace and the Mechanistic Universe." In *God and Nature: Historical Essays On the Encounter between Christianity and Science*, ed. David C. Lindberg and Ronald L. Numbers, 256–276. Berkeley: University of California Press, 1986.

Halper, E. "Aristotle's Solution to the Problem of Sensible Substance." *The Journal of Philosophy* 84 (1987): 666–672.

Hatfield, Gary. "Review Essay: The Importance of the History of Science for Philosophy in General." *Synthese* 11 (1995): 1–26.

Heath, T. L. *Mathematics in Aristotle*. Oxford: Clarendon Press, 1949. Reprinted, New York: Garland Publishing, 1980.

Heidel, W. A. "The Pythagoreans and Greek Mathematics." *American Journal of Philology* 61 (1940): 1–33. Reprinted in *Selected Papers*. New York: Garland Publishing, 1980.

Hine, William L. "Inertia and Scientific Law in Sixteenth-Century Commentaries on Lucretius." *Renaissance Quarterly* 48 (1995): 728–741.

Hoffmann, Philippe. "Les Catégories *où* et *quand* chez Aristote et Simplicius." In *Concepts et catégories dans la pensée antique: Etudes publiées sous la direction de Pierre Aubenque,* 217–245. Paris: Vrin, 1980.

Hussey, Edward. "Aristotle's Mathematical Physics: A Reconstruction." In *Aristotle's Physics: A Collection of Essays,* ed. Lindsay Judson, 213–242. Oxford: Clarendon Press, 1991.

Hyland, Drew A. "Why Plato Wrote Dialogues." *Philosophy and Rhetoric* 1 (1968): 38–50.

Inwood, Brad. "Chrysipus on Extension and the Void." *Revue Internationale de Philosophie* 45 (1991): 245–266.

Irwin, T. *Aristotle's First Principles.* Oxford: Clarendon Press, 1988.

Jaeger, W. *Aristotle: Fundamentals of the History of His Development.* Translated with the author's corrections and additions by Richard Robinson. Oxford: Clarendon Press, 1934.

Jones III, Joe F. "Intelligible Matter and Geometry in Aristotle." *Apeiron* 17 (1983): 94–102.

Judson, L. "Heavenly Motion and the Unmoved Mover." In *Self-Motion: From Aristotle to Newton,* ed. Mary Louise Gill and James G. Lennox, 155–171. Princeton: Princeton University Press, 1994.

Kahn, Charles H. "La *Physique* d'Aristote et la tradition grecque de la philosophie naturelle." In *La Physique d'Aristote et les conditions d'une science de la nature,* ed. F. De Gandt et P. Souffrin; trans. F. De Gandt, 41–52. Paris: Vrin, 1991.

"The Place of the Prime Mover in Aristotle's Teleology." In *Aristotle on Nature and Living Things: Philosophical and Historical Studies Presented to David M. Balme on His Seventieth Birthday,* ed. A. Gotthelf, 183–205. Pittsburgh: Mathesis Publications, 1985.

Konstan, David. "Points, Lines, and Infinity: Aristotle's *Physics* Zeta and Hellenistic Philosophy." In *Proceedings of the Boston Area Colloquium in Ancient Philosophy,* ed. John J. Cleary. Vol. 3, 1–32. New York: University Press of America, 1988.

Kosman, L. A. "Aristotle's Definition of Motion." *Phronesis* 14 (1969): 40–62.

"Aristotle's Prime Mover." In *Self-Motion: From Aristotle to Newton,* ed. Mary Louise Gill and James G. Lennox, 135–153. Princeton: Princeton University Press, 1994.

"Substance, Being, and Energeia." In *Oxford Studies in Ancient Philosophy,* ed. Julia Annas. Vol. II, 121–149. Oxford: Oxford University Press, 1984.

Kronz, Frederick M. "Aristotle, the Direction Problem, and the Structure of the Sublunar Realm." *The Modern Schoolman* 67 (1990): 247–257.

Lang, Berel. "Presentation and Representation in Plato's Dialogues." *The Philosophical Forum: A Quarterly* 4 (n.s. 1972–73): 224–240.

Lang, Helen. "The Concept of Place: Aristotle's Physics and the Angelology of Duns Scotus." *Viator: Medieval and Renaissance Studies* 14 (1983): 245–266.

"Points, Lines and Infinity: A Response to the Paper of David Konstan." In *Proceedings of the Boston Area Colloquium in Ancient Philosophy,* ed. John J. Cleary. Vol. 3, 33–43. New York: University Press of America, 1988.

Aristotle's Physics and Its Medieval Varieties. Albany: State University of New York Press, 1992.

"The Structure and Subject of *Metaphysics* Λ." *Phronesis* 38 (1993): 257–280.

Review of *Aristotle and Philoponus on Light*, by Jean De Groot. *Ancient Philosophy* 14 (1994): 190–192.

"Why the Elements Imitate the Heavens: *Metaphysics* IX, 8, 1050b28–34." *Ancient Philosophy* 14 (1994): 335–354.

"Aristotle's *Physics* IV, 8: A Vexed Argument in the History of Ideas." *The Journal of the History of Ideas* 56 (1995): 353–376.

"Inclination, Impetus, and the Last Aristotelian." *Archives internationales d'histoire des sciences* 46 (1996): 221–260.

"Thomas Aquinas and the Problem of Nature in *Physics* II, 1." *History of Philosophy Quarterly* 13 (1996): 411–432.

"Topics and Investigations: Aristotle's *Physics* and *Metaphysics*." *Philosophy and Rhetoric* 29 (1996): 416–435.

Le Blond, J. M. *Logique et méthode chez Aristote: Etude sur la recherche des principes dans le physique aristotélicienne*. Paris: Vrin, 1939.

Lear, J. "Aristotle's Philosophy of Mathematics." *Philosophical Review* 91 (1982): 161–192.

Aristotle: The Desire to Understand. Cambridge: Cambridge University Press, 1988.

Lettinck, Paul. *Aristotle's* Physics *and Its Reception In the Arabic World: With an Edition of the Unpublished Parts of Ibn Bajja's Commentary on the Physics*. Vol. 7 of *Aristoteles Semitico-Latinus*, founded by H. J. Drossaart Lulofs. H. Daiber and R. Kruk, gen. eds. Leiden: E. J. Brill, 1994.

Lewis, Eric. "Commentary on O'Brien." In *Proceedings of the Boston Area Colloquium in Ancient Philosophy*, ed. John J. Cleary and W. Wians. Vol. 11, 87–100. Washington D.C.: University Press of America, 1997.

Lewis, Frank A. *Substance and Predication in Aristotle*. Cambridge: Cambridge University Press, 1991.

Lloyd, G. D. R. "Right and Left in Greek Philosophy." In *Methods and Problems in Greek Science: Selected Papers*, 27–48. Cambridge: Cambridge University Press, 1991.

Lynch, John P. *Aristotle's School: A Study of a Greek Educational Institution*. Berkeley, Los Angeles, and London: University of California Press, 1972.

Machamer, Peter. "Aristotle on Natural Place and Natural Motion." *ISIS* 69 (1978): 377–387.

Macierowski, E. M., and R. F. Hassing. "John Philoponus on Aristotle's Definition of Nature: A Translation from the Greek with Introduction and Notes." *Ancient Philosophy* 8 (1988): 73–100.

Malcolm, John. "On Avoiding the Void." In *Oxford Studies in Ancient Philosophy*, ed. Julia Annas. Vol. IX, 75–94. Oxford: Oxford University Press, 1991.

Mansion, Augustin. "La Genèse de l'oeuvre d'Aristote d'apres les travaux récents." *Revue neoscholastique de philosophie* 29 (1927): 307–341; 423–466.

Introduction à la physique aristotélicienne. 2d ed. Revue et augmentée. Paris: Vrin, 1946.

Mansion, S. "Tò σιμόν et la définition physique." In *Naturphilosophie bei Aristoteles und Theophrast: Verhandlungen des 4. Symposium Aristotelicum veranstaltet in Göteborg, August 1966*, ed. I. Düring, 124–132. Heidelberg: Verlag, 1969.

Matthen, M., and R. J. Hankinson. "Aristotle's Universe: Its Form and Matter." *Synthese* 96 (1993): 417–435.

McCue, James F. "Scientific Procedure in Aristotle's *De Caelo.*" *Traditio* 18 (1962): 1–24.

McGuire, J. E. "Philoponus on *Physics* II: Φύσις, Δύναμις, and the Motion of the Simple Bodies." *Ancient Philosophy* 5 (1985): 241–267.

Mendell, Henry. "Topoi on Topos: The Development Aristotle's Concept of Place." *Phronesis* 32 (1987): 206–231.

Menn, Stephen. "The Origin of Aristotle's Concept of Ἐνέργεια: Ἐνέργεια and Δύναμις." *Ancient Philosophy* 14 (1994): 73–114.

Meyer, Susan S. "Self-Movement and External Causation." In *Self-Motion: From Aristotle to Newton,* ed. Mary Louise Gill and James G. Lennox, 65–80. Princeton: Princeton University Press, 1994.

Modrak, D. K. W. "Aristotle on the Difference between Mathematics and First Philosophy." In *Nature, Knowledge, and Virtue: Essays in Memory of Joan Kung,* ed. Terry Penner and Richard Kraut, issue of *Apeiron* 22 (1989): 121–139.

———. "Aristotle's Epistemology: One or Many Theories?" In *Aristotle's Philosophical Development: Problems and Prospects,* ed. W. Wians, 151–170. London: Rowman and Littlefield, 1996.

Moraux, Paul. *D'Aristote à Bessarion: trois exposés sur l'histoire et la transmission de l'aristotélisme grec.* Quebec: Les Presses de l'Université Laval, 1970.

Morrison, Donald. "Some Remarks on Definition in *Metaphysics* Z." In *Biologie, logique et métaphysique chez Aristote: Actes du séminaire C. N. R. S. - N. S. F. Oléron 28 juin - 3 juillet 1987,* ed. et publiés par Daniel Devereux et Pierre Pellegrin, 130–144. Paris: Editions du C.N.R.S., 1990.

Morrow, Glenn R. "Qualitative Change in Aristotle's *Physics.*" In *Naturphilosophie bei Aristoteles und Theophrast: Verhandlungen des 4. Symposium Aristotelicum veranstaltet in Göteborg, August 1966,* ed. I. Düring, 154–167. Heidelberg: Verlag, 1969.

Mueller, Ian. "Aristotle on Geometrical Objects." *Archiv Für Geschichte der Philosophie* 52 (1970): 156–171.

O'Brien, Dennis. "Aristotle's Theory of Movement." In *Proceedings of the Boston Area Colloquium in Ancient Philosophy,* ed. John J. Cleary and W. C. Wians. Vol. 11, 47–86. Washington D.C.: University Press of America, 1997.

Owen, G. E. L. "Aristotelian Mechanics." In *Aristotle on Nature and Living Things: Philosophical and Historical Studies Presented to David M. Balme on His Seventieth Birthday,* ed. A. Gotthelf, 227–245. Pittsburgh: Mathesis Publications, 1985.

———. "Aristotle: Method, Physics, and Cosmology." In *Logic, Science, and Dialectic: Collected Papers in Greek Philosophy,* ed. M. Nussbaum, 151–164. Ithaca: Cornell University Press, 1986.

———. "The Platonism of Aristotle." In *Logic, Science, and Dialectic: Collected Papers in Greek Philosophy,* ed. M. Nussbaum, 200–220. Ithaca: Cornell University Press, 1986.

Owens, Joseph. *The Doctrine of Being in the Aristotelian Metaphysics.* 3d ed. Toronto: Pontifical Institute of Medieval Studies, 1978.

———. "The Teleology of Nature in Aristotle." In *Aristotle: The Collected Papers of Joseph Owens,* ed. John R. Catan, 136–147. Albany: State University of New York Press, 1981.

Paulus, J. "La Théorie du premier moteur chez Aristote." *Revue de philosophie* 33 (1933): 259–294; 394–424.

Pegis, A. C. "St. Thomas and the Coherence of the Aristotelian Theology." *Mediaeval Studies* 35 (1973): 67–117.

Pellegrin, P. "The Platonic Parts of Aristotle's *Politics*." In *Aristotle's Philosophical Development: Problems and Prospects*, ed. W. Wians, 347–357. London: Rowman and Littlefield, 1996.

Pellicer, André. *Natura: Etude sémantique et historique du mot Latin*. Paris: Presses Universitaires de France, 1966.

Pfeiffer, Rudolf. *History of Classical Scholarship: From the Beginnings to the End of the Hellenistic Age*. Oxford: Clarendon Press, 1968.

Pines, S. "Philosophy, Mathematics, and the Concepts of Space in the Middle Ages." In *The Interaction between Science and Philosophy*, ed. Y. Elkana, 75–90. The Van Leer Jerusalem Foundation Series. Atlantic Highlands, N.J.: Humanities Press, 1974.

Polansky, R. "*Energeia* in Aristotle's *Metaphysics* IX." *Ancient Philosophy* 3 (1983): 160–170.

Potts, Timothy C., and C. C. W. Taylor. "States, Activities, and Performances." Part II, *Proceedings of the Aristotelian Society*, supp. vol. 39 (1965): 85–102.

Preus, Anthony. "Man and Cosmos in Aristotle: *Metaphysics* Λ and the Biological Works." In *Biologie, logique et métaphysique chez Aristote: Actes du séminaire C. N. R. S. - N. S. F. Oléron 28 juin - 3 juillet 1987*, ed. et publiés par Daniel Devereux et Pierre Pellegrin, 471–490. Paris: Editions du C.N.R.S., 1990.

Rist, John M. *The Mind of Aristotle: A Study in Philosophical Growth*. Toronto: University of Toronto Press, 1989.

Ross, W. D. "The Development of Aristotle's Thought." In *Aristotle and Plato in the Mid-fourth Century: Papers of the Symposium Aristotelicum Held at Oxford in August, 1957*, ed. by I Düring and G. E. L. Owen, 1–17. Göteborg: Elanders Boktryckeri Aktiebolag, 1960.

Samburaky, S. *The Concept of Place in Late Neoplatonism: Texts with Translation, Introduction, and Notes*. Jerusalem: The Israel Academy of Sciences and Humanities, 1982.

Sandbach, F. H. *Aristotle and the Stoics*. Supp. vol. xi. Cambridge: The Cambridge Philological Society, 1985.

Schofield, Malcolm. "Explanatory Projects in *Physics*, 2.3 and 7." In *Oxford Studies in Ancient Philosophy*, ed. Julia Annas. Supp. vol.: *Aristotle and the Later Tradition*, ed. Henry Blumenthal and Howard Robinson, 29–40. Oxford: Clarendon Press, 1991.

Sedley, David. "Two Conceptions of Vacuum." *Phronesis* 27 (1982): 175–193.

"Philoponus' Conception of Space." In *Philoponus and the Rejection of Aristotelian Science*, ed. R. Sorabji, 140–153. Ithaca: Cornell University Press, 1987.

Seeck, Gustav Adolf. "Licht-schwer und der Unbewegte Beweger (DC IV 3 and Phys. VIII, 4)." In *Naturphilosophie bei Aristoteles und Theophrast: Verhandlungen des 4. Symposium Aristotelicum veranstaltet in Göteborg, August 1966*, ed. I. Düring, 210–223. Heidelberg: Verlag, 1969.

Sharples, R. W. "The Unmoved Mover and the Motion of the Heavens in Alexander of Aphrodisias." *Apeiron* 17 (1983): 62–66.

"Eudemus' *Physics*: Change, Place, and Time." In *Eudemus of Rhodes*, ed. W. W. Fortenbaugh and I. Bodnar. Rutgers University Studies in Classical Humanities, Vol. XI. New Brunswick and London: Transaction Publishers, in press.

Solmsen, F. *Aristotle's System of the Physical World: A Comparison with His Predecessors.* Ithaca: Cornell University Press, 1960.

———. "Platonic Influences in the Formation of Aristotle's Physical System." In *Aristotle and Plato in the Mid-fourth Century: Papers of the Symposium Aristotelicum Held at Oxford in August, 1957*, ed. I. Düring and G. E. L. Owen, 213–235. Göteborg: Elanders Boktryckeri Aktiebolag, 1960.

Sorabji, R. *Matter, Space, and Motion: Theories in Antiquity and Their Sequel.* Ithaca: Cornell University Press, 1988.

———. "Theophrastus on Place." In *Theophrastean Studies: On Natural Science, Physics, and Metaphysics, Ethics, Religion, and Rhetoric.* Vol. III: *Studies in Classical Humanities*, 139–166. New Brunswick: Transaction Books, 1988.

———. "Infinite Power Impressed: The Transformation of Aristotle's Physics and Theology." In *Aristotle Transformed: The Ancient Commentators and Their Influence*, ed. R. Sorabji, 181–198. Ithaca: Cornell University Press, 1990.

———. Introduction to Simplicius, *On Aristotle's Physics 4.1–5, 10–14.* Trans. by J. O. Urmson, 1–14. Ithaca: Cornell University Press, 1992.

Sprague, Rosamond Kent. "The Four Causes: Aristotle's Exposition and Ours." *The Monist* 52 (1968): 298–300.

Strange, Steven K. "Plotinus on the Nature of Eternity and Time." In *Aristotle in Late Antiquity*, ed. Lawrence P. Schrenk, 22–53. Washington, D.C.: The Catholic University of America Press, 1994.

Summers, James W. "Aristotle's Concept of Time." *Apeiron* 18 (1984): 59–71.

Thorp, John. "Aristotle's *Horror Vacui.*" *The Canadian Journal of Philosophy* 20 (1990): 149–166.

Trifogli, Cecilia. "Giles of Rome on Natural Motion in the Void." *Mediaeval Studies* 54 (1992): 136–161.

Urmson, J. O. *The Greek Philosophical Vocabulary.* London: Gerald Duckworth, 1990.

Verdenius, W. J. "Critical and Exegetical Notes on *De Caelo.*" In *Naturphilosophie bei Aristoteles und Theophrast: Verhandlungen des 4. Symposium Aristotelicum veranstaltet in Göteborg, August 1966*, ed. I. Düring, 268–284. Heidelberg: Verlag, 1969.

Vogel, C. J. de. "The Legend of the Platonizing Aristotle." In *Aristotle and Plato in the Mid-fourth Century: Papers of the Symposium Aristotelicum Held at Oxford in August, 1957*, ed. I. Düring and G. E. L. Owen, 248–256. Göteborg: Elanders Boktryckeri Aktiebolag, 1960.

Von Staden, Heinrich. "Jaeger's "Skandalon der historischen Vernunft": Diocles, Aristotle, and Theophrastus." In *Werner Jaeger Reconsidered: Proceedings of the Second Oldfather Conference, Held on the Campus of the University of Illinois at Urbana-Champaign, April 26–28, 1990*, ed. W. M. Calder III, 227–265. Atlanta: Scholars Press, 1992.

Wardy, Robert. *The Chain of Change: A Study of Aristotle's Physics VII.* Cambridge: Cambridge University Press, 1990.

Waterlow, S. *Nature, Change, and Agency in Aristotle's Physics: A Philosophical Study.* Oxford: Clarendon Press, 1982.

Wedin, Michael V. "Aristotle on the Mind's Self-Motion." In *Self-Motion: From Aristotle to Newton*, ed. Mary Louise Gill and James G. Lennox, 81–116. Princeton: Princeton University Press, 1994.

Weisheipl, James A. "Motion in a Void: Aquinas and Averroes." In *St. Thomas Aquinas, 1274–1974: Commemorative Studies,* ed. A. Maurer. Vol. I, 467–488. Toronto: Pontifical Institute of Medieval Studies, 1974. Reprinted in *Nature and Motion in the Middle Ages,* ed. William E. Carroll, 121–142. Washington, D.C.: Catholic University of America Press, 1985.

"The Principle *Omne Quod Movetur Ab Alio Movetur* in Medieval Physics." *ISIS* 56 (1956): 26–45. Reprinted in *Nature and Motion in the Middle Ages,* ed. William E. Carroll, 75–97. Washington, D.C.: Catholic University of America Press, 1985.

"The Specter of *Motor Coniunctus* in Medieval Physics." In *Studi sul XIV secolo in memoria di Anneliese Maier,* ed. A. Maiery and A. P. Bagliani, 81–104. Rome: Edizioni di Storia e Letterature, 1981. Reprinted in *Nature and Motion in the Middle Ages,* ed. William E. Carroll, 99–120. Washington, D.C.: Catholic University of America Press, 1985.

Wians, W. Introduction to *Aristotle's Philosophical Development: Problems and Prospects,* ed. W. Wians, ix–xiv. London: Rowman and Littlefield, 1995.

Wieland, W. "The Problem of Teleology." In *Articles on Aristotle.* Vol. 1: *Science,* ed. J. Barnes, M. Schofield, and R. Sorabji, 141–160. London: Duckworth, 1975.

Wildberg, Christian. *John Philoponus' Criticism of Aristotle's Theory of Aether.* Peripatoi: Philologisch-historische studien zum aristotelismus in Verbindung mit H. J. Drossaart Lulolfs, R. Weil Herausgegeben von Paul Moraux. Vol. 16. Berlin and New York: Walter de Gruyter, 1988.

"Two Systems in Aristotle?" In *Oxford Studies in Ancient Philosophy,* ed. Julia Annas. Vol. 7, 193–202. Oxford: Oxford University Press, 1989.

Witt, Charlotte. *Substance and Essence in Aristotle: An Interpretation of Metaphysics VII–IX.* Ithaca: Cornell University Press, 1989.

"The Evolution of Developmental Interpretations of Aristotle." In *Aristotle's Philosophical Development: Problems and Prospects,* ed. W. Wians, 67–82. London: Rowman and Littlefield, 1996.

Wolff, Michael. *Fallgesetz und Massebegriff: Zwei wissenschaftshistorische Untersuchungen zur Kosmologie des Johannes Philoponus.* Vol. 2 in *Quellen und Studien zur Philosophie; Herausgegeben von Günther Patzig, Erhard Scheibe, Wolfgang Wieland.* Berlin: Walter de Gruyter and Co., 1971.

"Philoponus and the Rise of Preclassical Dynamics." In *Philoponus and the Rejection of Aristotelian Science,* ed. R. Sorabji, 84–120. Ithaca: Cornell University Press, 1987.

Wolfson, Harry A. *Crescas' Critique of Aristotle: Problems of Aristotle's Physics in Jewish and Arabic Philosophy.* Vol. 6 in Harvard Semitic Series. Cambridge: Harvard University Press, 1929.

"The Plurality of Immovable Movers in Aristotle, Averroes, and St. Thomas." In *Studies in the History of Philosophy and Religion,* ed. I. Twersky and G. H. Williams. Vol. 1, 1–21. Cambridge: Harvard University Press, 1973. Originally published in *Harvard Studies in Classical Philosophy* 63 (1958): 233–253.

Zimmerman, F. "Philoponus' Impetus Theory in the Arabic Tradition." In *Philoponus and the Rejection of Aristotelian Science,* ed. R. Sorabji, 121–129. Ithaca: Cornell University Press, 1987.

Subject and Name Index

308

Name Index

312   INDEX

Index of Aristotelian Texts

Milton Keynes UK
Ingram Content Group UK Ltd.
UKHW041207201024
449640UK00025B/8